PARALLEL JOURNEYS

Parallel Journeys

Eurasian History Through Travelers' Eyes (400 BCE–1936 CE)

With Introductions and
New Translations from the Chinese

Anthony J. Barbieri-Low

Cambria Sinophone World Series
General Editor: Wendy Larson (University of Oregon)

Cambria Sinophone Translation Series
General Editor: Kyle Shernuk (Georgetown University)
Advisor: Christopher Lupke (University of Alberta)

CAMBRIA
PRESS

Amherst, New York

Requests for permission should be directed to permissions@cambriapress.com, or
mailed to: Cambria Press, 100 Corporate Parkway, Suite 128, Amherst, New York 14226, USA.

Library of Congress Cataloging-in-Publication Data

Names: Barbieri-Low, Anthony J. (Anthony Jerome), 1967- editor
Title: Parallel journeys : Eurasian history through travelers' eyes (400 BCE-1936 CE) /
with introductions and new translations from the Chinese by Anthony J. Barbieri-Low.
Other titles: Eurasian history through travelers' eyes (400 BCE-1936 CE)

Description: Amherst, New York : Cambria Press, [2025] |
Series: Cambria sinophone world series |
Includes bibliographical references and index.
| Summary: "Parallel Journeys: Eurasian History through Travelers' Eyes (400 BCE-1936 CE)
is a comparative anthology of travel accounts by more than two dozen Eastern and Western
travelers spanning 2,400 years of Eurasian history. Structured around paired narratives-from
Chinese envoys, monks, and scholars to European merchants, missionaries, and journalists-
this volume explores diplomacy, commerce, religion, and cultural exchange through critical
introductions and new English translations from the Chinese. The anthology highlights
how these parallel journeys shaped perceptions of self and other across time, offering
insight into the evolving geography of the "Orient" and "Occident." Featuring rare and
newly translated Chinese texts alongside Western classics, Parallel Journeys contributes to
global history, travel literature, and cross-cultural studies by illuminating shared motifs and
divergent worldviews in historical accounts of long-distance travel"-- Provided by publisher.

Identifiers: LCCN 2025040928 |
ISBN 9781638573807 (library binding) | ISBN 9781638573821 paperback |
ISBN 9781638573951 epub | ISBN 9781638573814 pdf

Subjects: LCSH: East and West | Eurasia--Description and
travel | Eurasia--Description and travel--History--Sources

Classification: LCC G155.E75 P37 2025

LC record available at https://lccn.loc.gov/2025040928

*To the one who has been my travel
companion on the greatest journey of all*

TABLE OF CONTENTS

LIST OF FIGURES

LIST OF MAPS

Acknowledgments

This anthology began as a reader to accompany my undergraduate course, "Eurasian Currents," that I began teaching at the University of Pittsburgh in 2002. I compiled these accounts from Western and Eastern travelers with the support of an NRC Course Development Grant and a Faculty Fellowship from the University Center for International Studies at Pitt. I continued to use these accounts in my world history courses when I moved to UCSB in 2007, and I wish to thank the thousands of students who enrolled in these courses for their insights and feedback. Unfortunately, an anthology based on my reader was continually sidelined by other projects.

I must acknowledge my partner, Li Xiaorong, for urging me to revive this long-dormant project and to balance the male dominated travel accounts with more female perspectives. She also led me to Ellen Widmer, who is an expert on the Chinese female traveler, Shan Shili, whom I later incorporated. I also want to thank one of the anonymous outside readers for encouraging me to "fully commit" to my "parallel journeys" approach.

My undergraduate research assistant Zhang Yexu helped me construct the maps for the travel itineraries. I would also like to thank my History Department colleagues, John W. I. Lee and Adam Sabra, for reading drafts of chapters in their areas of specialty, and to Utathya Chattopadhyaya, for testing my translations of Faxian and Xuanzang in his survey course on South Asia.

PARALLEL JOURNEYS

INTRODUCTION

The notion of foreign travel can conjure up feelings of a romantic adventure but can also stoke fears of the unknown, with anxieties about physical or moral perils. But travel to distant lands can also be transformative, for an encounter with the foreign can lead to introspection and personal growth.

The ancient historian Diodorus Siculus (fl. ca. 60–30 BCE) once wrote that the greatest pleasure in reading history lay in vicariously experiencing momentous events and distant places from the safety of one's own home. In that spirit, I invite you to embark with me on a virtual journey across the expansive continent of Eurasia through this anthology and hopefully set off on your own personal voyage of intellectual discovery and personal transformation.

The connected history of Eurasia is the story of the movement and entanglement of objects, technologies, ideas, animals, pathogens, and people over vast stretches of space and time. While I shall deal with the adaptation of art and technology across ancient Eurasia in a future study, this book focuses on the ambassadors, pilgrims, merchants, missionaries, and journalists who traveled across this vast continent over the last 2,400 years. It presents critical introductions and substantial excerpts from the accounts of more than two dozen remarkable individuals and their encounters with foreign cultures. Among these are the writings of famous travelers to the East like William of Rubruck, Maro Polo, Matteo Ricci, Lord George Macartney, and Edgar Snow, paired alongside the lesser-known reports of the West by their Chinese contemporaries

Rabban Sauma, Zhao Rugua, Xie Qinggao, Guo Songtao, and Shan Shili. (To facilitate their potential use as primary sources to accompany world history classes, briefer excerpts of 500 to 1,000 words are highlighted within each account). The knowledge that these persons transmitted or carried home contributed to a complex process of discovery, witnessing, and reflecting that was crucial for defining the self and the other. Such knowledge could be employed for internal cohesion or to justify external domination, though both were pivotal in the integration of Eurasia over the past two millennia.

I had been inspired by Donald F. Lach's *Asia in the Making of Europe* (1965–1993) and by Jonathan D. Spence's more accessible *The Chan's Great Continent* (1998). These landmark scholarly works highlighted the importance of the discovery and knowing of Asia in shaping modern European civilization and its self-image. However, these works told only one side of the story, for they inadvertently perpetuated a false notion, one inherited from the Age of Discovery and from the likes of Adam Smith and Georg Hegel. This fallacy of Western exceptionalism, activism, and progress stripped China of its agency, reducing it to a stagnant monolith and a passive, uncharted territory waiting to be discovered, explored, and exploited by the West. A more balanced historical assessment acknowledges that Chinese travelers were nearly as active in the shared processes of discovering and knowing, adapting, and rejecting, as their European and American counterparts. These shared mechanisms were some of the key interactions that drove the history of Eurasia. Some earlier literary anthologies, like Jeannette Mirsky's *The Great Chinese Travelers* (1964), did anthologize several East Asian pioneers, but what was needed was a comparative framework that integrates the travelers from East and West in a balanced, global historical narrative.

Looking for methodological inspiration in the ancient past, we see that the Roman-period Greek author Plutarch, in his monumental collection of nearly four dozen biographies, *Parallel Lives,* had pioneered a new genre of comparative historical biography. He juxtaposed the life of one renowned Greek figure, such as Alexander the Great, with that of a comparable Roman personage, like Julius Caesar. This approach enabled Plutarch to draw parallels and contrasts between these men's virtues and moral failings and connect those to their personal destiny. Through this method, Plutarch also skillfully connected Greek and Roman history by identifying recurring historical patterns evident in his biographies.

In my book *Ancient Egypt and Early China* (2021), I demonstrated the enduring value of a Plutarchian-style comparative biography when I juxtaposed the life of the heretic pharaoh Akhenaten (r. ca. 1353–1326 BCE) with the Chinese emperor Wang Mang (r. 9–23 CE). Pairing these two seemingly incomparable figures allowed me to see past the themes that had transfixed earlier generations of scholars to envision the broader picture of the internal dynamics of reform within decaying empires.

Accordingly, the method of this literary anthology will be to juxtapose the account of a Western and an Eastern traveler during each time period, structured under a specific theme in a global, connected history. The value of comparative biography, or in the "parallel journeys" approach of this anthology, resides in the critical insights revealed by the juxtaposition of outwardly dissimilar cases. The comparison helps us realize shared structural processes in operation during a particular era in global history, while also highlighting prominent variations in cultural practice and historical trajectories. In other words, the "conversation" between the two travelers allows us to see both in a new light.

For example, in chapter 7, I have drawn into parallel two apparently mismatched 13th-century accounts: that of the Southern Song trade official Zhao Rugua (1225), who never voyaged outside China, and the renowned book of Marco Polo (1298), a man who traversed the breadth of Eurasia by both land and sea. By reading their accounts in parallel, however, we recognize that both men were attempting to present a comprehensive picture of the inhabited world. They each revealed a vast continent of foreign countries connected by trade and diplomacy, where marvelous and wonderful things still inhabited the margins of the earth. Furthermore, by critically examining each of these worlds, we understand how fundamentally the Mongol conquests reoriented the configuration of Eurasia, for Zhao's world in the early 13th century was still a Muslim-centered universe, dominated by the Abbasids in Baghdad, while Polo's late 13th–century world revolved around the great khanates of the Mongols in China, Central Asia, and Persia, which had conquered Baghdad in 1258 and leveled many of the great trading cities of Central Asia.

In building out this construction of "parallel journeys," it has not always been possible, or in some cases even preferable, to present a perfect pairing of travelers, one going East and the other headed West. Sometimes, our travel accounts, particularly those from the ancient and medieval era, appear less like unbroken parallel lines and more like short line-segments. However, these can be combined to complement one another and thereby present a general trajectory. For example, chapter 4 features the two secondhand accounts of Pliny the Elder (23–79 CE) and Yu Huan (fl. ca. 239–265 CE), but the chapter introduction also presents complementary information from the works of the geographer Claudius Ptolemy (fl. ca. 100 CE) and the Chinese historian Fan Ye (398–446) to fill in gaps in the narrative. In chapter 5, the East Asian side of the parallel travelers is fulfilled by the complementary duo of the monks Faxian and Xuanzang. Read together as one composite "Chinese pilgrim," they are even more compelling and insightful, for the former reveals the dynamics of Indian Ocean trading voyages, while the latter delivers an ethnography of seventh-century India. Similarly, chapter 13 has been divided into two parts. The first part combines the travel accounts of two Chinese intellectuals, a male scholar (Liang Qichao) who traveled to America and a female scholar (Shan Shili) who went to Europe, together providing a more comprehensive picture in terms of geographic coverage and gender balance. The second part is reserved for their counterpart, the American journalist Edgar Snow, and his encounter with Mao Zedong.

CHAPTER HIGHLIGHTS

Chapter 1, "Monsters at the Edge of the World," opens the anthology with a complementary pair of texts from the fourth century BCE. These accounts represent an early literate expression of a universal human cultural construct, the imagining of monsters that reside beyond the bounds of the civilized realm. From the Greek speaking world, we have the *Indica* (early 4th century BCE) of Ctesias of Cnidus, a physician to the Persian court who wrote from a comfortable distance about the peoples, creatures, and products of India, then believed to be the edge of the inhabited world. His account combines real locations, peoples, and animals of India, alongside monstrous races like his Dog-Headed People. The *Guideways Through Mountains and Seas* is a mythogeography of the mountainous areas within the civilized lands of China and of the wildernesses beyond the seas. These outlying wilds are inhabited by an array of inhuman or transgressive peoples (like the Country of Women), strange hybrid creatures, and powerful deities like the Spirit Mother of the West.

Chapter 2, "Imperial Ambassadors," takes us beyond the hazy lands of myth and hearsay to explore the new world of international diplomacy that developed in the wake of Alexander's conquests in the West and the Chinese imperial expansion in the East, juxtaposing the reports of two diplomatic envoys. Megasthenes, the Seleucid ambassador to Chandragupta Maurya's court in India, composed his own *Indica* around 300 BCE, detailing India's geography, history, flora, fauna, customs, and administration. He provides reasonably accurate information on institutions like the caste system, as well as a thrilling account of elephant taming, but he still spins tales of monstrous races in remote parts of India, such as the Mouthless Ones who subsist only on aromas. Megasthenes's account is paired with the report of the Han imperial envoy Zhang Qian, who was sent to the far west of China around 135 BCE to find allies for the fight against the Xiongnu nomads. Though frustrated in his mission to secure allies, he returned with a wealth of strategic information about the lands to the Far West, revealing a whole new world of foreign countries to China and paving the way for impactful cultural and economic exchanges.

Chapter 3, "Monsoon Winds and the Lure of Indian Luxuries," juxtaposes two travel itineraries from the early Indian Ocean world, where seasonally shifting monsoon winds facilitated the movement of people and commodities, tightening the web of Eurasian integration. The *Navigating Around the Red Sea* text, written by a Greek-speaking merchant-sailor from Egypt, details ports along three itineraries, including one across the open ocean to India, describing profitable trade items like silk and spices as well as the maritime hazards one might encounter. The *Voyage to Huangzhi* is a Chinese itinerary starting from coastal South China and voyaging to India and Sri Lanka. It describes state-sponsored trading missions, where the Chinese exchanged gold bullion and silk for Indian gems and imported Roman glass.

Chapter 4, "Hazy Notions of Far-Off Empires," parallels the description by the Roman naturalist Pliny the Elder concerning the Seres ("Silk People") who lived in the Far East, alongside an account penned by the Chinese historian Yu Huan concerning the powerful Far Western kingdom of Da Qin ("Greater Qin"), identified by modern scholars as the Roman Empire. Neither of these authors actually traveled to the distant kingdoms they described, for they relied on the reports of merchants and envoys to compile their accounts. They each combined some accurate knowledge about things like trade products and routes of contact, together with legends, utopian fantasies, and mistaken or garbled information, such as "Silk is combed off the leaves of trees" or "Romans say they were originally a branch of the Chinese people."

Chapter 5, "The Decentering Worldviews of Universal Religions," delves into the early medieval period in Eurasia, when the universal religions of Buddhism and Christianity formed symbiotic relationships with imperial formations and religious pilgrimage sometimes fueled long-distance travel. On the Chinese side, this chapter presents the complementary accounts of two renowned Chinese Buddhist pilgrims. The monk Faxian traveled to India in 399 BCE, seeking authentic scriptures and visiting holy sites. The excerpted portions from his *Record of Buddhist Kingdoms* (414 CE) includes his retelling of the Buddhist legend of "King Ashoka's Hell on Earth" and the harrowing tale of his own return voyage to China. Two centuries later, the monk Xuanzang traveled nearly the same route on his own pilgrimage. I have translated his "General Account of India" from his larger book, *Record of the Western Regions of the Great Tang Dynasty* (646 CE). This is an informative ethnography of seventh-century India, filled with scholarly observations of religion and education but also including descriptions of social practices that would astonish his readers, such as naked ascetics, endogamous castes, and judicial ordeals. The accounts of these Eastern pilgrims are paralleled in this chapter by the account of an Alexandrian-born Christian merchant known as Cosmas Indicopleustus. His *Christian Topography* (543–550 CE) provides information on the Indian Ocean trade, the spread of Christianity, silk trade with China, and an entertaining story I have called "Sopatrus and the Contest of Coins."

Chapter 6, "Collisions and Connections Between New Empires," examines interactions between Tang China and the Abbasid Caliphate, two empires that collided and connected at the zenith of their power and extent. Du Huan was a Chinese officer who had been captured in battle and lived in the Arabic capital for a decade. His *Record of a Journey* (762 CE) provides the earliest first-person Chinese account of Islamic teachings and practices, such as the five daily prayers, the veiling of women in public, and abstention from pork and alcohol. To offer a complimentary perspective, Du's account is juxtaposed with the compilation of Arabic sailor accounts from Sulaymān the Merchant and others called *Accounts of China and India* (851 CE). This compilation provides a detailed sailing itinerary of the main Indian Ocean route from the Persian Gulf to Canton. It also includes the earliest eyewitness accounts of the governance, law, commerce, and customs of China to be found in any Middle Eastern or Western work, including the salacious (but accurate) information that the Chinese ate unclean animals, frequented prostituted slave boys in Buddhist temples, and wiped their bottoms with paper after defecating.

Chapter 7, "Spanning Eurasia During the Mongol Century," incorporates the parallel journeys of two remarkable travelers during the pivotal 13th century, when Mongol incursions into Central Asia and Europe broke down Muslim barriers to direct contact between Europe and China. The monk William of Rubruck entered Mongol-controlled lands to preach and convert, eventually reaching the court of the Great Khan Möngke. His *Travel Account* (1255) provides one of the most detailed early descriptions of Mongol society, domestic architecture, food, customs, and religion. A few decades after William visited the Mongol court, a Chinese-born Nestorian Christian pilgrim named Rabban Sauma visited Constantinople, Italy, and France on a diplomatic mission from the Mongol Il-Khan of Persia (1287–1288 CE), making him the first person from China to leave a record of visiting Western Europe. He narrates his debate with the cardinals in Rome over the nature of Christ and the Trinity, his encounters with Christian relics, and offers a brief mention of late 13th–century Paris and its growing scholarly community.

Chapter 8, "Descriptions of the World," presents two parallel accounts that paint comprehensive pictures of the known world of the late Middle Ages. Zhao Rugua was the supervisor of maritime trade in Quanzhou, a bustling Song Chinese port with a Muslim merchant diasporic community. His *Account of Foreign Countries* (1225), based on earlier texts and Zhao's interviews with foreign merchants, describes 58 foreign lands and their exotic products. In the translated excerpts, Zhao records wondrous African animals like the giraffe and zebra, retells the story of Joseph in Egypt, and scandalizes his readers with tales of two "Countries of Women." The illustrious Marco Polo traveled to China with his father and uncle on a mission from Kublai Khan (1271–1275) then spent another 17 years in China fulfilling various imperial commissions. Polo's *Description of the World* (1298) presented the "wonders of the East" in a whole new fashion, revealing to Europe an enormous empire of great prosperity. His description of the world's largest city of Quinsai (modern Hangzhou), excerpted in this chapter, is one of the centerpieces of his book.

Chapter 9, "Epic Voyages During the Global Age of Discovery," broadens the usual timeframe of the Age of Discovery to incorporate the earlier voyages of Zheng He (1405–1433), alongside the better-known European voyages. It excerpts the account entitled *Wondrous Sightseeing Along the Great Ocean's Shores*, by Zheng He's interpreter, Ma Huan, which includes his descriptions of the kingdom of Calicut in India and his personally meaningful visit to Mecca. This is paired with the account of the Portuguese smuggler Galeote Pereira, entitled "Some Things Known About China" (1561), which provides some of the first detailed descriptions of Chinese governance and customs since the days of Marco Polo, including the earliest accurate account of the civil service exam system. Pereira's harrowing description of the Chinese penal system had a lasting impact on European impressions of Chinese brutality.

Chapter 10, "Gathering Empirical Knowledge for the Early Modern World," features accounts from two 17th- and 18th-century travelers who each spent decades living within a foreign culture, so they provide much more accurate empirical information than those of previous travelers. The Jesuit missionary Matteo Ricci lived in China for nearly 30 years, mastered the Chinese language, and greatly impressed the native literati with his knowledge of Western cartography and mathematics. The excerpts from his journals showcase his nuanced understanding of Chinese language, government, and customs. Father Ricci is paired with the intrepid Cantonese sailor Xie Qinggao, who traveled the world aboard Portuguese trading ships and collaborated on an account of his voyages called *Maritime Records* (1820), which discusses many countries in the Indian Ocean and Atlantic World. This chapter excerpts his fascinating account of the marriage customs, social greetings, and Easter rituals in Portugal, fresh-water delivery systems in 18th-century London, and the earliest American steamboats.

Chapter 11, "Political Ambassadors," explores early modern diplomacy through the paired written accounts of two envoys. Lord George Macartney was sent by the British East India Company and the government to the Qing court (1792–1793) on an embassy to negotiate favorable trading conditions and a permanent ambassador in Peking. Macartney famously refused to perform a kowtow of submission before the Qing emperor, and excerpts from his journal show that Macartney realized that Britain could obtain its demands through force from a decaying Chinese empire but that he preferred the gentler approach of diplomacy. Almost a century later, Guo Songtao was sent by the Qing court to be the first ambassador to Great Britain (1877–1888). Guo's letter from London expresses his admiration for Western technologies like the railroad and telegraph but also praises Western principles of governance that provided a strong foundation for a modern state. An appendix to chapter 11 reproduces captioned illustrations by Macartney's draftsman, William Alexander, which demonstrate the disdainful attitude toward China that had developed in parts of Europe during the 18th century.

Chapter 12, "Cultural Ambassadors," parallels the accounts of two unofficial travelers during the 19th century. Eliza Jane Bridgman was the first female American Protestant missionary to preach in China. Her gender granted her access to the cloistered realm of Chinese women that had been inaccessible to earlier male travelers. In her book, *Daughters of China* (1853), she denounces practices like foot binding, female infanticide, and concubinage, but she also actively promotes the education of women as a means of social and spiritual liberation. Wang Tao was the first Confucian scholar to spend considerable time living in Europe, and he used his observations to argue for reform in China. His travel diary, *Illustrated Record of My Jottings on Carefree Travel* (1890), includes admiration for efficient Western technologies like trains and telegraphs but also praise for institutions like patents that gave Great Britain a competitive advantage. His rapturous description of a dance festival in Scotland is another highlight of his travelogue.

Chapter 13, "Witnesses to the Contradictions of Modernity," showcases three travel accounts from the first third of the twentieth century. The first part of the chapter presents travel accounts of two Chinese intellectuals from the turn of the 20th century, one sojourning in America, the other in Europe. Liang Qichao's *Notes on Travels in the New World* (1903) was the most influential account of America by a Chinese traveler. Liang wanted to see the American republic in action and evaluate its political model for adoption in a post-imperial China. In his travelogue, Liang expressed some misgivings about American democracy and felt it would be difficult to transplant elsewhere. He criticized the racism and inequality he witnessed in America but reserved his harshest critique for his own Chinese people, whom he felt were unsuited and unprepared for self-governance. Shan Shili was a Chinese female scholar who accompanied her ambassador husband to Europe. Her *Travelogue of the Guimao Year* (1903) delivers her observations traveling across Eurasia by rail, including her unique take on social and industrial evolution and the role of Chinese women in building a modern state. Her *Writings During My Husband's Retirement* (1910) compiles her scholarly essays on subjects like Marco Polo and the Jewish Ghetto in Rome, which ends with a prophetic warning about failing to heed the lessons of anti-Semitism.

The second part of the chapter pairs these Chinese travelers with the American journalist Edgar Snow's and his book *Red Star over China* (1938). Snow recounts his time behind the lines of "Red China," when he interviewed Mao Zedong and other Communist leaders at their base in the northwest, shortly after the Long March. During his decade-long sojourn in China, Snow had witnessed the glaring inequalities and corruption in Chinese society that gave rise to the Communist movement. He portrays Mao rather sympathetically as a rustic but cultured popular hero. Snow's account is prophetic, for it was far from certain at that time that Mao would win out during the Chinese Civil War and become the paramount leader of China.

CARTOGRAPHIC APPENDICES

The travel reports in this anthology influenced the development of both Eastern and Western cartography. However, popular views of the world were quite conservative and lagged behind the mapmakers' innovations. To illustrate this, I have included three important world maps that demonstrate cartographic revolutions resulting from distant travel, each placed in an appendix to the relevant chapter (with links to higher resolution online versions). The *Catalan Atlas* (1375) advanced medieval cartography by presenting a more faithful rendering of the world, combining precise European navigation charts with information from Marco Polo about East and South Asia (appendix to chapter 8). The *Record of Military Preparations* chart originally represented a nautical scroll map that was drawn up for Zheng He's seventh voyage (appendix to chapter 9). It incorporated knowledge gained from the previous six voyages about Southeast Asia, India, the Persian Gulf, and East Africa. Matteo Ricci's *Complete Diagram of the Myriad Countries of the Earth* (1602) was the first world map to combine the planar projections of the Western cartographic tradition, information about the New World, and traditional Chinese geographic knowledge (appendix to chapter 10).

THE GEOGRAPHY OF EURASIAN TRAVEL

The concepts of what constituted the Orient or the Occident evolved over time, as longer distance journeys broadened the horizons of the inhabited world. For residents of the ancient Mediterranean or Near East, the Oriental East originally signified India, Persia, and Afghanistan. By the Roman period, this was extended to include the land of the Seres ("Silk People," i.e., China) and, by the time of Marco Polo and the later Iberian voyagers, it incorporated the Japanese and Philippine archipelagos as well.

For those living in early imperial China, the "Western Regions" originally signified the Gansu Corridor, Tarim Basin, and Turfan Depression but was extended after Zhang Qian's forays to include Bactria, Sogdiana,

India, Persia, and Mesopotamia. By the early third century CE, the concept of the Far West was enlarged to include Egypt and Italy. Subsequent Chinese journeys during medieval and early modern times expanded this notion to include East Africa, Western Europe, and the Americas.

Though East Asia was never totally isolated from Western contact and cultural exchange, formidable geographic barriers and the immense distances involved (8,000 km from Rome to Beijing) limited the easy or direct movement of people and goods. Directly west of China lay the vast Gobi and Taklamakan Deserts, followed immediately by the lofty Pamir Mountains. Communication toward the southwest with Burma was hindered by mountains and malarial jungles and toward India by the Tibetan Plateau and Himalayas. Historically, three established avenues of contact traversed the Eurasian continent: two overland networks and one by sea through the Indian Ocean. Each traveler in this anthology utilized one or more of these routes.

The central overland route, known today as the Silk Road or Silk Routes, was a network of land itineraries passing through a diverse terrain that went from the Chinese imperial capital along alternate routes to ports in western India, the mouth of the Persian Gulf, or all the way to the Black Sea and eastern Mediterranean (see maps 1–5). These were important avenues for cultural transmission across Eurasia (e.g., Hellenism, Buddhism, Nestorian Christianity, and Islam), but even at their height, they never rivalled the Indian Ocean route in commercial trade volume.

Commencing westward from the Chinese capital, a traveler passed through the Gansu Corridor, then, after passing through the oasis of Dunhuang, he or she skirted the Tarim Basin, either through a string of northern oases (like Kucha) or southern ones (like Khotan), exiting via Kashgar. After hazarding high mountain passes over the Pamirs, travelers entered Ferghana and Sogdiana (Samarkand). At this point, various alternate routes traveled south through Bactria and toward the Indian subcontinent or across the Iranian Plateau to Hormuz. Those traveling even farther west could go through places like Bukhara and Merv, ending at Palmyra in Syria. Many travelers in our anthology traversed at least part of this route, including Zhang Qian (chapter 2), Faxian and Xuanzang (chapter 5), Du Huan (chapter 6), Rabban Sauma (chapter 7), and Marco Polo (chapter 8).

The northern steppe route, which the Mongols utilized for their invasion of Eastern Europe, ran from Mongolia across the Kipchak and Kazakh steppes to Sarai on the Volga River and then on to the emporium of Soldaia (Sudak) on the Black Sea. It was a vast ocean of grass, interrupted only by the Altai and Ural Mountains. This route was traversed by William of Rubruck (chapter 7).

The Indian Ocean route is sometimes called the Maritime Silk Road. Seamen at opposite ends of Eurasia discovered how to take advantage of the seasonally shifting monsoon winds by the second century BCE to arrive in India, and some sailors from the Roman world traversed the entire ocean from the Red Sea to South China by the second century CE (chapters 3 and 4). The entirety of the Indian Ocean sea lanes are most clearly described by the Arab sailors in chapter 6 (see map 4). Starting from Red Sea ports in Egypt in summer or from the mouth of the Persian Gulf, sailors could arrive on India's western coast by September. Ships stopped in Sri Lanka, then crossed the Bay of Bengal, arriving in western Malaysia about a month later. They then sailed through the Straits of Malacca, stopping in Sumatra or Java, then traveled due north to Champa (central Vietnam), ending at the emporium of Canton in China about six months after departure (see map 2 and map 4). Marco Polo used this route to return to Persia (map 6), and Ma Huan sailed these waters with Zheng He's armadas (map 8).

The pace of travel and communication across Eurasia was dependent on the technology of transport and the rhythms of the natural world. And with advances in technology, the Eurasian world appears to shrink, and the pace of communication quickened. However, movement overland was always hindered by fragmented political conditions. Limited to travel on foot and horseback and waylaid by his enemies for a decade, Zhang Qian expended nearly twelve years to reach Uzbekistan during the second century BCE (chapter 2). It took the 60-year-old Buddhist pilgrim Faxian over five years to walk to Central India from China in 399 CE (chapter 5). During the 13th century, the overweight monk William of Rubruck (chapter 7) required only four months to travel from the Holy Land to the Mongol capital of Karakorum (6,000 km), while Marco Polo and his father and uncle consumed a full four years (1271–1275) to reach China via the central land route a couple decades later (chapter 8). At times, these arduous land journeys were aided by postal stations, caravanserai, or tokens of safe passage granted by rulers, but such periods were exceptional. In contrast, the Chinese female scholar Shan Shili required just three weeks to cross all of Eurasia on the Trans-Siberian Railway in 1903 (chapter 13).

The swiftness of the ocean route between China and the West was dictated by the timing of the monsoon winds in the Indian Ocean. No amount of government facilitation or nautical efficiency could help to accomplish this one-way journey in less than six months before the advent of steamships. Departing from Portuguese Atlantic-facing ports added considerably to the duration of this voyage, since one had to round the Cape of Good Hope in southern Africa first. It took Matteo Ricci nine months to sail this route to India in 1578. Sailing across the Pacific to Asia was equally time consuming, as Lord George Macartney (chapter 11) required ten months to travel from Portsmouth to China in 1792–1793, though when Eliza Jane Bridgman (chapter 12) sailed from the east coast of North America to China on the clipper ship *Horatio* in 1844–1845, it took her only 131 days. Wang Tao's steamship from Hong Kong to Marseilles required only a little more than 40 days in 1867 (chapter 12). Our anthology closes in 1936, on the eve of transpacific flight, a technological advance that would dramatically shrink the world.

THE ACCOUNT OF DISTANT LANDS AS A LITERARY GENRE

The surviving written texts from this cross-cultural literary genre were produced within different social and political contexts, and their transmission would remain rare until the advent of printing and commercial book culture in the West and East. Some early accounts survive in only a single manuscript exemplar. Many of the pieces in this anthology were commissioned by the royal or ecclesiastical authority that had sent out the envoy on their journey. Other men and women were motivated to write out of private interest, either to consolidate their reputation or to enlighten a particular audience of readers.

The long-distance travel genre varies depending on the author's social class and occupation, influencing the aspects of the other society the traveler notices and their interpretation. For example, ancient diplomatic envoys like Megasthenes or Zhang Qian (chapter 2) or more recent ambassadors like Lord Macartney or Guo Songtao (chapter 11) were interested in what we might refer to today as military intelligence. They wanted to know about geography, population, subsistence, technology, and military strength, but they were also attuned to customs and political trends that might make conflicts more or less likely.

Premodern religious pilgrims like the Buddhists Faxian and Xuanzang (chapter 5) or Christian envoys and pilgrims like William of Rubruck and Rabban Sauma (chapter 7) were more attentive to fine distinctions in doctrine and practice and to sites and relics of religious power. More recent missionaries like the Jesuit Matteo Ricci (chapter 10) or the Protestant Eliza Jane Bridgman (chapter 12) still viewed the world in terms of the faithful versus non-believers but showed more sympathy for the social plight of the unconverted.

Merchants, on the other hand, were outfitted with a very different set of eyes. Men like the anonymous Greek Indian Ocean trader (chapter 3), the Alexandrian former merchant Cosmas Indicopleustes (chapter 5), and the Arab merchants who sailed to China (chapter 6) recorded details that would be advantageous to men of their occupation, like sailing itineraries, ports of call, valuable products, exchange rates, and tariffs, but they also noted local customs, laws, and political conditions, for these could impact trade. Some travelers, such as Marco Polo (chapter 8) and Ma Huan (chapter 9), held dual identities as both merchant and official envoy, which made them equally attentive to both commerce and politics.

William Alexander (appendix to chapter 11), the sole artist featured in the anthology, was appropriately observant of scenes of natural beauty, ships, and architecture, but he often seemed more focused on cultural differences in dress, customs, rituals, and punishments. A scholar and fictional author like Wang Tao (chapter 12) was a keen observer of technology, governance, and social customs, but he also looked upon people as character studies.

Chinese journalists and social critics like Liang Qichao and Shan Shili (chapter 13) were driven by a desire to modernize China, but they were incisive in their observations of western Europe and America, distinguishing between those aspects worth emulating and those which exposed the institutionalized inequality underlying Western modernity. The American journalist Edgar Snow's lionizing account of the early Communist Party in China was less critical or self-reflective by comparison (chapter 13).

The travel account genre also evolves historically in the broadest sense, with transformations reflecting different geopolitical relations and economic conditions. In Europe and the Mediterranean, the earliest framework for these accounts is the "Marvels of the East," where the mysterious lands of India, and later China, were seen as home to fantastic creatures and monstrous races. First witnessed in the works of Ctesias (chapter 1), Megasthenes (chapter 2), and Pliny the Elder (chapter 4), this framework persisted throughout the Middle Ages. Medieval Arabic accounts of China and India in this collection (chapter 6) also reported on some strange and wonderful things, but they were more pragmatic and empirical about China, noting commercial duties, eating habits, and debt regulations.

The style of European travel accounts would be fundamentally transformed with the arrival of the Portuguese and Spanish in East Asia during the 16th century. Men like Galeote Pereira (chapter 9) described the governance, law, and customs of China in general and admiring terms but also maintained an eye toward the possibilities of economic gain, religious conversion, and military conquest. Subsequently, Matteo Ricci and other Jesuits lived in China for decades and immersed themselves in the literature and culture of the elite, providing highly detailed and nuanced information about Chinese governance, language, customs, and handicrafts (chapter 10). Some Jesuits sent glowing letters back to Europe, describing a civilized empire run by philosophers, which fostered an idealized view of China.

The accounts written by English travelers and envoys starting in the middle of the 18th century formed a harsh counterreaction to these Jesuit reports. They reflected the growing strength of the English and their imperialist and Orientalist viewpoint. As seen in the journals of Lord George Macartney and the drawings of his draughtsman William Alexander (chapter 11 and appendix), the English were highly critical of Chinese governance and customs, viewing the Qing Empire as backward, corrupt, and effeminate. After the Opium Wars, the travel accounts of English and American missionaries in China were still condescending in some ways (chapter 12), but more in a paternalistic, patronizing fashion. Though they still viewed China as backward, they sympathized with the plight of the poor and that of bound-footed, uneducated women. Edgar Snow's journalistic travel account from 1936 (chapter 13) partially continues this 19th-century missionary style, still focusing on "China's plight." However, he sympathizes with peasants and the urban proletariat, portraying Communist revolutionaries like Mao Zedong as democratic heroes.

At the Chinese end of Eurasia, accounts of foreign lands also commenced with a "marvels" genre, as seen in the mythogeographical text *Guideways Through Mountains and Seas* (chapter 1). But while this illustrated genre remained popular as a minor current for two millennia, it was supplanted in the mainstream by the cataloguing-style report on foreign lands pioneered by Zhang Qian and solidified by subsequent Chinese historians (chapters 2 and 4). These reports catalogue foreign countries along an itinerary starting near China and moving outward, describing their location, geography, climate, customs, religion, dress, and natural products. Even the accounts of the Buddhist pilgrims Faxian and Xuanzang (chapter 5) still adhere to this cataloguing style, though they focus more on religious practice and holy sites. This mainstream tradition continues through the Tang-period writings of Du Huan (chapter 6) and those of Ma Huan (chapter 9) from the 15th century. Even Xie Qinggao's *Maritime Records* (chapter 10) from 1820 largely observed this genre, though he occasionally infused his account with the Qing tradition of the "investigation of things," such as in his exploration of the water delivery systems in London.

The situation would fundamentally change after the shock of the Opium Wars and the Chinese realization of the precarious imbalance in military and industrial technology between the Far East and the West. The diary and letters penned by the first Chinese ambassador to England, Guo Songtao (chapter 11) and Wang Tao's travelogue (chapter 12) reflect this shift, as both men frequently bemoaned Chinese backwardness and encouraged the adoption of Western technologies like railways and the telegraph. This new written tradition of "admiring the West" became more sophisticated and critical at the turn of the century, with the "journalistic" travel account style of both Liang Qichao and Shan Shili (chapter 13), for they found much to admire but also made their readers aware of the darker side of modernity.

When viewed as a collective whole, the travel accounts of distant lands reveal several motifs and themes in common, and the genre can be viewed as a subtype of the mythological "hero's journey." The archetypal traveler makes a fateful decision to undertake a perilous journey to a distant land or is commissioned to depart by some authority. The journey is long and arduous, full of misfortunes and physical or spiritual trials. Our traveler is full of resolve and determination but is not necessarily a Herculean specimen of a man or woman. On their journey, they must cross over some vast liminal space, whether that be the Indian Ocean, the Taklamakan Desert, the Eurasian Steppe, or the lofty Pamir Mountains. After transiting this liminal barrier, they are "transported into another world," which were the exact words used by William of Rubruck when he entered the Mongol domains (chapter 7).

In this new world, the traveler is overcome with a sense of strangeness as they encounter the ultimate other. This foreign culture might seem entirely backward or could feel uncanny and unsettling due to its strange familiarity. This confrontation with the other inevitably leads to comparisons with one's own culture and sometimes prompts an examination or redefinition of the self. Some travelers employ their description of the foreign land to criticize their own culture's shortcomings, while others glorify or idealize the other land as a utopia.

On the other hand, the encounter with the other can also lead to the opposite phenomenon, sharp criticisms and condescending dismissals of the barbaric customs and backward nature of the foreign people and their government. Such attitudes are an outgrowth of the ideologies associated with empire and serve to bolster exclusion, conquest, and colonialism. We can clearly see such attitudes in the journal of Lord Macartney (chapter 11) and in the critiques of Chinese punishments, foot binding, homosexuality, and infanticide authored by Western travelers (chapters 9, 10, and 12). Finally, as in the mythological hero's journey, the traveler frequently undergoes some fundamental transformation because of their odyssey.

NOTE ON THE TRANSLATIONS

I have provided new English translations for all the accounts of Chinese travelers. While several of these texts (e.g., Zhao Rugua, Faxian, Xuanzang, and Ma Huan) have been translated in their entirety before, most of those translations are over a century old and quite outdated and inaccurate. My translations are based on the best modern Chinese annotated editions, with consultation of recent French, German, and Japanese scholarly translations. Selected studies and alternate translations are listed in the chapter bibliographies. Some of the Chinese accounts, such as those of Shan Shili, Xie Qinggao, and Wang Tao, have never been translated before, beyond some passages found in scholarly articles. For all those works originally in Greek, Latin, Syriac, Franco-Italian, Italian, and Arabic, I have either licensed recent translations or reprinted those in the public domain, while updating the spelling and correcting any errors.

NOTE ON ROMANIZATION

For the translated Chinese accounts in this anthology, I have almost always transliterated words using the pinyin system of romanization for standard Mandarin. The only two exceptions are in the *Maritime Records* of Xie Qinggao (chapter 10) and a couple instances in the *Notes on Travels in the New World* of Liang Qichao (chapter 13), where I have used the Yale system of Cantonese romanization, since this spelling better reveals which Portuguese or English words that these Cantonese men were trying to sound out in Chinese. For the accounts of Arabic travelers in Tang China (chapter 6), I have included the standard Arabic transliteration of the words, followed by the pinyin and Chinese graphs I believe they were trying to represent. For Western accounts, like those of Galeote Pereira (chapter 9), Lord Macartney (chapter 11), and Eliza Jane Bridgman (chapter 12), I have retained their own idiosyncratic romanizations of the Chinese place names, people, and objects they encountered, followed by standard pinyin and Chinese characters. In the case of Edgar Snow's

Red Star over China (chapter 13), copyright restrictions prevented me from updating his older Wade-Giles romanization to pinyin (e.g., Mao Tse-tung to Mao Zedong), but I have provided, on first appearance, the standard romanization in brackets.

CHAPTER 1

MONSTERS AT THE EDGE OF THE WORLD (CA. 400–300 BCE)

Since humans first committed their thoughts and feelings to writing, they imagined that the mountains and wildernesses beyond their civilized world were inhabited by monstrous races, baleful spirits, and strange creatures. These distant lands were seen as the domains of strange beings with supernatural powers and human/animal hybrid bodies who engaged in transgressive behaviors that excluded them from incorporation into human society. Travel into this unknown realm generated considerable fear of bodily and spiritual harm.

Cultural historians have argued that monsters are an essential construct of the human mind, a projection of fear of the unknown and a simultaneous revulsion against and an attraction to tabooed and transgressive behavior. The monsters who reside at the edge of the world are a cultural necessity, however, for they serve to define the limits of the civilized realm, reinforce group identity and cohesion, and articulate, through contrast, what it means to be human, for the self is always constructed through reference to the other.

So there must always be monsters out there, for we cannot fully envision ourselves as human without contemplating the boundaries of the inhuman. When these far-off lands are eventually conquered and domesticated to become part of the civilized world, the monsters will always reappear at the new margin of the known world. And when the entire terrestrial globe has become fully known, they are reimagined to dwell in the abyss of the ocean or in the depths of outer space.

This anthology opens with two parallel journeys to the edge of the world from the fourth century BCE, which demonstrates that this psychological and mythical projection of monsters was a practice shared among people from both ends of Eurasia during this time period. In later centuries, after explorers and diplomats have blazed a path through these distant lands, we shall see that the monsters, and people with monstrous behavior, don't really disappear; they are only pushed back to other as yet unknown lands or seas.

SELECTED STUDIES

Wittkower, Rudolf. "Marvels of the East: A Study in the History of Monsters." In *Allegory and the Migration of Symbols*. Thames and Hudson, 1977.

Wright, Alexa. "Monstrous Strangers at the Edge of the World: The Monstrous Races." In *The Monster Theory Reader*, edited by Jeffrey Andrew Weinstock. University of Minnesota Press, 2020.

INTRODUCTION TO CTESIAS OF CNIDUS

In his seminal work of history, Herodotus (ca. 484–ca. 425 BCE) briefly describes the Persian domains near the Indus River (book 3, 98–106) and relates some secondhand knowledge of other Indian tribes and marvels like his infamous gold-digging ants. In the mental topography of the Greeks of Herodotus's time, the lands directly east and south of the Indus were inhabited by creatures of unusual size and by dark-skinned cannibals, but everywhere beyond that was an uninhabitable tract of scorching sand, for this was the end of the world.

The *Indica* by Ctesias of Cnidus (ca. 440–ca. 390 BCE) is the oldest Western account of India to survive, albeit only in fragments. Written a generation after Herodotus, Ctesias's account predates the invasion of the Indus River valley by Alexander the Great and the subsequent report on India by the Seleucid ambassador Megasthenes, excerpted in chapter 2. The *Indica* parallels the anonymous Chinese text *Guideways Through Mountains and Seas* (Shanhai jing 山海經), excerpted in this chapter, in terms of its time of composition and in thematic content, for it is a curious mixture of real locations, peoples, and animals, alongside exaggerated creatures and fantastic monsters who inhabit the fringes of the world.

Ctesias was born in the Greek city of Cnidus (in present-day southwestern Turkey) into a family of doctors, sometime after 450 BCE. He served for as many as 17 years as a physician at the Persian court, attending Artaxerxes II (r. 405–359 BCE) and his mother, Parysatis (d. ca. 395 BCE). He gained fame for healing the wound the king suffered at the Battle of Cunaxa (401 BCE). He was sent back to the Greek world in 398 BCE to relate a communication between Artaxerxes II and the Spartans, and he remained in the Aegean for the remainder of his life, where he wrote several works on the East, including the *Persica*, a work on Persian history closer to a historical novel in character, and the *Indica*, excerpted here.

It is clear that Ctesias never personally traveled to India, for he remained at the side of the Persian king as his physician, traveling around the empire. His information about India came from stories told by Indian and Bactrian merchants or envoys who traveled to the Persian court and from his personal observations of plants, products, and animals brought from India. Some of his descriptions of fantastic animals could have been derived from representations he saw on textiles or other artworks.

When Ctesias talks about "India," he is speaking only about the area of the Indus River, for he was apparently not aware of the Gangetic civilization. During his time, direct knowledge of the East among the Greeks extended only to the core areas of the Persian Empire, where Greeks served as specialists and mercenaries. Their indirect knowledge extended up into Bactria (a Persian satrapy).

Many of the wonders of India that Ctesias describes are fairly accurate, like his description of parakeets that can mimic human speech. His account of certain plants like the bamboo and the camphor tree are also reliable, as is his description of the lac insect used to make shellac. Some of his more fantastic animals are probably exaggerations of real beasts. For example, his unicorn is probably based on the rhinoceros, and his fearsome "man-eater" predator, the *martichoras,* was probably inspired by the Bengal tiger. Others, like the Dog-Headed People (*kalystrioi*) are typical strange races from the edge of the world but may also have been influenced by stories from Indian mythology. His men with eyes in their chests have a parallel in the headless rebel Punished by Heaven (Xingtian 刑天), who was said in the Chinese *Guideways Through Mountains and Seas* to have occupied lands to the far west of China. Both might derive from the same Indian or Tibetan myth.

The *Indica* of Ctesias had a strong influence on the Western imagination of the exotic East for millennia. His accounts of the fantastic animals and monstrous races of India continued to be transmitted in textual and visual form throughout the Roman period and into the Middle Ages, illustrated in bestiaries and on the margins of world maps, and they even colored the later supposedly more reliable first-hand travel accounts of the East by Odoric of Pordenone (d. 1331 CE).

No manuscript of Ctesias's *Indica* survives, and we know the text only through quotations and summaries in other works, above all the lengthy epitome by Photios I (ca. 810–ca. 893 CE), patriarch of Constantinople, in his collection of extracts from ancient works called *Bibliotheca.* The excerpts below were taken from the *Bibliotheca* and from a quote in the *Natural History* of Pliny, as translated by J. W. McCrindle in his *Ancient India as Described by Ktêsias the Knidian* (Trübner, 1882) (7–13, 15–30, 31–

32, 34, 61), with updates to the English spelling and Greek transliteration. The introductory subject headings are my own.

Primary Source, Alternate English Translation, and Selected Studies

McCrindle, J. W. *Ancient India as Described by Ktêsias the Knidian.* Trübner, 1882.

Nichols, Andrew. *Ctesias on India: Introduction, Translation, and Commentary.* Bristol Classical Press, 2011.

Waters, Matt. *Ctesias' Persica and its Near Eastern Context.* University of Wisconsin Press, 2017.

Excerpts from the *Indica*, by Ctesias of Cnidus (early fourth c. BCE)

Geography and Climate of India

Ctesias reports of the river Indus that, where narrowest, it has a breadth of 40 stadia [7.4 km], and where widest of 200 [37 km]; and of the Indians themselves that they almost outnumber all other men taken together.... He states that there are no men who live beyond the Indians and that no rain falls in India but that the country is watered by its river.

Fantastic Animals, Plants, and Natural Features

He notices the *pantarba*, a kind of seal stone, and relates that when seal stones and other costly gems to the number of 477 that belonged to the Bactrian merchant had been flung into the river, this *pantarba* drew them up to itself, all adhering together.

He notices also the elephants that demolish walls; the kind of small apes that have tails four cubits long [1.85 m]; the roosters that are of extraordinary size; the kind of bird called the *bittakos* [probably the plum-headed parakeet] and which he thus describes: it has a tongue and

voice like the human, is of the size of a hawk, has a red bill, is adorned
with a beard of a black color, while the neck is red like cinnabar, it talks
like a man in Indian, but if taught Greek can talk in Greek also.

He notices the fountain that is filled every year with liquid gold, out
of which are annually drawn a hundred earthen pitchers filled with the
metal. The pitchers must be earthen since the gold when drawn becomes
solid, and to get it out the containing vessel must needs be broken in
pieces. The fountain is of a square shape, eleven cubits in circumference
[5.08 m], and a fathom in depth. Each pitcherful of gold weighs a talent
[approx. 26 kg]. He notices also the iron found at the bottom of this
fountain, adding that he had in his own possession two swords made
from this iron, one given to him by the king of Persia [Artaxerxes II],
and the other by Parysatis, the mother of that same king. This iron, he
says, if fixed in the earth, averts clouds and hail and thunderstorms, and
he avers that he had himself twice seen the iron do this, the king on both
occasions performing the experiment.

We learn further that the dogs of India are of very great size, so that
they fight even with the lion; that there are certain high mountains
having mines which yield the sardonyx, the onyx, and other seal stones;
that the heat is excessive, and that the sun appears in India to be ten
times larger than in other countries; and that many of the inhabitants
are suffocated to death by the heat. Of the sea in India, he says, that it
is not less than the sea in Greece [Aegean]; its surface however for four
finger breadths downward is hot, so that fish cannot live that go near the
heated surface, but must confine themselves always to the depths below.

He states that the river Indus flows through the level country, and
through between the mountains, and that what is called the Indian reed
[bamboo] grows along its course, this being so thick that two men could
scarcely encompass its stem with their arms, and of a height to equal the
mast of a merchant ship of the heaviest burden. Some are of a greater
size even than this, though some are of less, as might be expected when
the mountain it grows on is of vast range. The reeds are distinguished

by sex, some being male, others female. The male reed has no pith, and is exceedingly strong, but the female has a pith.

He describes an animal called the *martichoras* found in India. Its face is like a man's—it is about as big as a lion, and in color red like cinnabar. It has three rows of teeth—ears like the human—eyes of a pale-blue like the human, and a tail like that of the land scorpion, armed with a stinger more than a cubit long [46 cm]. It has besides stingers on each side of its tail, and, like the scorpion, is armed with an additional stinger on the crown of its head, wherewith it stings anyone who goes near it, the wound in all cases proving mortal. If attacked from a distance, it defends itself both in front and in rear—in front with its tail, by uplifting it and darting out the stingers, like shafts shot from a bow, and in rear by straightening it out. It can strike to the distance of a hundred feet, and no creature can survive the wound it inflicts save only the elephant. The stingers are about a foot in length, and not thicker than the finest thread. The name *martichora* means in Greek "man-eater," and it is so called because it carries off men and devours them, though it no doubt preys upon other animals as well. In fighting it uses not only its stingers but also its claws. Fresh stingers grow up to replace those shot away in fighting. These animals are numerous in India and are killed by the natives who hunt them with elephants, from the backs of which they attack them with darts.

Indian Worship and Climate
He describes the Indians as extremely just and gives an account of their manners and customs. He mentions the sacred spot in the midst of an uninhabited region which they venerate in the name of the Sun and the Moon. It takes one a 15 days' journey to reach this place from Mount Sardo. Here for the space of 5 and 30 days the Sun every year cools down to allow his worshipers to celebrate his rites and return home unscorched by his burning rays. He observes that in India there is neither thunder nor lightning nor rain, but that storms of wind and violent hurricanes which sweep everything before them, are of frequent occurrence. The

morning sun produces coolness for one half of the day, but an excessive heat during the other half, and this holds good for most parts of India.

It is not, however, by exposure to the sun that the people are swarthy, but by nature, for among the Indians there are both men and women who are as fair as any in the world, though such are no doubt in a minority. He adds that he had himself seen two Indian women and five men of such a fair complexion.

Pygmies

He writes that in the middle of India are found the swarthy men called Pygmies, who speak the same language as the other Indians. They are very diminutive, the tallest of them being but two cubits in height [92 cm], while the majority are only one and a half [69 cm]. They let their hair grow very long—down to their knees, and even lower. They have the largest beards anywhere to be seen, and when these have grown sufficiently long and copious, they no longer wear clothing, but, instead, let the hair of the head fall down their backs far below the knee, while in front are their beards trailing down to their very feet. When their hair has thus thickly enveloped their whole body, they bind it round them with a girdle, and so make it serve for a garment. Their sexual organs are thick, and so large that they reach to their ankles. They are moreover snub-nosed, and otherwise ill-favored. . . . Of the Pygmies, 3,000 men attend the king of the Indians, on account of their great skill in archery. They are eminently just and have the same laws as the Indians. They hunt hares and foxes not with dogs but with ravens and kites and crows and vultures. In their country is a lake 800 stadia [148 km] in circumference, which produces an oil like our own. If the wind be not blowing, this oil floats upon the surface, and the Pygmies going upon the lake in little boats collect it from amidst the waters in small tubs for household use. They also use oil of sesame and nut oil, but the lake oil is the best. The lake has also fish.

Gold, Griffins, and Fat-Tailed Sheep

There is much silver in their part of the country, and the silver mines though not deep are deeper than those in Bactria. Gold also is a product of India. It is not found in rivers and washed from the sands like the gold of the river Pactolus [near ancient Sardis in Turkey], but is found on those many high-towering mountains which are inhabited by the Griffins, a race of four-footed birds, about as large as wolves, having legs and claws like those of the lion, and covered all over the body with black feathers except only on the breast where they are red. On account of those birds the gold with which the mountains abound is difficult to be got.

The sheep and the goats of the Indians are bigger than asses and generally produce young by four and by six at a time. The tails grow to such a size that those of the females must be cut off before the rams can get at them. India does not however produce the pig, either the tame sort or the wild. Palm trees and their dates are in India three times the size of those in Babylon, and we learn that there is a certain river flowing with honey out of a rock, like the one we have in our own country.

Truth-Telling Potion

The justice of the Indians, their devotion to their king, and their contempt of death are themes on which Ctesias loves to expatiate. He notices a fountain having this peculiarity, that when any one draws water from it, the water coagulates like cheese, and should you then detach from the solid lump a piece weighing about three obols [2.16 g], and having pulverized this, put the powder into common water, he to whom you give this potion blabs out whatever he has done, for he becomes delirious, and raves like a madman all that day. The king avails himself of this property when he wishes to discover the guilt or innocence of accused persons. Whoever incriminates himself when undergoing the ordeal is sentenced to starve himself to death, while he who does not confess to any crime is acquitted.

The Indians are not afflicted with headache, or toothache, or eye inflammation, nor have they mouth sores or ulcers in any part of their body. The age to which they live is 120, 130, and 150 years, though the very old live to 200.

Poisonous Snakes

In their country is a serpent a span long [23 cm], in appearance like the most beautiful purple with a head perfectly white but without any teeth. The creature is caught on those very hot mountains whose mines yield the sardonyx stone. It does not sting, but on whatever part of the body it casts its vomit, that place invariably putrefies. If suspended by the tail, it emits two kinds of poison, one like amber which oozes from it while living, and the other black, which oozes from its carcass. Should about a sesame seed's bulk of the former be administered to anyone, he dies the instant he swallows it, for his brain runs out through his nostrils. If the black sort be given, it induces consumption but operates so slowly that death scarcely ensues in less than a year's time.

Fantastic Plants and Animals

He mentions an Indian bird called the *dikairon,* a name equivalent in Greek to "just." It is about the size of a partridge's egg. It buries its dung under the earth to prevent its being found. Should it be uncovered nevertheless, and should a person at morning tide swallow so much of it as would about equal a grain of sesame, he falls into a deep unconscious sleep from which he never awakes, but dies at the going down of the sun.

In the same country grows what is called the *parebon* [possibly the sacred fig], as a plant about the size of the olive, found only in the royal gardens, producing neither flower nor fruit, but having merely fifteen roots, which grow down into the earth, and are of considerable thickness, the very slenderest being about as thick as one's arm. If a span's length [23 cm] of this root be taken, it attracts to itself all objects brought near it —gold, silver, copper, stones and all things else except amber. If, however, a cubit's length [46 cm] of it be taken, it attracts lambs and birds, and it is in fact with this root that most kinds of birds are caught. Should you wish to turn water solid, even a whole *chous* [3.27 l] of it, you have but to throw into the water not more than an obol's weight [0.72 g] of this root, and the thing is done. Its effect is the same upon wine which, when condensed by it, can be held in your hand like a piece of wax, though it melts the next day. It is found beneficial in the cure of bowel disorders.

Through India there flows a certain river, not of any great size, but only about two stadia in breadth [370 m], called in the Indian tongue, Hyparchos, which means in Greek, "the bearer of all things good." This river for thirty days in every year floats down amber, for in the upper part of its course where it flows among the mountains, there are said to be trees overhanging its current which for 30 days at a particular season in every year continue dropping tears like the almond tree and the pine tree and other trees. These tears on dropping into the water harden into gum. The Indian name for the tree is *siptachora*, which means when rendered into Greek, "sweet." These trees then supply the Indians with their amber. And not only so, but they are said to yield also berries, which grow in clusters like the grapes of the vine and have stones as large as the hazelnuts of the Pontus [southern Black Sea coast].

* * * * * * * * * * * * * * * **BRIEF EXCERPT** * * * * * * * * * * * * * * * *

The Dog-Headed People

Ctesias writes that on the mountains just spoken of there live men having heads like those of dogs, who wear the skins of wild beasts and do not use articulate speech but bark like dogs and thus converse so as to be understood by each other. They have larger teeth than dogs and claws like those of dogs, only larger and more rounded. They inhabit the mountains and extend as far as the Indus River. They are swarthy, and like all the other Indians, they are extremely just men. With the Indians they can hold intercourse, for they understand what they say, though they cannot, it is true, reply to them in words, still by barking and by making signs with their hands and their fingers like the deaf and the dumb, they can make themselves understood. They are called by the Indians *kalystrioi*, which means in Greek "dog-headed" (*cynocephali*). Their food is raw flesh. The whole tribe numbers not less than 120,000 men [figure 1].

Figure 1. "The Dog-Headed People of India."

Source: Marco Polo, *Livre des merveilles.* Bibliothèque nationale de France, gallica.bnf.fr, Département des Manuscrits, Français 2810, 76 verso.

Near the sources of this river there grows a certain purple flower [possibly the fire-flame bush], which is used for dying purple, and is not inferior to the Greek sort, but even imparts a far more florid hue. In the same parts there is a wild insect about the size of a beetle, red like cinnabar, with legs excessively long [probably the lac insect]. It is as soft as the worm called *scolex* and is found on the trees which produce amber, eating the fruits of those trees and destroying them, as in Greece the woodlouse ravages the vine-trees. The Indians grind these insects to a powder and therewith dye such robes, tunics, and other vestments as they want to be of a purple hue. Their dyestuffs are superior to those used by the Persians.

The Dog-Headed People living on the mountains do not practice any of the arts but subsist by the produce of the chase. They slaughter the prey and roast the flesh in the sun. They rear however great numbers of sheep and goats and asses. They drink the milk of the sheep and the whey which is made therefrom. They eat moreover the fruit of the *siptachora*—the tree which produces amber, for it is sweet. They also dry this fruit, and pack it in hampers as the Greeks do raisins. The same people construct rafts, and freight them with the hampers as well as with the flowers of the purple plant, after cleansing it, and with 260 talents weight of amber [6,723 kg], and a like weight of the pigment which dyes purple, and 1,000 talents [25,860 kg] more of amber. All this cargo, which is the season's produce, they convey annually as tribute to the king of the Indians. They take also additional quantities of the same commodities for sale to the Indians, from whom they receive, in exchange loaves of bread and flour and cloth which is made from a tree-grown stuff [cotton]. They sell it also for swords such as they use in hunting wild beasts, and bows and javelins, for they are deadly marksmen both in shooting with the bow and in hurling the javelin. As they inhabit steep and pathless mountains, they cannot possibly be conquered in war, and the king moreover once every five years sends them as presents 300,000 arrows and as many javelins, 120,000 shields and 50,000 swords.

These Dog-Headed People have no houses but live in caves. They hunt wild beasts with the bow and the spear and run so fast that they can overtake them in the chase. Their women bathe once a month at the time of menstruation, and then only. The men do not bathe at all but merely wash their hands. Three times a month, however, they anoint themselves with an oil made from milk and wipe themselves with skins. Skins denuded of the hair, and made thin and soft, constitute the dress both of the men and their wives. Their richest men however use cotton raiment, but the number of such men is small. They have no bed but sleep on a litter of straw or leaves. That man is considered the richest who possesses the most sheep, and all their wealth consists of property

of this sort. Both men and women have, like dogs, tails above their
buttocks but larger and more hairy. They copulate like quadrupeds
in doggy-style, and to copulate otherwise is thought shameful.
They are just, and of all men are the longest-lived, attaining the
age of 170, and some even of 200 years.

* *

The Men Without Anuses

Beyond these again are other men who inhabit the country above the
sources of the river, who are swarthy like the other Indians, do no work,
and neither eat grain nor drink water, but rear a good many cows and
goats and sheep, and drink their milk as their sole sustenance. Children
are born among them with the anus closed up, and the contents of the
bowels are therefore voided, it is said, as urine, this being something
like curds, though not at all thick, but very foul like feces. When they
drink milk in the morning and take another draught at noon, and then
immediately after eat a certain sweet-tasted root of indigenous growth
which is said to prevent milk from coagulating in the stomach, this
root towards evening acts as an emetic, and they vomit up everything
quite readily.

The Unicorn

Among the Indians, Ctesias proceeds, there are wild asses as large as
horses, some being even larger. Their head is of a dark red color, their
eyes blue, and the rest of their body white. The Unicorn has a horn on
its forehead, a cubit in length [46 cm]. The filings of this horn, if given
in a potion, are an antidote to poisonous drugs. This horn for about two
palm breadths upwards from the base is of the purest white, where it
tapers to a sharp point of a flaming crimson, and, in the middle, is black.
These horns are made into drinking cups, and such as drink from them
are attacked neither by convulsions nor by the sacred disease [epilepsy].
Indeed, they are not even affected by poisons, if either before or after

swallowing them they drink from these cups wine, water, or anything else. While other asses, moreover, whether wild or tame, and indeed all other solid-hoofed animals, have neither anklebones nor gall in the liver, these Unicorns have both. Their anklebone is the most beautiful of all I have ever seen and is, in appearance and size, like that of the ox. It is as heavy as lead, and of the color of cinnabar both on the surface, and all throughout. It is exceedingly fleet and strong, and no creature that pursues it, not even the horse, can overtake it.

On first starting, it scampers off somewhat leisurely, but the longer it runs, it gallops faster and, faster till the pace becomes most furious. These animals therefore can only be caught at one particular time—that is when they lead out their little foals to the pastures in which they roam. They are then hemmed in on all sides by a vast number of hunters mounted on horseback, and being unwilling to escape while leaving their young to perish, stand their ground and fight, and by butting with their horns and kicking and biting kill many horses and men. But they are in the end taken, pierced to death with arrows and spears, for to take them alive is in no way possible. Their flesh being bitter is unfit for food, and they are hunted merely for the sake of their horns and their anklebones.

The Crocodile

He states that there is bred in the Indus River a worm like in appearance to that which is found in the fig, but seven cubits more or less in length [approx. 3.22 m], while its thickness is such that a boy ten years old could hardly clasp it within the circuit of his arms. These worms have two teeth—an upper and a lower, with which they seize and devour their prey. In the daytime they remain in the mud at the bottom of the river, but at night they come ashore, and should they fall in with any prey as a cow or a camel, they seize it with their teeth, and having dragged it to the river, there devour it. For catching this worm a large hook is employed, to which a kid or a lamb is fastened by chains of iron. The worm being landed, the captors hang up its carcass, and placing vessels underneath it, leave it thus for thirty days. All this time oil drops from it, as much being got as would fill ten Attic *kotylae* [approx. 2.7 l]. At the end of the thirty days, they throw away the worm, and preserving the oil, they take it to the king of the Indians, and to him alone, for no subject is allowed to get a drop of it. This oil, like fire, sets everything ablaze over which it is poured, and it consumes not alone wood but even animals. The flames can be quenched only by throwing over them a great quantity of clay, and that of a thick consistency.

Camphor Trees

Moreover, there are certain trees in India as tall as the cedar or the cypress, having leaves like those of the date palm, only somewhat broader, but having no shoots sprouting from the stems. They produce a flower like the male laurel, but no fruit. In the Indian language they are called *karpion* [Sanskrit: *karpūras*, "camphor"], but in Greek, "scented-rose" [*muroroda*]. These trees are scarce. There oozes from them an oil in drops, which are wiped off from the stem with wool, from which they are afterward wrung out and received into alabaster boxes of stone. The oil is in color of a faint red, and of a somewhat thick consistency. Its smell is the sweetest in all the world and is said to diffuse itself to a distance of five stadia [925 m] around. The privilege of possessing this perfume belongs only to the king and the members of the royal family. A present of it was sent by the king of the Indians to the king of the Persians, and Ctesias alleges that he saw it himself, and that it was of such an exquisite fragrance as he could not describe, and he knew nothing whereunto he could liken it.

The Hoary Race

On those Indian mountains where the Indian reed [bamboo] grows, there is a race of men whose number is not less than 30,000 and whose wives bear offspring only once in their whole lifetime. Their children have teeth of perfect whiteness, both the upper set and the under, and the hair both of their head and of their eyebrows is from their very infancy quite hoary, and this whether they be boys or girls. Indeed, every man among them till he reaches his 30th year has all the hair on his body white, but from that time forward it begins to turn black, and by the time they are sixty, there is not a hair to be seen upon them but what is black. These people, both men and women alike, have eight fingers on each hand, and eight toes on each foot. They are a very warlike people, and 5,000 of them armed with bows and spears follow the banners of the king of the Indians. Their ears, he says, are so large that they cover their arms as far as the elbows while at the same time they cover all the back, and the one ear touches the other.

More Monstrous Races

Ctesias mentions also a race of men called Monocoli, who, though they had but a single leg, could hop upon it with wonderful agility, and that they were also called Sciapodes, because that when they lay on their back in very hot weather, they shaded themselves from the sun with their feet. They lived not very far from the Troglodytes [cave-dwellers]. To the west of these, he adds, lived men without a neck, and who had their eyes placed in their shoulders.

. . .

Ctesias, thus writing and romancing, professes that his narrative is all perfect truth, and, to assure us of this, insists that he has recorded nothing but what he either saw with his own eyes, or learned from the testimony of credible eyewitnesses. He adds moreover that he has left unnoticed many things far more marvelous than any he has related, lest anyone who had not a previous knowledge of the facts might look upon him as a complete teller of tales.

Introduction to the *Guideways Through Mountains and Seas*

The *Guideways Through Mountains and Seas* is a schematic mythogeography of the lands within the inhabited Chinese world and the vast wildernesses beyond, extending to the four cardinal directions. Since the earliest recensions, the text was probably accompanied by illustrations. It is a composite text, incorporating layers with different dates of composition. The oldest stratum of the text, which probably dates to the fourth–third centuries BCE, describes a finite, inhabited world that is rectangular and girdled by holy mountains that supported the rounded vault of the sky. The younger layers in the text (third–first centuries BCE) extend this world by envisioning that inhabited rectangle to be surrounded by four encircling seas, and beyond those seas to lie four vast wildernesses extended to four directional limits. This evolution probably reflects the influence of the geographical speculations of Zou Yan (ca. 350–ca 270 BCE) and some indirect knowledge of far-off lands. Chinese knowledge of the far West at this time was probably limited to peripheral awareness of the Bactria area in northern Afghanistan. Awareness of India, Iran, and Mesopotamia would await the reports of Zhang Qian around 125 BCE (chapter 2).

The Northern, Eastern, and Southern Seas of this world had concrete meaning, while the "Western Sea" in the text seems to refer to Lake Qinghai as well as to the "flowing sands" of the deserts in Gansu. In this anthology, the general term "Western Sea," will be employed by later Chinese travelers to refer to the Caspian Sea, the Black Sea, the Indian Ocean, or even the Mediterranean.

The first five chapters of the *Guideways* record a series of itineraries through over four hundred sacred mountains, beginning in the center and radiating out towards the four directions, noting minerals, medicinal plants, creatures, and deities that could be encountered at each. Human cities and political divisions are never mentioned. Much like the *Christian Topography* of Cosmas Indicopleustes (chapter 5), the *Guideways* is a spiritual topography of the world, which subordinates the actual landscape to an overarching schema of how the world should be organized. There are a few recognizable rivers and mountains along its itineraries, but the vast majority could never be mapped onto a real geography of China. Several real animals are mentioned in the *Guideways*, as well as some exaggerated creatures inspired by real animals, but most of the strange hybrid and monstrous creatures belong to the realm of the mythic imagination. Some scholars have argued that these five books constitute a repository of knowledge for shamans, who were the traditional healers and spirit intermediaries in ancient China. The later chapters of the book vary in structure and include some brief mythic narratives of gods, rebels, and famous shamans. These outlying lands are much more hazardous than the central domains and include mention of dangerous animals, powerful demons, and dozens of foreign tribes with strange bodies and transgressive practices. A focus throughout the text is on the power of knowing the names of deities and monsters, for to know a demon's name gives one power to placate or mollify its potential harmful aspects.

One powerful being called the Spirit Mother of the West (Xiwangmu 西王母) is mentioned three times in the *Guideways*. She is said to reside in a cave on the slopes of a holy mountain in the Far West, an axis mundi that communicates with heaven. The two most extensive mentions of her are excerpted below. She assumes a hybrid, human/animal form, with the ferocious traits of predator felines. Her attributes are similar to those of plague demonesses in pre-Buddhist Tibetan mythology, for she is said to control natural disasters and destructive astrological forces. Recent scholarship by Knauer even suggests that her cult and iconography were inspired by the Anatolian mother goddess, Cybele.

Over the course of the Warring States and Han periods, the demonic Spirit Mother of the West seen in the *Guideways* gradually became domesticated. She first evolved into an alluring sovereign who entices Chinese monarchs, then turned into a beneficent goddess of personal salvation who dispenses the elixir of immortality to worthy supplicants, holding court in the West alongside her fantastic animal helpers. She is even given a complimentary male consort to tame her archaic wild aspects. And while her more demonic aspects became gradually subdued, her domain also appears to have moved farther away from central China. In the oldest layer of the text, she resides within the inhabited world of

China that is encircled by mountains. In the next oldest layer, she resides in a cave on Mt. Kunlun, which is located in the wilderness "beyond the seas" to the West, envisioned to be in the nearest part of Chinese Central Asia. When the ambassador Zhang Qian journeys great distances through the previously unknown lands to the west, he does not encounter the heralded domain of the Spirit Mother. His report to the emperor now relocates her realm to the west of Tiaozhi (Mesopotamia), as told to him by elders in the Parthian Empire (see chapter 2).

In early Chinese texts like the *Guideways*, the Country of Women (Nüzi Guo 女子國) represents a projection of elite male anxiety about gender inversion and transgressive female power, parallel to the tales of the Amazons in Greek myth. The women there are said to reside communally (or as lesbian couples) and to become pregnant by immersing themselves in water. In the passage translated here, this "monstrous" land is located roughly within the gateway to Chinese Central Asia, in present-day Qinghai. Once that territory had become conquered and domesticated by Han Chinese colonists, the Country of Women vanishes, only to reappear in later texts at a much further remove. For example, the Buddhist pilgrim Xuanzang (chapter 5) places it west of Fulin (Constantinople), on an island in the Mediterranean or Black Sea, as does the Chinese sojourner during the Abbasid Caliphate, Du Huan (chapter 6), suggesting a conflation by confirmation with the Greek Amazon myth. In several Ming novels, including *Journey to the West*, a Country of Women continued to be a feature of the West. In the 19th century, Chinese travelers like the scholar Wang Tao (chapter 12) and Guo Songtao's vice-ambassador Liu Xihong (chapter 11) allowed this tradition of a Country of Women in the far West to color their views of England, where they misunderstood chivalry and deference to women for true gender inversion. In England, of course, Queen Victoria did reign over a great empire. Liang Qichao (chapter 13), in his account of his trip to America, was more incisive, recognizing that while American women were paid courtesy by men on streetcars and such, they still did not have the right to vote.

In a similar fashion, the land of the Amazons of Greek myth was first placed in Asia Minor, then pushed farther away, near the shores of the Black Sea, before it disappeared from view, only to reappear in the minds of explorers during the Age of Exploration as a land of female warriors in South America. Once the terrestrial world was fully explored, a threatening kingdom of women reappeared on the planet Venus in the campy 1958 sci-fi flick *Queen of Outer Space.*

Primary Source, Alternate European Translations, and Selected Studies

Birrell, Anne., trans. *The Classic of Mountains and Seas.* Penguin, 1999.

Dorofeeva-Lichtmann, Vera. "Mapping a 'Spiritual' Landscape: Representation of Terrestrial Space in the *Shanhaijing.*" In *Political Frontiers, Ethnic Boundaries, and Human Geographies in Chinese History,* edited by Nicola di Cosmo and Don J. Wyatt. Routledge, 2003.

Goldin, Paul R. "On the Meaning of the Name *Xi Wangmu,* Spirit-Mother of the West." *Journal of the American Oriental Society* 122, no. 1 (2002): 83–85.

Knauer, Elfriede R. "A Study of the Influence of Western Prototypes on the Iconography of the Taoist Deity." In *Contact and Exchange in the Ancient World,* edited by Victor H. Mair. University of Hawai'i Press, 2006.

Mathieu, Rémi. *Étude sur la mythologie et l'ethnologie de la Chine ancienne.* 2 vols. Collège de France, Institut des hautes études chinoises, 1983.

Strassberg, Richard E. *A Chinese Bestiary: Strange Creatures from the Guideways Through Mountains and Seas.* University of California Press, 2002.

Yuan Ke 袁珂. *Shanhaijing jiaozhu* 山海經校注. Shanghai guji chubanshe, 1980.

EXCERPTS FROM THE *GUIDEWAYS THROUGH MOUNTAINS AND SEAS*, CA. FOURTH–FIRST CENTURIES BCE

* * * * * * * * * * * * * * * BRIEF EXCERPT * * * * * * * * * * * * * * * *

Guideways Through the Western Mountains

Three hundred and fifty Chinese miles farther west, there is a peak called Mount Jade. This is where the Spirit Mother of the West [Xiwangmu 西王母] resides. The Spirit Mother of the West has the general appearance of a human but with a leopard's tail and the fangs of a tigress, and she is quite good at howling [like a ferocious feline]. In her disheveled hair, she wears the *sheng* ["warp beam"] headdress-crown. She controls the natural calamities emanating from Heaven and the ill-omened star of the Five-Destructions.

There is a creature here that has the form of a dog but the spots of a leopard. It has horns like those of a bull, and it is named the Vigorous Whelp. Its call is like the bark of a dog, and when it is sighted, that means the country will have a bountiful harvest. A type of bird is found here that has a form like that of a long-tailed pheasant but with flame-red plumage. It is called meet-with-destruction [*xingyu* 勝遇; a pun on *xingyu* 腥魚 "raw fish"]. It eats fish and its cry is like that of a deer. When it appears, a country shall endure great flooding.[1]

Guideway Through the Western Regions Beyond the Seas

The region beyond the sea, from the southwestern corner to the northwestern corner.

The *miemeng* bird [maybe the fairy-bluebird] resides north of the Bound-Breast Country [possibly near Myanmar or the Malaysian peninsula]. The bird has blue-green plumage with crimson tail feathers.

Mount Dayun is 300 fathoms tall [approx. 480 m] and is located north of the habitat of the *miemeng* bird.

The wilderness of Dayue; the King of the Xia Dynasty named Qi pantomimed to the "Dance of Nine Successions" here. Arriving in a carriage pulled by two dragons, with a three-layered cloud canopy, he held a feathered fan in his left hand and clasped a jade disc in his right. Suspended from his belt was a semicirclet of jade. This was located north of Mount Dayun, but some say it was in the wilderness of Dayi.

The Country of the Three-Bodied People is situated north of the territory of King Qi of the Xia [possibly upper Mekong on Tibetan Plateau]. The people have one head but three torsos.

Single-Forearm Country lies to its north. The people have one forearm, one eye, and only one nostril. There is a brown horse here with the stripes of a tiger. It has one eye and only one foreleg.

Single-Upperarm Country is located to its north. The people have one arm and three eyes, some with whites and some fully pigmented. They ride on dappled horses. There are birds there with two heads, with crimson and yellow plumage. The birds remain by the sides of these people.

The rebel named Punished by Heaven came here to contend with the Yellow Emperor over the godhead. The Yellow Emperor cut off his head and buried it under Mount Changyang. So Punished by Heaven used his nipples to serve as eyes and his navel as a mouth, and performed a martial dance while brandishing shield and battle axe.

Shamaness Sacrifice and Shamaness Blood Stained are located north of here, residing between two rivers. Shamaness Blood Stained grasps a rhinoceros horn rhyton, while Shamaness Sacrifice holds a sacrificial altar table.

* *

The *ci* owl and the *zhan* falcon reside north of the domain of Shamaness Sacrifice. Their plumage is blue-green and yellow, and whichever kingdom they fly over shall be ruined. The *ci* owl has a face like a human and lives atop mountains. Some say that this is the region in which the *wei* birds and the blue-green and yellow orioles gather in flocks.

The Country of Grown Men lies north of the *wei* bird's territory. The people there wear robes and caps and carry swords.

The Corpse of Shamaness Ugly: While she was still alive, she was roasted to death [to ritually end a drought], under the ten suns that came out together in the sky [nine of which were later shot down by Archer Yi]. The corpse is located to the north of the Country of Grown Men. Though she employed her right sleave to screen her face, the ten suns still shined on above. Shamaness Ugly resided at the top of a mountain.

The land of Shaman Xian is located to the north of the Corpse of Shamaness Ugly. In his right hand he grasps a blue-green snake, and in his left hand he holds a scarlet snake. At the site of Mount Dengbao, groups of shamans who followed him ascend and descend the mountain, [communing with heavenly spirits and searching for herbs].

The Bingfeng creature dwells to the east of the land of Shaman Xian. Its outer appearance is like a swine, but it has a head at the forepart and another on the rear. They are black.

The Country of Women is located north of the land of Shaman Xian. The women live together in pairs, and the place is surrounded by water [in which they immerse themselves to conceive children]. Some say that they all reside communally in a single household.

The Kingdom of the Yellow Emperor is located here at the foothills of Mount Qiong [north of present-day Lake Qinghai]. Even those without longevity still live 800 years. It is located to the north of the Country of Women. The people have human faces but the bodies of serpents. Their tails coil above their heads.

Mount Qiong lies to the north of here. The people don't dare shoot arrows toward the west, for they fear disturbing the burial mound of the Yellow Emperor. The mound is square in shape, and four serpents are coiled around it for protection.

In the wilderness of Zhuyao, the Simurgh bird sings freely, and the Phoenix bird dances unreservedly. The common people eat the eggs of the Phoenix, drink sweet dew from heaven, and are able to pursue anything that they desire. The hundred species of beasts live in harmony there in herds, to the north of the four serpents [coiled around the mound of the Yellow Emperor]. The men grasp eggs with both hands when they eat them. The two mythical birds precede the people and guide them.

Dragonfish live there on hills to the north and are shaped like carp. Some say it is a type of crustacean. This is what the gods and sages rode when they traveled through the Nine Wildernesses. Some say that soft-shelled turtles live to the north of the wilderness of Zhuyao. They are actually fish and resemble carp.

The country of White Folk [possibly inspired by the Tocharian-speaking peoples in the Tarim Basin] lies to the north of the Dragonfish territory. Their bodies are white, and they wear their auburn hair draped over their shoulders. There is a creature here in the country of White Folk called a "yellow-mount" [possibly inspired by the camel]. It has the overall form of a fox, but along its spine there are fleshy horns [humps]. Those who ride it are granted a longevity of 2,000 years.

The Country of Circumspection is located to the north of the White Folk. There is a tree here named *luotang*. In ancient times, when a sage would ascend the throne, people could come here to take the bark from this tree to fashion clothing.

The country of the Long-Thigh People is located to the north of the territory of the *luotang* tree. They wear their hair draped down. Some people called them the Long-Calf People.

The god of the western region, Rushou [who overlooks where the sun sets], has a snake coming out his left ear and rides a pair of dragons.[2]

Guideways Through the Great Wilderness of the West

In the midst of the vast wilderness of the West there is Mount Long [possibly near present-day Ledu, Qinghai], which is where the sun and the moon set. There are marshy rivers there, which are called the "three quagmires." This is where the Kunwu people get their food.

There is a creature wearing blue-green robes who employs her sleeve to shield her face. She is called the Corpse of Shamaness Ugly.

There is the Country of Women.

. . .

South of the Western Sea [probably referring to Lake Qinghai], adjacent to the sea of Flowing Sands [referring to the Tenggeli Desert in Ningxia and Badan Jilin Desert in Gansu], beyond the Scarlet River and before the Black River [in Qilian, Qinghai], there is a great mountain named the Mound of Kunlun. There is a deity here that has a human face but a tiger's body. It has stripes and a tail, both flecked with white. This is where it lives.

At the base of this Mount Kunlun we find the Abyss of Weak Water, which surrounds it. Beyond, there is a Mountain of Flames [a volcano]. If you throw objects at it, they instantly catch fire. There is a being there who wears a *sheng* ["warp beam"] headdress-crown and has the fangs of a tiger and the tail of a leopard, and lives in a cave. She is called the Spirit Mother of the West. This mountain harbors all the myriad creatures of nature.[3]

Notes

1. Yuan Ke 袁珂, *Shanhaijing jiaozhu* 山海經校注 (Shanghai guji chuban-she, 1980), 50. Translation by the author.
2. Yuan Ke, *Shanhaijing jiaozhu*, 207–228. Translation by the author.
3. Yuan Ke, *Shanhaijing jiaozhu*, 400, 407. Translation by the author.

CHAPTER 2

IMPERIAL AMBASSADORS
(300–115 BCE)

When Alexander the Great (356–323 BCE) crossed into Asia and campaigned across the breadth of the once mighty Persian Empire, from Asia Minor to Afghanistan and the northern Indian subcontinent, he briefly welded together the world's most expansive empire and fostered a cross-fertilization of people, technologies, and ideas between East and West. When King Ying Zheng of Qin (259–210 BCE) conquered the last of his rival kingdoms in 221 BCE, he fashioned the first multiethnic territorial empire in East Asia.

Upon Alexander's sudden death in Babylon (323 BCE), his unwieldy empire was carved up by family members and his former generals into rival Hellenistic kingdoms, meaning that West Asia and Central Asia reverted to the kind of international system of peer kingdoms, such as had flourished before the founding of the Persian Empire. These included the Ptolemaic Kingdom in Egypt (305–30 BCE), the Pergamene Kingdom (282–129 BCE) in Anatolia, and the Seleucid Kingdom (312–63 BCE) in West Asia and Central Asia, along with its offshoot, the Greco-Bactrian Kingdom (256– ca. 120 BCE). When Seleucus I Nicator (r. 305–281 BCE) invaded India in 305 or 304 BCE, duplicating Alexander's audacious foray, he ran into the recently founded empire of Chandragupta Maurya (r. ca. 321–ca. 297 BCE), with its formidable elephant-equipped armies, necessitating a negotiated treaty (Treaty of the Indus) and a delicate withdrawal.

After the Qin empire's (221–207 BCE) similarly rapid demise, it was replaced by the longer-lived Han dynasty (206 BCE–220 CE), which operated in a bipolar geopolitical situation with its equally powerful rival to the north, the Xiongnu Confederation. When the Han emperors sent envoys to the west in search of allies in their long-running standoff with the Xiongnu (135 BCE), they unexpectedly stumbled into their own version of an international order, among states of which they had been unaware.

For both the Seleucid monarchs and the Han emperors, diplomatic engagements with distant realms like the Mauryan Empire in India and the Yuezhi Kingdom in Bactria required a reorientation of their archaic worldviews. Lands that were formerly thought to be inhabited by monstrous races and shadowy deities were now populated by real men, still quite foreign, but also militarily formidable and potentially hostile. This new world of diplomacy required exchanges of envoys, gift giving, treaties, marriage alliances, and, when diplomacy failed, open conflict. The initial motivation to reach these distant lands may have been military conquest or alliance building, but those initial forays soon fostered a trade in exotic luxuries that bound East and West even tighter together in an interconnected net (see chapter 3). Monsters still inhabited the fringes of these worlds, since the human mind requires monsters to live there to delineate the civilized realm, but the monster-laden edges were necessarily pushed a few thousand miles farther back.

INTRODUCTION TO MEGASTHENES

Under the Treaty of the Indus, negotiated around 304 BCE, Seleucus I agreed to cede all his eastern satrapies to Chandragupta, who in exchange, gave the Seleucid monarch 500 war elephants. By some accounts, the monarchs also arranged some sort of marriage alliance. Megasthenes (ca. 350–ca. 290 BCE), who may have helped Seleucus negotiate that treaty, then served as his ambassador to the Mauryan court at Pāṭaliputra (near present-day Patna; see map 1). He journeyed there from the Seleucid satrapy of Arachosia, with its capital near present-day Kandahar, Afghanistan. So unlike Ctesias, who resided only at the Persian court and delivered secondhand tales of India (chapter 1), Megasthenes made repeated visits there and may have been in residence for extended periods. Around 300 BCE, he wrote the *Indica*, a three-section book about the Mauryan realm. It was originally intended as a political report to his monarch but also contained observations on India's geography, history, flora, fauna, and customs. Megasthenes's *Indica* is now lost, but portions of it survive in quotations and paraphrases in the writings of classical authors like Arrian, Strabo, Diodorus Siculus, and Pliny.

Megasthenes gives the first relatively accurate account of the geography of the subcontinent, informing his readers that it was bounded on the east by an ocean and did not fade off into Terra Incognita as had been thought during the time of Herodotus. Megasthenes's *Indica* also gives the West its first account of the Indian caste system. His seven castes do incorporate all four of the classical *varna* of India, but he subdivides some and folds others together. His fascinating account of Indian philosophers does make an accurate distinction between the Brahmanes (Brahmins) and Garmanes (*śramaṇa*, "ascetics"), but he conflates Buddhists, Jains, and Ajivikas within the second category.

Like other travelers in this anthology, Megasthenes occasionally engages in utopian projection, indirectly criticizing his own society. Like the later Chinese Buddhist pilgrims Faxian and Xuanzang (chapter 5), Megasthenes was impressed by the Indians' justice system and native honesty. It was also remarkable to him that "all Indians are free, and not one of them is a slave." This was not a true statement, for India had many forms of servitude, but they were different in nature from Greek chattel slavery, since Indian slaves were still considered persons and were not employed in industry. The most vivid section of Megasthenes's account is his report on the capture and taming of elephants for military use. Those elephants obtained in the Treaty of the Indus would prove crucial to Seleucus I's victory at Ipsus in 301 BCE and would encourage the Ptolemaic rulers in Egypt to obtain their own elephants from the East African coast, further developing Indian Ocean trade routes (see chapter 3).

But alongside these accurate new facts about Indian politics, society, and animals, Megasthenes continues to tell tales of the monstrous races and wondrous animals of India, such as the Dog-Headed People reported by Ctesias, mouthless people who subsist on aromas, and the infamous gold-digging ants. But the location of these wonders has shifted. For Ctesias, the Indus River valley was the civilized center, and the Ganges was the land of the strange, whereas for Megasthenes, who resided in the cultured capital of Pāṭaliputra on the Ganges, monstrous races were now relegated to the far south or to the northwestern mountains.

SOURCE TEXT, ALTERNATE ENGLISH TRANSLATIONS, AND SELECTED STUDIES

Bosworth, A. B. "The Historical Setting of Megasthenes' *Indica*." *Classical Philology* 91 (1996): 113–127.

Kosmin, Paul. *The Land of the Elephant Kings: Space, Territory, and Ideology in the Seleucid Empire*. Harvard University Press, 2014.

McCrindle, John Watson. *Ancient India as Described by Megasthenês and Arrian*. Trübner, 1877.

Rackham, H., W. H. S. Jones, and D. E. Eichholz, trans., *Pliny, Natural History*. Harvard University Press, 1938–1963.

Stoneman, Richard. *Megasthenes' Indica: A New Translation of the Fragments with Commentary.* Routledge, 2021.

EXCERPTS FROM THE *INDICA,* BY MEGASTHENES

Geography of India

According to Eratosthenes and Megasthenes, who lived with Sibyrtius [late fourth century BCE], the satrap of Arachosia [see map 1], who as he himself tells us often visited Sandrocottus [Chandragupta Maurya, r. ca. 321–ca. 297 BCE], the king of the Indians, India forms the largest of the four parts into which South Asia is divided, while the smallest part is that region that is included between the Euphrates and our own [Mediterranean] sea. The two remaining parts, which are separated from the others by the Euphrates and the Indus and lie between these rivers are scarcely of sufficient size to be compared with India, even should they be taken both together. The same writers say that India is bounded on its eastern side, right onward to the south, by the Great Ocean; that its northern frontier is formed by the Caucasus Range as far as the junction of that range with Mt. Taurus; and that the boundary towards the west and northwest, as far as the Great Ocean, is formed by the Indus River. A considerable portion of India consists of a level plain.[1]

Map 1. Zhang Qian's first embassy to the West and Megasthenes's embassy to the Mauryan court.

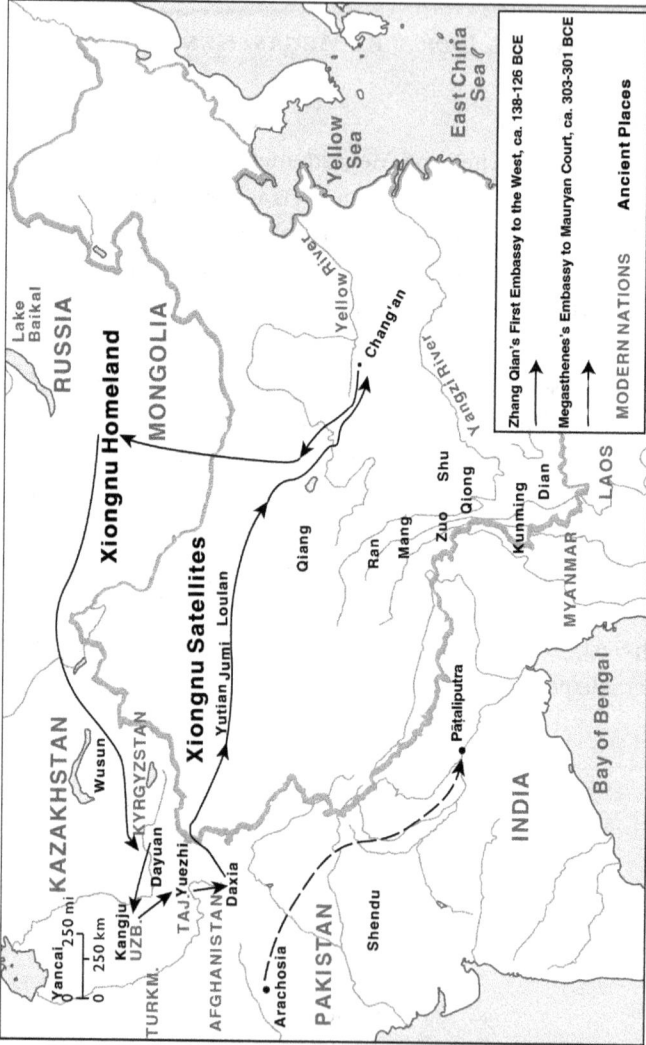

Source: Map created by Zhang Yexu and the author.

Alexander's Invasion of India and Earlier Incursion of Dionysus

This same Megasthenes then informs us that the Indians neither invade other men nor do other men invade the Indians, for Sesostris the Egyptian [a legendary conflation of pharaohs Senwosret III and Ramesses II], after having overrun the greater part of Asia and advanced with his army as far as Europe, returned home. Idanthyrsus the Scythian [ca. 513 BCE] issuing from Scythia, subdued many nations of Asia, and carried his victorious arms even to the borders of Egypt. And Semiramis, the Assyrian queen [a semi-mythical sovereign based on Shammuramat, wife of Shamshi-Adad V, r. 824–811 BCE], planned an expedition against India, but died before she could execute her design. Thus, Alexander the Great was the only conqueror who actually invaded the country.

And regarding Dionysus, many traditions are current to the effect that he also made an expedition into India and subjugated the Indians before the days of Alexander. But of Hercules, tradition does not say much. Of the expedition, however, which Dionysus led, the city of Nysa is no mean monument, while Mt. Meros is yet another, and the ivy which grows thereon, and the practice observed by the Indians themselves of marching into battle with drums and cymbals, and of wearing a spotted dress such as was worn by the Bacchanals of Dionysus. On the other hand, there are but few memorials of Hercules.[2]

* * * * * * * * * * * * * * * **BRIEF EXCERPT** * * * * * * * * * * * * * * * *

Indian Caste System

In India, the whole people are divided into about seven castes. Among these are the wise men, who are not so numerous as the others but hold the supreme place of dignity and honor, for they are not required to do any bodily labor at all, or of contributing from the produce of their labor anything to the common stock, nor indeed is any duty absolutely binding on them except to perform the sacrifices offered to the gods on behalf of the state. If anyone, again, has a private sacrifice to offer, one of these sages shows him the proper mode, as if he could not otherwise make an acceptable offering to the gods. To this class the knowledge of divination among the Indians is exclusively restricted, and none but a sage is allowed to practice that art. They predict about such matters as the seasons of the year and any calamity that may befall the state, but the private fortunes of individuals they do not care to predict, either because divination does not concern itself with trifling matters, or because to take any trouble about such is deemed unbecoming. But if anyone fails three times to predict truly, he incurs, it is said, no further penalty than being obliged to be silent for the future, and there is no power on earth able to compel that man to speak who has once been condemned to silence.

These sages go naked, living during winter in the open air to enjoy the sunshine and during summer, when the heat is too powerful, in meadows and low grounds under large trees, the shadow of which Nearchus [Alexander's admiral, d. 300 BCE] says extends to five *plethora* [approx. 153 m] in circuit, adding that even 10,000 men could be covered by the shadow of a single tree. They live upon the fruits that each season produces and on the bark of trees, the bark being no less sweet and nutritious than the fruit of the date palm.

After these, the second caste consists of the tillers of the soil, who form the most numerous class of the population. They are neither furnished with arms nor have any military duties to perform, but they cultivate the soil and pay tribute to the kings and the independent cities. In times of civil war, the soldiers are not allowed to molest the farmers or ravage their lands; hence, while the former are fighting and killing each other as they can, the latter may be seen close at hand tranquilly pursuing their work, perhaps ploughing, or gathering in their crops, pruning the trees, or reaping the harvest.

The third caste among the Indians consists of the herdsmen, both shepherds and cowherds, and these live neither in cities nor in villages, but they are nomadic and live on the hills. They too are subject to tribute, and this they pay in cattle. They scour the country in pursuit of fowl and wild beasts.

The fourth caste consists of handicraftsmen and retail dealers. They must perform gratuitously certain public services and pay tribute from the products of their labor. An exception, however, is made in favor of those who fabricate the weapons of war, and not only so, but they even draw pay from the state. In this class are included shipbuilders, and the sailors employed in the navigation of the rivers.

The fifth caste among the Indians consists of the warriors, who are second in number to the husbandmen but lead a life of supreme freedom and enjoyment. They have only military duties to perform. Others make their arms, and others supply them with horses, and they have others to attend on them in the camp, who take care of their horses, clean their arms, drive their elephants, prepare their chariots, and act as their charioteers. As long as they are required to fight, they fight, and when peace returns, they abandon themselves to enjoyment, the pay that they receive from the state being so liberal that they can with ease maintain themselves and others besides.

The sixth class consists of those called superintendents. They spy out what goes on in the countryside and town, and report everything to the king where the people have a king, and to the magistrates where the people are self-governed. It is not lawful to make any false reports to them—but indeed no Indian is accused of lying.

The seventh caste consists of the councilors of state, who advise the king, or the magistrates of self-governed cities, in the management of public affairs. This is the smallest caste in terms of numbers, but it is distinguished by superior wisdom and justice and hence enjoys the prerogative of choosing governors, chiefs of provinces, deputy governors, superintendents of the treasury, generals of the army, admirals of the navy, controllers, and commissioners who superintend agriculture.

The custom of the country prohibits intermarriage between the castes. For instance, the farmer cannot take a wife from the artisan caste nor the artisan a wife from the agricultural caste. Custom also prohibits anyone from exercising two trades or from changing from one caste to another. One cannot, for instance, become a farmer if he is a herdsman or become a herdsman if he is an artisan. It is permitted that only the sage can be drawn from any caste; for the life of the sage is not an easy one, but the hardest of all.[3]

* *

Royal Capital of Pāṭaliputra

According to Megasthenes, the mean breadth of the Ganges is 100 stadia [approx. 18.5 km], and its least depth 20 fathoms. At the meeting of this river and another is situated Palibothra [the Mauryan capital of Pāṭaliputra, near present-day Patna], a city 80 stadia in length [14.8 km] and 15 [2.8 km] in breadth. It is of the shape of a parallelogram and is girded with a wooden wall, pierced with loopholes for the discharge of arrows. It has a ditch in front for defense and for receiving the sewage of the city. The people in whose country this city is situated is the most distinguished in all India and are called the Prasioi [Sanskrit: Prācyas; "Easterners"]. The king, in addition to his family name, must adopt the surname Palibothros, as Sandrocottus [Chandragupta], for instance, did, to whom Megasthenes was sent on an embassy.[4]

Funerary Customs and Absence of Slavery

It is further said that the Indians do not rear monuments to the dead but consider the virtues that men have displayed in life and the songs in which their praises are celebrated sufficient to preserve their memory after death. . . . The same writer [Megasthenes] tells us further this remarkable fact about India that all the Indians are free, and not one of them is a slave. The Spartans and the Indians are here so far in agreement. The Spartans, however, hold the Helots as slaves, and these Helots do servile labor, but the Indians do not even use aliens as slaves, and much less a countryman of their own.[5]

Indian Elephants

The Indians hunt all wild animals in the same way as the Greeks, except the elephant, which is hunted in a mode altogether peculiar, since these animals are not like any others. The mode may be thus described: The hunters having selected a level tract of arid ground dig a trench all round it, enclosing as much space as would suffice to encamp a large army. They make the trench with a breadth of five fathoms and a depth of four. But the earth that they throw out in the process of digging they heap up in mounds on both edges of the trench and use it as a wall. Then they make huts for themselves by excavating the wall on the outer edge of the trench, and in these they leave loopholes, both to admit light and to enable them to see when their prey approaches and enters the enclosure. They next station some three or four of their best-trained female elephants within the trap, to which they leave only a single passage by means of a bridge thrown across the trench, the framework of which they cover over with earth and a great quantity of straw to conceal the bridge as much as possible from the wild animals, which might else suspect treachery. The hunters then go out of the way, retiring to the cells which they had made in the earthen wall. Now the wild elephants do not go near inhabited places in the daytime, but during the nighttime, they wander about everywhere and feed in herds, following as leader the one who is biggest and boldest, just as cows follow bulls. As soon, then, as they approach the enclosure and hear the cry and catch scent of the females, they rush at full speed in the direction of the fenced ground and being arrested by the trench move round its edge until they come across the bridge, along which they force their way into the enclosure. The hunters, meanwhile, perceiving the entrance of the wild elephants, hasten, some of them, to tear down the bridge, while others, running off to the nearest villages, announce that the elephants are within the trap. The villagers, on hearing the news, mount their most spirited and best-trained elephants and as soon as mounted ride off to the trap. Although they ride up to it, they do not immediately engage in a conflict with the wild elephants but wait till these are sorely pinched by hunger and tamed by thirst. When they think their strength has been enough weakened, they rebuild the bridge and ride into the enclosure, when a fierce assault is made by the tame elephants upon those that have been entrapped. Then, as might be expected, the wild elephants, through loss of spirit and faintness from hunger, are overpowered. On this, the hunters, dismounting from their elephants, bind with fetters the feet of the wild ones, now by this time quite exhausted. Then they instigate the tame ones to beat them with repeated blows, until their sufferings wear them out and they fall to the ground. The hunters, meanwhile, standing near them, slip nooses over their necks and mount them while yet lying on the ground. To prevent them shaking off their riders or doing mischief otherwise, they make with a sharp knife an incision all round their neck, and fasten the noose round in the incision. By means of the wound thus made, they keep their head and neck quite steady, for if they become restive and turn round, the wound is chafed by the action of the rope. They shun, therefore, violent movements and, knowing that they have been vanquished, suffer themselves to be led in fetters by the tame ones.

Such as are too young or through the weakness of their constitution not worth keeping, their captors allow to escape to their old haunts, while those that are retained they lead to the villages, where at first they give them green stalks of bamboo and grass to eat. The creatures, however, having lost all spirit, have no wish to eat. But the Indians, standing round them in a circle, soothe and cheer them by chanting songs to the accompaniment of the music of drums and cymbals, for the elephant is of all brutes the most intelligent. Some of them, for instance, have taken up their riders when slain in battle and carried them away for burial; others have covered them, when lying on the ground, with a shield; and others have borne the brunt of battle in their defense when fallen. There was one even that died of remorse and despair because it had killed its rider in a fit of rage.

I have myself actually seen an elephant playing on cymbals, while other elephants were dancing to his strains. A cymbal had been attached to each foreleg of the performer, and a third to what is called his trunk, and while he beat in turn the cymbal on his trunk, he beat in proper time those on his two legs. The dancing elephants all the while kept dancing in a circle, and as they raised and curved their forelegs in turn they too moved in proper time, following as the musician led. . . . [6]

Gold-Guarding Ants of India

Megasthenes insists that this tradition about the ants is strictly true, that they are gold diggers, not for the sake of the gold itself, but because by instinct they burrow holes in the earth to lie in, just as the tiny ants of our own country dig little holes for themselves, only those in India being larger than foxes, make their borrows proportionately larger. But the ground is impregnated with gold and the Indians thence obtain their gold.[7]

Monstrous Races

Megasthenes states that on the mountain named Nulus, there are people with their feet turned backward and with eight toes on each foot, while on many of the mountains, there is a tribe of human beings with dogs' heads who wear a covering of wild beasts' skins, whose speech is a bark, and who live on the produce of hunting and fowling, for which they use their nails as weapons. He says that they numbered more than 120,000 when he published his work.

Megasthenes tells of a race among the nomads of India that has only holes in the place of nostrils, like snakes, and are bandy legged. They are called the Sciritae. At the extreme boundary of India to the east, near the source of the Ganges, he puts the tribe of the Mouthless Ones, who have no mouth and a body hairy all over. They dress in cottonwool and live only on the air they breathe and the scent they inhale through their nostrils. They have no food or drink except the different odors of the roots and flowers and wild apples, which they carry with them on their longer journeys so as not to lack a supply of scent. He says they can easily be killed by a rather stronger odor than usual.[8]

Indian Philosophers

Speaking of the philosophers, Megasthenes says that such of them as live on the mountains are worshippers of Dionysus, who show as proofs the wild vine, which grows in their country only, and the ivy, and the laurel, and the myrtle, and the boxtree, and other evergreens, none of which are found beyond the Euphrates, except a few in parks, which it requires great care to preserve. They observe also certain customs which are Bacchanalian. Thus, they dress in muslin, wear the turban, use perfumes, array themselves in garments dyed of bright colors, and their kings, when they appear in public, are preceded by the music of drums and gongs. But the philosophers who live on the plains worship Hercules.

Megasthenes makes a different division of the philosophers, saying that they are of two kinds, one of which he calls the Brahmanes [Sanskrit: Brāhmaṇa, "Brahmin" or Vedic priest], and the other the Garmanes [Sanskrit: śramaṇa, "ascetic"]. The Brahmanes are best esteemed, for they are more consistent in their opinions. From the time of their conception in the womb, they are under the guardian care of learned men, who go to the mother and, under the pretense of using some incantations for the welfare of herself and her unborn babe, in reality give her prudent hints and counsels. The women who listen most willingly are thought to be the most fortunate in their children. After their birth, the children are under the care of one person after another, and as they advance in age each succeeding master is more accomplished than his predecessor. The philosophers have their abode in a grove in front of the city within a moderate-sized enclosure. They live in a simple style and lie on beds of rushes or deer skins. They abstain from animal food and sexual pleasures and spend their time in listening to serious discourse and in imparting their knowledge to such as will listen to them. The hearer is not allowed to speak, or even to cough, and much less to spit, and if he offends in any of these ways, he is cast out from their society that very day as being a man who is wanting in self-restraint. After living in this manner for 37 years, each individual retires to his own property, where he lives for the rest of his days in ease and security. They then array themselves in fine muslin and wear a few trinkets of gold on their fingers and in their ears. They eat flesh, but not that of animals that are employed in labor. They abstain from hot and highly seasoned food. They marry as many wives as they please, with a view to have numerous children, for by having many wives, greater advantages are enjoyed, and since they have no slaves, they have more need to have children around them to attend to their wants.

The Brahmanes do not communicate a knowledge of philosophy to their wives, lest they should divulge any of the forbidden mysteries to the profane if they became depraved, or lest they should desert them if they became good philosophers, for no one who despises pleasure and pain, as well as life and death, wishes to be in subjection to another, but this is characteristic both of a good man and of a good woman.

Death is with them a very frequent subject of discourse. They regard this life as, so to speak, the time when the child within the womb becomes mature, and death as a birth into a real and happy life, for those who engage in philosophy. On this account they undergo much discipline as a preparation for death. They consider nothing that befalls men to be either good or bad, to suppose otherwise being a dream-like illusion, else how could some be affected with sorrow, and others with pleasure, by the very same things, and how could the same things affect the same individuals at different times with these opposite emotions?

Their ideas about physical phenomena, the same author [Megasthenes] tells us, are very crude, for they are better in their actions than in their reasonings, inasmuch as their belief is in great measure based upon fables. Yet on many points, their opinions coincide with those of the Greeks, for like them, they say that the world had a beginning and is liable to destruction and is in shape spherical and that the god who made it, and who governs it, is diffused through all its parts.

They hold that various first principles operate in the universe and that water was the principle employed in the making of the world. In addition to the four elements, there is a fifth agency, from which the heaven and the stars were produced. The earth is placed in the center of the universe. Concerning generation, and the nature of the soul, and many other subjects, they express views like those maintained by the Greeks. They wrap up their doctrines about immortality and future judgment, and similar topics, in allegories, after the manner of Plato.

. . .

Of the Garmanes, Megasthenes tells us that those who are held in most honor are called the Forest Dwellers. They live in the woods, where they subsist on leaves of trees and wild fruits, and wear garments made from the bark of trees. They abstain from sexual intercourse and from wine. They communicate with the kings, who consult them by messengers regarding the causes of things, and who through them worship and supplicate the deity. Next in honor to the Forest Dwellers are the physicians, since they are engaged in the study of the nature of man. They are simple in their habits but do not live in the fields. Their food consists of rice and barley meal, which they can always get for the mere asking or receive from those who entertain them as guests in their houses. By their knowledge of pharmacology, they can make marriages fruitful and determine the sex of the offspring. They effect cures rather by regulating diet than by the use of medicines. The remedies most esteemed are ointments and plasters. All others they consider to be in a great measure pernicious in their nature. This class and the other class practice fortitude, both by undergoing active toil and by the endurance of pain, so that they remain for a whole day motionless in one fixed attitude. Besides these there are diviners and sorcerers, and adepts in the rites and customs relating to the dead, who go about begging both in villages and towns. . . . Women pursue philosophy with some of them but abstain from sexual intercourse.[9]

INTRODUCTION TO ZHANG QIAN

The Han imperial envoy Zhang Qian's 張騫 (d. 114 BCE) incredible 13-year odyssey from the heart of China into the depths of Central Asia not only unfolds as a remarkable human drama but also ushers in a dramatic change in the relations between the Far East and the Western world. Before Zhang Qian's trip, official Chinese knowledge of what lay to the west of their world was meager and unreliable. Most Chinese believed that those distant and hard-to-reach lands were inhabited by uncivilized tribes of strange humans, grotesque monsters, or divine beings (chapter 1).

The motivation behind Zhang Qian's first embassy was to seek allies in the Han empire's conflict with the Xiongnu, the powerful confederation of nomadic warriors who had held the empire hostage since its inception, requiring massive payments under a treaty system deceptively called Peace and Brotherhood (*heqin* 和親). This onerous treaty had been renewed under each successive reign of the Han, but Emperor Wu (r. 140–87 BCE) had designs to abrogate the treaty and provoke an all-out conflict that could change the balance of power. It was for this reason that Zhang Qian was sent west to seek out the Great Yuezhi, former enemies of the Xiongnu, who had been driven westward after a great defeat. Though he was frustrated in his goal of convincing the Yuezhi to become allies of the Han in an attack on the Xiongnu, the unexpected gains of the embassy were far more important in the long term.

The narrative of Zhang Qian's journey was related by the Han historian Sima Qian 司馬遷 (ca. 145–ca. 86 BCE). Sima Qian's account incorporates verbatim Zhang Qian's detailed written report to Emperor Wu on the places he personally visited, such as the Ferghana Valley, Sogdiana, and Bactria, as well as those which he only heard about, such as the Parthian Empire, the Kingdom of Characene in southern Mesopotamia, Alexandria in Egypt, and northwest India.

Today, we would classify the information in Zhang Qian's report as diplomatic and military intelligence. He details the location of each kingdom and their strength under arms, as well as their subsistence base (farmers vs. pastoralists) and their favorability toward the Han. For Parthia, which Zhang Qian did not visit in person but to which he later sent sub-envoys, he was surprised to learn of Parthian coinage that displayed the face of the current king (unheard of for Chinese coinage) and horizontal writing on leather, probably in Pahlavi script. He also deduced an alternative route from China to Bactria, when he witnessed that Chinese products from Sichuan had been brought overland to Bactria by merchants trading in India (a land where, he noted, elephants were ridden into battle). And while Emperor Wu was attracted by the exotic products he could obtain from these newly discovered foreign lands, he was more interested in how he could cajole these polities into recognizing his suzerainty, thereby enhancing his legitimacy as a universal monarch.

And like Megasthenes's experience in India, Zhang Qian's reports of the Far West forced the historian Sima Qian to reconsider the traditional location of the mythical Mount Kunlun, the source of the Yellow River and the abode of the deity called the Spirit Mother of the West. When they were not sighted by Zhang, their supposed location was now shifted much farther west, with the Spirit Mother now said to reside somewhere near Mesopotamia.

After Zhang Qian's opening, China was never again out of diplomatic contact with the countries of Central and South Asia. Contacts were frequent during the Han with the Parthian Empire, and Parthian ambassadors even toured the empire at the side of Emperor Wu. Relations with India increased in subsequent centuries, as the lure of Buddhist wisdom attracted Chinese pilgrims in search of scriptures (see chapter 5). Chinese knowledge of the Roman Empire and later Byzantine Empire remained scanty (chapters 4 and 6), based mostly on corrupted secondhand accounts, and no contact with Western European monarchs was attempted until after the Mongol conquests (see chapter 7). But throughout the Tang dynasty, relations with Sassanid Persia and the later Abbasid Caliphate were extremely active over both land and sea routes (see chapter 6).

But it was the resolute traveler Zhang Qian who first opened China's door to the wider world.

SOURCE TEXT, ALTERNATE ENGLISH TRANSLATIONS, AND RELEVANT RECENT STUDIES

Hulsewé, A. F. P., and Michael Loewe. *China in Central Asia.* Brill, 1979.

Sima Qian 司馬遷, *Shi ji* 史記 (Zhonghua shuju, 1959), 123.3157–69, 3179.

Watson, Burton. *Records of the Grand Historian: Han Dynasty II.* Columbia University Press, 1993.

THE TRAVELS OF ZHANG QIAN, EXCERPTED FROM THE "ACCOUNT OF DAYUAN," BY SIMA QIAN

Sima Qian's Historical and Biographical Narrative

The traces of the country of Dayuan [present-day Ferghana Valley, split between Tajikistan, Uzbekistan, and Kyrgyzstan], were first seen from the time of Zhang Qian [d. 114 BCE]. Zhang Qian was a man from Chenggu County in Hanzhong Commandary [in present-day Shaanxi]. During the Jianyuan era [140–135 BCE], he became a palace gentleman. At this time, the Son of Heaven [Emperor Wu, r. 140–87 BCE] had made inquiries among those Xiongnu who had surrendered, and all had said that the Xiongnu had defeated the king of the Yuezhi and had taken his skull and turned it into a drinking vessel. The Yuezhi had fled but harbored constant resentment against the Xiongnu, considering them their enemy, but had no allies with whom to strike against them. At that time, the Han desired to undertake the project of wiping out the northern barbarians, and so when the emperor heard these words, he, thereupon, desired to send out an envoy. The route would necessarily pass through Xiongnu territory, so he put out a muster call for those who were willing and able to serve as an envoy.

In his capacity as a court gentleman, Zhang Qian answered the muster call and was made envoy to the Yuezhi. Along with the Xiongnu slave Ganfu, of the Tangyi clan, the party set out from Longxi [in present-day Gansu] and passed through Xiongnu territory. The Xiongnu captured them and transported them to the court of the Chanyu [the Xiongnu leader, whose court was in present-day Mongolia; see map 1]. The Chanyu detained him and said, "The Yuezhi reside to the north of me. Why would the Han be able to send an envoy to go there? If I wanted to send an envoy to the Yue people [south of the Han, in present-day Guangdong], would the Han submit to my request?" He detained Zhang Qian for more than 10 years and gave him a wife by whom he had children, but all along, Zhang Qian did not let go of his staff of authority, given by the emperor.

While he resided among the Xiongnu, Zhang Qian was treated with increasing liberality, so he took advantage of this to escape along with his attendants and headed toward the land of the Yuezhi. He traveled westward for several tens of days and finally reached Dayuan. The people of Dayuan had heard of the bountiful wealth of the Han, and they had wanted to open up communication but had been unable.

When they saw Zhang Qian, they were delighted, and the King of Dayuan asked him, "Where do you wish to go?"

Zhang Qian replied, "I have been sent by the Han as an envoy to the Yuezhi, but I had my path closed off by the Xiongnu. Now, I have escaped, and I only ask that the King of Dayuan employ men to serve as my guides to send me on my way. If I am truly able to make it there and return to the Han, the amount of riches with which the Han would reward Your Majesty would be beyond words!"

The ruler of Dayuan considered this acceptable and sent out Zhang Qian, mobilizing for him guides and interpreters till he reached Kangju [Sogdiana, in present-day Uzbekistan]. The state of Kangju conveyed him onward till he reached the Great Yuezhi [whose court was then located north of the Amu Darya River]. The king of the Great Yuezhi had already been killed by the Xiongnu, so they had set up his heir apparent as the new king. The Great Yuezhi had already made subjects of the people of Bactria [in northern Afghanistan, at the time a weakened Hellenistic kingdom], and occupied that territory. The land was fertile and full of abundance, and bandits were scarce. They were determined to live a peaceful and joyous existence, and moreover, they considered the Han as quite distant, so they harbored no particular desire to seek revenge on the Xiongnu. Zhang Qian went from the court of the Great Yuezhi into the land of Bactria, but in the end, he was not able to grab the Yuezhi by the "waist and neck" to become a Han ally.

He remained for more than a year and then returned, skirting along the southern mountains [of the Tarim Basin]. He wished to return through the territory of the Qiang [a Tangut people, living in present-day Qinghai], but again he was captured by the Xiongnu. After Zhang Qian was detained for more than a year, the Chanyu died [126 BCE], and the Luli Prince of the Left attacked the heir apparent of the Chanyu and set himself up as Chanyu, throwing the Xiongnu state into disorder. Zhang Qian, along with his Xiongnu wife and Ganfu, were able to escape together and return to Han territory. The Han conferred on Zhang Qian the title of supreme palace counsellor, and Ganfu was named Gentleman who Supports the Emissary.

As a man, Zhang Qian was strong and powerful, and he was magnanimous and trusting of others. The barbarians were very fond of him. Ganfu of the Tangyi clan was formerly a Xiongnu subject. He was skilled at shooting arrows, and in times when their provisions were sorely exhausted, he could always shoot birds or beasts to provide them with nourishment. At the start, Zhang Qian had set off with more than 100 persons, but after he had left for 13 years, only two of those persons were able to make it back. Those countries that Zhang Qian personally visited included Dayuan [Ferghana], the Great Yuezhi, Daxia [Bactria], and Kangju [Sogdiana]. He also transmitted knowledge of five or six other large countries in the vicinity and compiled these into a report delivered orally to the emperor, which said:

Kingdom of Dayuan

Dayuan is located southwest of Xiongnu territory, directly west of the Han domain, distant from the Han by perhaps ten thousand Chinese miles [approx. 4,000 km]. Their tradition is to cultivate the land, plowing their fields and planting rice and wheat. They also have wine made from grapes. Many of them are skilled at horsemanship, and their horses sweat blood. Its ancestor was the offspring of a Heavenly Horse. They have walled cities with suburbs and live in houses. They have more than 70 walled towns of various sizes under them, with a total population of around several hundred thousand people. Their weaponry consists of the bow and spear, and they shoot while riding on horseback. To the north of Dayuan is Kangju; to the west is the Great Yuezhi; to the southwest is Daxia; to the northeast is the Wusun; to the east is Jumi [probably somewhere near Khotan] and Yutian [Khotan, in Xinjiang]. West of Yutian, all the rivers flow to the west, emptying into the Western Sea [probably Aral Sea], to the east of Yutian, the rivers flow eastward, emptying into the Salt Marsh [Lop Nur, in eastern Xinjiang]. The waters of the Salt Marsh flow hidden underground. South of Yutian emerges the source of the Yellow River. There is much jade stone in Yutian, and the Yellow River flows onward into the Middle Kingdom. The towns of Loulan and Gushi also have walls and overlook the Salt Marsh. The Salt Marsh is distant from Chang'an [the Han capital, present-day Xi'an] by about 5,000 Chinese miles [approx. 2,000 km]. The right quadrant [i.e. western area] of lands under Xiongnu control consists of the territory east of the Salt Marsh up until the Great Wall fortifications at Longxi and to the south abuts the territory of the Qiang, where they block the road coming from Han territory.

Kingdom of Wusun

The Wusun are located northeast of Dayuan by about 2,000 Chinese miles [approx. 800 km, in the Ili Valley of Kazakhstan]. They are nomadic, following their herds, and have the same customs as the Xiongnu. They have tens of thousands who can draw the bowstring and are daring in war. Formerly, they had submitted to the Xiongnu, but then they became powerful, and though they were controlled by the Xiongnu through "loose reins," they were not willing to travel for the court audiences of the Chanyu of the Xiongnu.

The Polity of Kangju

Kangju [Tashkent area in Sogdiana] is located northwest of Dayuan by about 2,000 Chinese miles [approx. 800 km]. It is a nomadic country, and their customs are largely shared with the Yuezhi. They have 80,000 or 90,000 skilled archers and are neighbors of Dayuan. The country is relatively small, and to the south, they are under the loose political control of the Yuezhi, and in the east, under the influence of the Xiongnu.

The Polity of Yancai

Yancai is located about 2,000 Chinese miles [approx. 800 km] northwest of Kangju. It is a nomadic country whose customs are largely the same as Kangju. They have more than 100,000 men who can control the bowstring. It borders a great marsh, without any shore in sight, for this is the Northern Sea [Aral Sea or possibly the Caspian].

Kingdom of the Great Yuezhi

The Great Yuezhi are located west of Dayuan by about 2–3,000 Chinese miles [approx. 800–1,200 km]. Their court resides north of the Amu Darya River. To their south is the land of Daxia [Bactria]; to their west is Anxi [Parthian Empire]; to their north is Kangju [Sogdiana]. It is a nomadic country, and the people move about following their herds. Their customs are similar to those of the Xiongnu. They have perhaps 100,000 to 200,000 men who can control the bowstring. In former times, they were powerful and took the Xiongnu lightly, but then when Maodun ascended to the position of Chanyu of the Xiongnu, he attacked and defeated the Yuezhi, to the point that his son, the Chanyu called Laoshang, killed the king of the Yuezhi and made his skull into a drinking vessel. Originally, they resided in the area east of Dunhuang and west of the Qilian Mountains [present-day Gansu and Qinghai], but when they were defeated by the Xiongnu, they moved far away, past Dayuan, moving west to strike at Daxia and subjugating them. Then, they made their capital north of the Amu Darya River, making this the king's court. Small numbers of remaining Yuezhi who could not migrate took refuge among the Qiang people of the southern mountains and are called the Lesser Yuezhi.

* * * * * * * * * * * * * * BRIEF EXCERPT * * * * * * * * * * * * * * *

Kingdom of Anxi

Anxi [Persian: Aršak, i.e., Arsacid dynasty of Parthia] is located around several thousand Chinese miles west of the Great Yuezhi. They are sedentary agriculturalists, plowing their fields and planting rice and wheat and cultivating grapes for wine. Their cities and towns are like those of Dayuan. There are also several hundred large and small towns subordinate to them, and their territory covers thousands of square miles. It is an extremely large country. Where it borders the Amu Darya River, there is a market where ordinary people and merchants, traveling in carts or by boat, travel to neighboring countries, some several thousand miles distant. They use silver for their coinage, and the coins display a likeness of the king's face. When the king dies, they immediately change the coinage to reflect the face of the new king! To make written records, they paint words on strips of leather in horizontal lines. To their west lies the polity of Tiaozhi [Characene Kingdom at the head of the Persian Gulf, subordinate to Parthia]; to their north is Yancai.

Kingdom of Tiaozhi

Lixuan [Alexandria in Egypt] and Tiaozhi are located several thousand Chinese miles west of Anxi, bordering on the Western Sea [Persian Gulf and Red Sea]. Tiaozhi is hot and damp. The people plow their fields and plant crops of rice. There is a large bird there with eggs the size of a pot. The population is very large, and there are numerous small chieftains. And even though they are all subordinate to the rulers of Anxi, Tiaozhi is considered by them to be a foreign country. The people of this country are good at juggling and other circus tricks. Even though the elderly people of Anxi transmit the tradition that in Tiaozhi can be found the "Weak Water" and the Spirit Mother of the West, these have not yet been seen there.

The Land of Daxia

Daxia [Bactria] is located more than 2,000 Chinese miles [approx. 800 km] southwest of Dayuan, south of the Amu Darya River. Its people are sedentary farmers, and they have cities and houses. Their customs are largely the same as those of Dayuan. They do not have one supreme ruler, but each city and town has established a lesser king or chief. They are weak at arms and timid in battle but are very skilled at trading in the marketplace. When the Great Yuezhi migrated westward, they conquered Daxia and made all their people subjects. The population of Daxia is very numerous, perhaps more than a million persons. Their capital is called Lanshi [present-day Balkh]. It has markets where all sorts of products are bought and sold.

Country of Shendu

To the southeast of Daxia lies the country of Shendu [Indian subcontinent]. Zhang Qian stated: "When I was in Daxia, I saw bamboo canes from Qiong and cloth from Shu [both in present-day Sichuan]. When I asked them, 'Where did you get this?' the people of Daxia replied, 'Our merchants traveled and bought them in Shendu. Shendu is located southeast of Daxia by perhaps several thousand Chinese miles. Its people are sedentary farmers, with customs largely similar to those of Daxia, but it is very low, humid, and hot. Their people ride elephants in battle. The country borders on a great river [Ganges].' According to my calculation, Daxia is 12,000 miles [approx. 4,800 km] distant from the Han and is located to the southwest of the Han. Now, Shendu is located several thousand miles southeast of Daxia and has trade goods from Shu [Sichuan], so it must not be far from Shendu to our province of Shu. Now, when we send envoys to Daxia, passing through Qiang territory, they are blocked, and the Qiang people despise them. A little to the north, they would be captured by the Xiongnu. Passing through Shu seems a fitting shortcut, without bandits."

The Han Emperor's Reaction

The Son of Heaven [Emperor Wu] had now heard that Dayuan, Daxia, and the subordinate states of Anxi were all large countries, with many strange and wondrous things, sedentary farmers, and with largely the same enterprises as China, yet with weak armies and being covetous of Han wealth and products, and that to their north there were the states subordinate to the Great Yuezhi and Kangju, who were strong in arms but could be bribed with wealth and made to serve the advantages of the court. Moreover, if one were able to truly acquire these territories and make them adherents through the use of righteous moral persuasion, then this would make a broad realm of 10,000 square miles, where languages would have to go through nine levels of translation among people of widely divergent customs, and the awesomeness and virtue of the emperor could spread to all the land within the four seas. The emperor was pleased with this idea and approved of Zhang Qian's suggestions.

* *

Attempt to Open the Southwestern Route

He thereupon ordered Zhang to go through Shu and Jianwei Commanderies [present-day Sichuan] and mobilize secret missions to go out by four different routes, from Mang, from Ran, from Si, and from Qiong and Bo. Each traveled about 1,000–2,000 Chinese miles [approx. 400–800 km]. The northern route was blocked by the Di and the Zuo. The southern route was blocked by the Sui and the Kunming. The tribes of the Kunming do not have a chieftain, and they are fond of banditry and robbery. They immediately murdered or abducted the Han envoys, and so in the end, no one was able to get through. Though they did hear that to the west by 1,000 miles or more was an elephant-riding country called Dian Yue and that the goods that merchants of Shu illegally exported sometimes made it to there. Therefore, it was because the Han was searching for a road to Daxia that they first came into contact with the Kingdom of Dian [in present-day Yunnan]. Earlier, the Han wanted to make contact with the southwestern barbarians, but the expense was very great, and the routes were all blocked, so they ceased their efforts. It was when Zhang Qian had said that they could communicate with Daxia by this route that they had once again made an important affair of making contact with the southwestern barbarians.

Zhang Qian's Subsequent Career

When Zhang Qian was sent out to attack the Xiongnu as a colonel serving under the supreme general [Wei Qing], he always knew the locations for water and pasture, and the army was able to be without shortages or privation. Then, he was enfeoffed as the Marquess of Bowang. This year was the sixth year of the Yuanshuo reign period [123 BCE]. The following year, Zhang Qian was made colonel of the guards, and along with General Li Guang rode out from Youbeiping to attack the Xiongnu. The Xiongnu surrounded General Li's army, and the casualties were very high, but Zhang Qian had arrived late at the rendezvous point and warranted execution but was able to redeem his crime by surrendering his ranks and becoming a freedman. This year, the Han had sent out the general of the swift cavalry, Huo Qubing, with several tens of thousands of troops, to defeat the Xiongnu's western territories, as far as Qilian Mountain. The following year [121 BCE], the subordinate Hunye King of the Xiongnu led his people to surrender to the Han, and the region west of Jincheng and Hexi and from the Southern Mountains to the Salt Marsh was now devoid of all Xiongnu presence. Occasionally, an advance scout of the Xiongnu would arrive, but this was rare indeed. Two years later [119 BCE], the Han attacked and drove the Chanyu's court north of the Gobi Desert.

Zhang Qian's Second Embassy to the West

[In 116 BCE, Zhang Qian was sent west again, to make allies with the Wusun: Abridged]

Zhang Qian thereupon sent out [from Wusun] his assistant envoys on separate embassies to Dayuan, Kangju, the Great Yuezhi, Daxia, Anxi, Shendu, Yutian, Jumi, as well as the various neighboring countries. The Wusun mobilized guides and translators to accompany Zhang Qian on his return journey and gave to him several tens of ambassadors and several tens of horses to requite the gifts from the emperor. Because of this, the Wusun were able to catch a glimpse of the Han and know something of its extent and greatness. When Zhang Qian finally returned to China, he was conferred with the title of superintendent of state visits [115 BCE] and ranked among the nine ministers of state. A little over a year later [114 BCE], he died.

Consequences of Zhang Qian's Embassies

When the ambassadors from the Wusun had seen how numerous and wealthy the people of China were and returned and reported this to their country, they now treated the Han as much more formidable. More than a year later, the assistant envoys that Zhang Qian had sent through to Daxia and the other countries all arrived back in China, accompanied by numerous people from those lands. From that time onward, the countries of the northwest were in contact with the Han, and it was Zhang Qian who "chiseled the hole" that made it possible. All later envoys invoked the name of the Marquess of Bowang [Zhang Qian] as a pledge of faithfulness towards the foreign countries, and because of this, those countries trusted them.

Sima Qian's Concluding Observations

The Grand Historian remarks, "The *Basic Annals of Yu the Great* [a lost ancient text] records, 'The source of the Yellow River is in the Kunlun Mountains, which are over 2,500 Chinese miles high where the sun and the moon in turn go to hide when they are not shining. On their summit, one finds the Sweetwater Spring and the Jade Lake [the abode of the Spirit Mother of the West].' Yet after the time when Zhang Qian was first sent as an envoy to Daxia, men have traced the source of the Yellow River and have seen no such Kunlun Mountains as the *Basic Annals* records. Therefore, what the *Book of Documents* [a key classical text] states about the mountains and rivers of the nine provinces of China seems to be nearer the truth, while when it comes to the strange creatures recorded in the *Basic Annals of Yu the Great* or the *Guideways Through Mountains and Seas* [chapter 1], I would not even venture to speak about such untrustworthy things."[10]

NOTES

1. Excerpted in Arrian, *Anabasis* (5.6.2–3), translation by John McCrindle, *Ancient India as Described by Megasthenes and Arrian* (Trübner, 1877), with updates to the spelling and diction.
2. Excerpted in Arrian, *Indica* (5.4–12), translation by McCrindle, *Ancient India*, 208–213, with a few corrections from Stoneman, *Megansthenes' Indica* (41–42), and updates to Americanize the spelling.
3. Excerpted in Arrian, *Indica* (11.1–12.9), translation by McCrindle, *Ancient India*, 208–213, with a few corrections from Stoneman, *Megansthenes' Indica* (58–59), and updates to modernize the spelling and phrasing.
4. Excerpted from Strabo, *Geography* (book 15, chapter 1, sections 35–36), translation by McCrindle, *Ancient India* (66–67), with a few corrections from Stoneman, *Megansthenes' Indica* (61–62), and updates to Americanize the spelling and phrasing.
5. Excerpted in Arrian, *Indica* (chapter 10.1–8), translation by McCrindle, *Ancient India* (67–68), with a few corrections from Stoneman, *Megansthenes' Indica* (60–61), and updates to Americanize the spelling and diction.
6. Excerpted in Arrian, *Indica* (chapters 13–14), translation by McCrindle, *Ancient India* (213–216), with a few corrections from Stoneman, *Megansthenes' Indica* (64–66), and updates to Americanize the spelling and phrasing.
7. Excerpted in Arrian, *Indica* (chapter 15), translation by McCrindle, *Ancient India* (217–218), with a few corrections from Stoneman, *Megansthenes' Indica* (54), and updates to Americanize the spelling.
8. Excerpted in Pliny, *Natural History* (book 7, chapter 2, lines 23, 25–26), translation based on H. Rackham, W. H. S. Jones, and D. E. Eichholz, trans., *Pliny, Natural History* (Harvard University Press, 1938–1963), 2:521, 523 (with modifications and updates to the spelling).
9. Excerpted from Strabo, *Geography* (book 15, chapter 1, sections 58–60), translation by McCrindle, *Ancient India* (97–103), with a few corrections from Stoneman, *Megansthenes' Indica* (69–71), and updates to modernize the spelling and phrasing.
10. Sima Qian 司馬遷, *Shi ji* 史記 (Zhonghua shuju, 1959), 123.3157–69, 3179. Translation and headings by the author.

CHAPTER 3

MONSOON WINDS AND THE LURE OF INDIAN LUXURIES (100 BCE–100 CE)

In this chapter, we explore the world of maritime trade, plying the great Indian Ocean, driven by the monsoon winds that made long-distance ocean exchanges possible. The Ptolemaic Kingdom that ruled Egypt had begun to sponsor voyages down the Red Sea to capture elephants for battle and later to acquire ivory, tortoiseshell, slaves, and spices. Eventually, they discovered from a shipwrecked Indian merchant how to sail across the open ocean to India, gaining access to India's treasured gems and pepper. When the Romans annexed Egypt in 30 BCE, this trade greatly expanded, facilitating access for the Roman elites to Indian luxuries such as pearls, diamonds, and aromatics, as well as silks brought overland from China, all of which they conspicuously displayed in their competition for status.

A similar scenario played out in East Asia, where the Qin dynasty had conquered the Yue territory (present-day Guangdong and Vietnam) between 221 and 214 BCE, gaining access to the luxury products of the south, such as rhinoceros horn, tortoiseshell, ivory, pearls, and kingfisher feathers, mostly for court display. The subsequent Han dynasty (206 BCE– 220 CE) reconquered these Yue territories in 111 BCE and relied on the experienced Yue sailors (who had plied the South China Sea for centuries) to take their envoys across the ocean. These trade envoys switched to local merchant vessels in Malaysia, and eventually reached southeastern India and Sri Lanka, searching for the same types of gemstones and fine pearls which attracted Roman merchants. However, Roman and Han voyagers did not traverse the entire Indian Ocean and reach each other's shores during this period. This great route between East and West, called the Maritime Silk Road by scholars, would remain a pivotal avenue of contact for the next two millennia.

INTRODUCTION TO *NAVIGATING AROUND THE RED SEA*

The *Períplous tēs Erythrâs Thalássēs* (Navigating Around the Red Sea), also known by its Latin title, *Periplus Maris Erythraei*, is a unique and invaluable manuscript text, written between 40 and 70 CE, by a Greek speaking merchant-sailor living in Egypt under Roman rule. It details ports and trading products along three different itineraries, all starting from the Egyptian Red Sea ports of Myos Hormos and Berenicê. I have abridged the portions that describe the voyage down the east coast of Africa (where ivory, rhinoceros horn, and slaves were acquired in exchange for textiles and metal) and the itinerary from the Red Sea to the bustling ports of southern Arabia (where frankincense was the main attraction), picking up with the detailed account of trade with western India.

Sailing across the open Arabian Sea of the Indian Ocean was only made possible by the phenomenon of the shifting monsoon winds. These blow powerfully in the western Indian Ocean from the southwest during June to September then shift to a milder prevailing northwesterly from November to April. Timing one's sailing to accord with these wind patterns was crucial, for missing the proper window could result in being stranded for up to nine months in a foreign land. If a merchant timed the legs of his voyages perfectly, he could manage a round trip from Alexandria to western India and back in twelve months. If his merchandise started in Rome, that could be extended to as many as 20 months.

Roman goods were first sailed upstream on the Nile from Alexandria to Koptos (2 weeks) in May, then overland by camel to Myos Hormos (6–7 days) or Berenicê (11–12 days) on the Red Sea. The ships had to depart the Red Sea ports by mid-July to arrive at the southern Arabian ports (like Kanê, 30 days) by mid- to late August, in time to make the monsoon-driven jump across the ocean to India (two to four weeks), ideally arriving by September at the Indian ports of Barbarikon, Barygaza, or Muziris (see map 2). They traded for a few months along the Indian coast before beginning their return voyage by early December (to time the winds), arriving back in Arabia by February and Alexandria by April, where they faced a 25 percent import duty.

The *Navigating* author claims that using the monsoon winds to travel across the open ocean to India was discovered by a Greek pilot named Hippalos, but Indian and Arab sailors were probably already quite familiar with that route. The geographer Strabo tells us of another Greek sailor named Eudoxus of Cyzicus who sailed twice to India and back (ca. 118–107 BCE), the first time under orders of Ptolemy VIII of Egypt, guided by a shipwrecked seaman from India. With the Roman annexation of Egypt in 30 BCE, the volume of trade with Arabia and India accelerated, reaching its peak shortly after the time the *Navigating* text was written.

Map 2. Itineraries of *Navigating Around the Red Sea* and *Voyage to Huangzhi*.

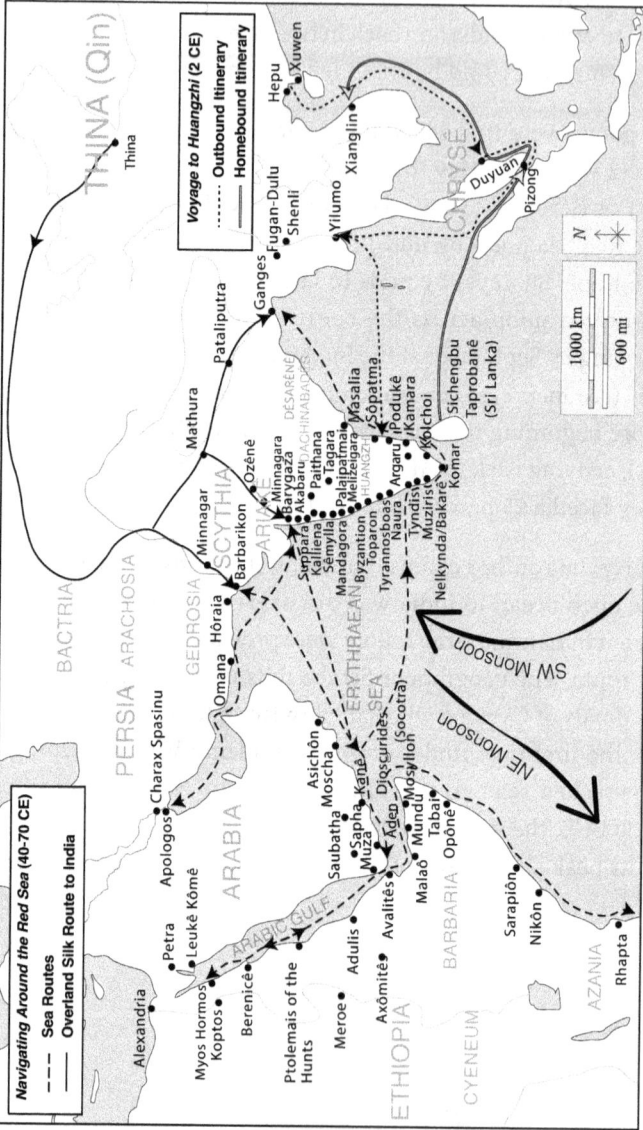

Source: Map created by Zhang Yexu and the author.

The *Navigating* account reveals a trade economy in the Indian Ocean that was largely driven by a core-periphery dynamic. The Roman world exported commodities like textiles, metals, wine, raw glass, and gold and silver coinage in exchange for ivory, tortoiseshell, pearls, gemstones, spices, aromatics, and Chinese silks, items that fed conspicuous consumption and display by the Roman elites and middle class. Even though the luxury trade was the most lucrative, merchants also traded in other commodities, since ships needed bulky ballast and no profitable opportunity could be wasted. It is only through this text that we learn of the southern extension of the Silk Road from China, which turned south near Bactria and made silk available to western merchants in all the major trading ports of India, bypassing the caravan trade through Parthian-controlled domains.

The *Navigating* author, like other merchant writers in this anthology such as the Arab sailors in chapter 6, showed great interest in trade products, exchange rates, and ports of call but were also curious about local customs, laws, and political conditions, for these influenced trade. Unlike religious pilgrims such as Faxian and Xuanzang (chapter 5), the *Navigating* author is mostly indifferent to religious practices and pilgrimage sites, and unlike diplomats such as Zhang Qian (chapter 2) or Lord Macartney (chapter 11), he cares little about military strength and technology unless they affected trade.

The *Navigating* author relates far less fantastic tales about India than either Ctesias (chapter 1) or Megasthenes (chapter 2), especially regarding the southwestern coast, with which he was quite familiar from experience. But when his account moves up the eastern coast of India and into the hinterland, where his information was no longer firsthand, he does mention barbaric men with "flattened noses" or cannibals with "horse-faces," but these might be hearsay accounts of northeast Asian peoples with more Mongolian features.

The *Navigating* author's knowledge is even hazier about the lands east of India, including the Malaysian Peninsula and Sumatra (Chrysê) and China (Thina), confirming that merchants from the Roman Empire probably didn't travel much beyond familiar ports in southeastern India before 166 CE. The *Navigating* text does contain the earliest reference in any Greco-Roman text to the Qin Empire (Thina), which is the origin of our word "China." The Qin dynasty fell in 206 BCE, so this name was already quite out of date when the *Navigating* text was written, but its power still reverberated westward.

PRIMARY SOURCE, ALTERNATE ENGLISH TRANSLATIONS, AND SELECTED STUDIES

Casson, Lionel. *The Periplus Maris Erythraei: Text with Introduction, Translation and Commentary.* Princeton University Press, 1989.

Cobb, Matthew Adam. *Rome and the Indian Ocean Trade from Augustus to the Early Third Century CE.* Brill, 2018.

Schoff, William H. *The Periplus of the Erythraean Sea: Travel and Trade in the Indian Ocean by a Merchant of the First Century.* Longmans, 1912.

EXCERPTS FROM *NAVIGATING AROUND THE RED SEA*

Ports on the Red Sea

Of the designated ports on the Erythraean Sea [incorporating the Red Sea, Gulf of Aden, and western Indian Ocean] and the market towns around it, the first is the Egyptian port of Myos Hormos [present-day Quseir al-Qadim; see map 2]. To those sailing down from that place, on the right hand, after 1,800 stadia [approx. 333 km], there is Berenicê [present-day Cape of Ras Banas]. The harbors of both are at the boundary of Egypt and are bays opening from the Erythraean Sea.

Ports along Red Sea, Somali Peninsula, Arabian Peninsula, and Persian Gulf
[Abridged]

Sindh and Gujarat Regions

Beyond this [Baluchistan] region, the continent making a wide curve from the east across the depths of the bays, there follows the coast district of Scythia [present-day Sindh Province, Pakistan], which lies above toward the north, the whole marshy, from which flows down the Sinthos River [Indus], the greatest of all the rivers that flow into the Erythraean Sea, bringing down an enormous volume of water, so that a long way out at sea, before reaching this country, the water of the ocean is fresh from it. Now as a sign of approach to this country to those coming from the sea, there are serpents coming forth from the depths to meet you, and a sign of the places just mentioned and in Persia are those called *graai* [Sanskrit: *grāha*, "rapacious sea animal"]. This river has seven mouths, very shallow and marshy, so that they are not navigable, except the one in the middle, at which, by the shore, is the market town Barbarikon. Before it there lies a small island, and inland behind it is the metropolis of Scythia, Minnagar ["Saka-town"]; it is subject to Parthian princes who are constantly driving each other out.

The ships lie at anchor at Barbarikon, but all their cargoes are carried up to the metropolis by the river, to the king. There are imported into this market a great deal of thin, unadorned clothing, figured linens, *chrysolithon* ["golden stone," probably peridot], coral, storax, frankincense, vessels of glass, silver plate, money, and a little wine. On the other hand, there are exported costus root, bdellium, *lykion* [probably cutch extract], nard oil, turquoise, and lapis lazuli; furs, cloth, and silk yarn of the "Silk People" [the Chinese]; and indigo. And sailors set out thither with the Indian winds, about the month of July, that is, Epeiph; it is more dangerous then, but through these winds, the voyage is more direct and sooner completed.

Beyond the Sinthos River there is another gulf, not navigable, running in toward the north; it is called Eirinon [Sanskrit: Iriṇa, "salt desert," present-day Rann of Kutch]; its parts are called separately the small gulf and the great; in both parts, the water is shallow, with shifting sandbanks occurring continually and a great way from shore, so that very often when the shore is not even in sight, ships run aground, and if they attempt to hold their course, they are wrecked. A promontory stands out from this gulf, curving around from Eirinon toward the east, then south, then west, and enclosing the gulf called Barakê [present-day Gulf of Kutch], which contains seven islands. Those who come to the entrance of this bay escape it by putting about a little and standing farther out to sea, but those who are drawn inside into the Gulf of Barakê are lost, for the waves are high and very violent, and the sea is tumultuous and foul and has eddies and rushing whirlpools. The bottom is in some places abrupt and in others rocky and sharp, so that the anchors lying there are parted, some being quickly cut off, and others chafing on the bottom. As a sign of these places to those approaching from the sea there are serpents, very large and black, for at the other places on this coast and around Barygaza [Gulf of Khambhat], they are smaller, and in color bright green, running into gold.

Beyond the Gulf of Barakê is the Gulf of Barygaza and the coast of the country of Ariakê ["of the Aryans"], which is the beginning of Manbanos's [the Saka ruler Nahapāna, r. first c. CE] kingdom and of all India. That part of it lying inland and adjoining Scythia is called Abêria, but the coast is called Syrastrênê [Kathiawar Peninsula]. It is a fertile country, yielding wheat and rice and sesame oil and clarified butter, cotton and the Indian cloths made therefrom, of the coarser sorts. Very many cattle are pastured there, and the men are of great stature and black in color. The metropolis of this country is Minnagara, from which much cotton cloth is brought down to Barygaza [present-day Broach, India]. In these places there remain even to the present time signs of the expedition of Alexander, such as ancient shrines, walls of forts and great wells. The sailing course along this coast, from Barbarikon to the promontory called Papikê [Kuda Point], opposite Barygaza, and before Astakapra [Kūkad, India], is of 3,000 stadia [approx. 555 km].

Beyond this there is another gulf exposed to the sea waves, running up toward the north, at the mouth of which there is an island called Baiônês [Piram Island]; at its innermost part there is a great river called Mais [Mahi River]. Those sailing to Barygaza pass across this gulf, which is 300 stadia in width [approx. 55.5 km], leaving behind to their left the island just visible from their tops toward the east, straight to the very mouth of the river of Barygaza, and this river is called Lamnaios [Narmada River].

This gulf is very narrow to Barygaza and very hard to navigate for those coming from the ocean; this is the case with both the right and left passages, but there is a better passage through the left. For on the right at the very mouth of the gulf there lies a shoal, long and narrow and full of rocks, called Hêrônê, facing the village of Kammôni, and opposite this on the left projects the promontory that lies before Astakapra, which is called Papikê, and is a bad anchorage because of the strong current setting in around it and because the anchors are cut off, the bottom being rough and rocky. And even if the entrance to the gulf is made safely, the mouth of the river at Barygaza is found with difficulty, because the shore is very low and cannot be made out until you are close upon it. And when you have found it, the passage is difficult because of the shoals at the mouth of the river.

Because of this, native fishermen in the king's service, stationed at the very entrance in well-manned large boats called *trappaga* and *kotymba*, go up the coast as far as Syrastrênê, from which they pilot vessels to Barygaza. And they steer them straight from the mouth of the bay between the shoals with their crews, and they tow them to fixed stations, going up with the beginning of the flood, and lying through the ebb at anchorages and in basins. These basins are deeper places in the river as far as Barygaza; which lies by the river, about 300 stadia [approx. 55.5 km] up from the mouth.

Now the whole country of India has very many rivers, and very great ebb and flow of the tides, increasing at the new moon, and at the full moon for three days, and falling off during the intervening days of the moon.

But about Barygaza it is much greater, so that the bottom is suddenly seen, and now parts of the dry land are sea, and now it is dry where ships were sailing just before. And the rivers, under the inrush of the flood tide, when the whole force of the sea is directed against them, are driven upward more strongly against their natural current, for many stadia.

For this reason, entrance and departure of vessels is very dangerous to those who are inexperienced or who come to this market town for the first time. For the rush of waters at the incoming tide is irresistible, and the anchors cannot hold against it, so that large ships are caught up by the force of it, turned broadside on through the speed of the current, and so driven on the shoals and wrecked, and smaller boats are overturned, and those that have been turned aside among the channels by the receding waters at the ebb are left on their sides, and if not held on an even keel by props, the flood tide comes upon them suddenly, and under the first head of the current, they are filled with water. For there is so great a force in the rush of the sea at the new moon, especially during the flood tide at night, that if you begin the entrance at the moment when the waters are still, on the instant there is borne to you at the mouth of the river, a noise like the cries of an army heard from afar; and very soon the sea itself comes rushing in over the shoals with a hoarse roar.

The country inland from Barygaza is inhabited by numerous tribes, such as the Aratrioi [the Ārattas in Punjab], the Arachosioi [in southern Afghanistan], the Gandharans [near Peshawar, Pakistan] and the people of Proklais [Chārsadda in Gandhara], in which is the city of Alexandria Bucephalus. Above these is the very war-like nation of the Bactrians [the Kushans], who are under their own king. And Alexander, setting out from these parts, penetrated to the Ganges, leaving aside Limyrikê [the Malabar coast] and the southern part of India; and to the present day ancient drachma coins are current in Barygaza, coming from this country, bearing inscriptions in Greek letters, and the devices of those kings who reigned after Alexander, namely Apollodotus [ca. 180–160 BCE] and Menander [r. 155–130 BCE].

Inland from this place [Barygaza] and to the east is the city called Ozênê [present-day Ujjain], formerly a royal capital; from this place are brought down all things needed for the welfare of the country about Barygaza and many things for our trade: onyx, agate, Indian cotton garments, and much ordinary cloth. Through this same region and from the upper country is brought the nard that comes through Proklais, that is, from Kattyburinê [Kashmir], Patropapigê [Hindu Kush], and Kabalitê [Kabul], and that nard brought through the adjoining country of Scythia, also costus and bdellium.

There are imported into this market town [of Barygaza], wine, Italian preferred, also Laodicean and Arabian; copper, tin, and lead; coral and peridot; thin clothing and inferior sorts of all kinds; bright-colored girdles a cubit wide; storax, yellow sweet clover, raw glass, realgar, antimony, gold and silver Roman coins, on which there is a profit when exchanged for the money of the country; and ointment, but not very costly and not much. And for the king, there are brought into those places very costly vessels of silver, singing boys, beautiful maidens for the harem, fine wines, thin clothing of the finest weaves, and the choicest ointments. There are exported from these places spikenard, costus, bdellium, ivory, onyx and agate, *lykion*, cotton cloth of all kinds, Chinese silk cloth, *molochinon* [fine cotton?] cloth, yarn, long pepper, and such other things as are brought here from the various market towns. Those bound for this market town from Egypt make the voyage favorably about the month of July, that is, Epeiph.

Maharashtra Region and Deccan Plateau

Beyond Barygaza, the adjoining coast extends in a straight line from north to south, and so this region is called Dachinabadês [Deccan], for *dachanos* [Sanskrit: *dakṣiṇa*] in the language of the natives means "southern." The inland country back from the coast toward the east comprises many desert regions and great mountains and all kinds of wild beasts—leopards, tigers, elephants, enormous serpents, *krokottas* [probably hyenas], and "dog-headed" creatures of many sorts—and many populous nations, as far as the Ganges.

Among the market towns of Dachinabadês there are two of special importance: Paithana [present-day Paithan], distant about 20 days' journey south from Barygaza, beyond which, about 10 days' journey east, there is another very great city, Tagara [present-day Tēr]. There are brought down to Barygaza from these places by wagons and through great tracts without roads, from Paithana onyx in great quantity, and from Tagara much common cloth, all kinds of cotton muslins and *molochinon* cloth, and other merchandise brought there locally from the regions along the [eastern] seacoast. And the whole course to the end of Limyrikê [Malabar Coast] is 7,000 stadia [approx. 1,295 km], but the distance is greater to the Coast Country [present-day Palk Straight, separating India and Sri Lanka].

The market towns of this region are, in order after Barygaza, Akabaru, Suppara [present-day Vasai], and the city of Kalliena [present-day Kalyān], which in the time of the elder Saraganos [either Sātakarṇi I or II, r. ca.70–25 BCE] became a lawful market town, but since it came into the possession of Sandanês, the port is much obstructed, and Greek ships landing there may chance to be taken to Barygaza under guard.

Karnataka Coast

Beyond Kalliena there are other market towns of this region: Sêmylla [present-day Chaul], Mandagora [Bankōt], Palaipatmai [Dābhol], Melizeigara [Jaigad], Byzantion [Vijaydurg], Toparon, and Tyrannosboas. Then there are the islands called Sêsekreienai [Vengurla Rocks], and that of the Aigidioi [Grand Island, Goa], and that of the Kaineitoi [Oyster Rocks], opposite the place called the Peninsula, and in these places there are pirates, and after this the White Island [Netrani Island]. Then come Naura [Mangalore] and Tyndis [Ponnāni], the first markets of Limyrikê, and then Muziris [Pattanam] and Nelkynda [Niranam], which are now of leading importance.

Malabar Coast

Tyndis is of the Kingdom of Kêrobotros [Keralaputra; Chera dynasty]; it is a village in plain sight by the sea. Muziris, of the same kingdom, abounds in ships sent there with cargoes from Arabia, and by the Greeks; it is located on a river, distant from Tyndis by river and sea 500 stadia, and up the river from the shore 20 stadia. Nelkynda is distant from Muziris by river and sea about 500 stadia and is of another kingdom, the Pandiôn [Pandya]. This place also is situated on a river, about 120 stadia [22.2 km] from the sea.

There is another place at the mouth of this river, the village of Bakarê [Purakkad], to which ships drop down on the outward voyage from Nelkynda, and anchor in the roadstead to take on their cargoes, because the river is full of shoals and the channels are not clear. The kings of both these market towns live in the interior. And as a sign to those approaching these places from the sea, there are serpents coming forth to meet you, black in color, but shorter, like snakes in the head, and with blood-red eyes.

They send large ships to these market towns [Muziris and Nelkynda] on account of the great quantity and bulk of pepper and malabathrum. There are imported here, in the first place, a great quantity of coin; peridot, thin clothing, not much; multicolored textiles, antimony, coral, raw glass, copper, tin, lead; wine, not much, but as much as at Barygaza; realgar and orpiment; and wheat enough for the sailors, for this is not dealt in by the merchants there. There is exported pepper, which is produced in quantity in only one region near these markets, a district called Kottanarikê [Kuṭṭanāḍu]. Besides this there are exported great quantities of fine pearls, ivory, Chinese silk cloth, nard from the Ganges, malabathrum from the places in the interior, transparent stones of all kinds, diamonds and sapphires, and tortoiseshell; that from Chrysê Island [Malaysian Peninsula and Sumatra], and that taken among the islands along the coast of Limyrikê [Laccadive Islands]. They make the voyage to this place in a favorable season who set out from Egypt about the month of July, that is, Epeiph.

Hippalos and the Monsoon Winds

This whole voyage as above described, from Kanê and Eudaimôn Arabia, they used to make in small vessels, sailing close around the shores of the gulfs. And Hippalos [ca. 116 BCE] was the pilot who by observing the location of the ports and the conditions of the sea first discovered how to lay his course straight across the ocean. For at the same time when with us the Etesian winds are blowing, on the shores of India the wind sets in from the ocean, and this southwest wind is called Hippalos, from the name of him who first discovered the passage across. From that time to the present day, ships start, some direct from Kanê and some from the Cape of Spices [Cape Guardafui, Somalia], and those bound for Limyrikê [Malabar Coast] throw the ship's head considerably off the wind, while those bound for Barygaza and Scythia [Barbarikon] keep along shore not more than three days and for the rest of the time hold the same course straight out to sea from that region, with a favorable wind, quite away from the land, and so sail outside past the aforesaid gulfs.

Kerala and Tamil Nadu Regions

Beyond Bakarê there is the Dark Red Mountain [present-day Varkala] and another district stretching along the coast toward the south called the Seaboard. The first place is called Balita; it has a fine harbor and a village by the shore. Beyond this there is another place called Komar, at which are the Cape of Komari [Cape Comorin] and a harbor; hither come those men who wish to consecrate themselves for the rest of their lives and bathe and dwell in celibacy and women also do the same, for it is told that a goddess [Durgā] once dwelt here and bathed.

From Komar toward the south this region extends to Kolchoi [present-day Korkai], where the pearl fisheries are—they are worked by condemned criminals—and it belongs to the Pandiôn Kingdom. Beyond Kolchoi there follows another district called the Coast Country [Palk Bay], which lies on a bay and has a region inland called Argaru [present-day Uraiyur]. At this place, and nowhere else, are bought the pearls gathered on the coast thereabouts, and from there are exported cotton muslins, those called Argaritides.

* * * * * * * * * * * * * * BRIEF EXCERPT * * * * * * * * * * * * * * * *

Coromandel Coast and Sri Lanka

Among the market towns of these countries and the harbors where the ships put in from Limyrikê and from the north [Ganges delta], the most important are, in order as they lie, first Kamara [mouth of Kāveri River], then Podukê [Arikamedu site, Puducherry], then Sôpatma [possibly Chennai], in which there are ships of the country coasting along the shore as far as Limyrikê and other very large vessels made of single logs bound together, called *sangara*, but those which make the voyage to Chrysê and to the Ganges are called *kolandiophônta* [Chinese: *kunlun bo* 崑崙舶, "SE Asian ships"], and are very large. There are imported into these places everything made in Limyrikê, and the greatest part of what is brought at any time from Egypt comes here, together with most kinds of all the things that are brought from Limyrikê and of those that are carried along this coast.

About the following region, the course trending toward the east, lying out at sea toward the west is the island Palaisimundu, called by the ancients Taprobanê [Sri Lanka]. The northern part is civilized, and the southern part trends gradually toward the west and almost touches the opposite shore of Azania [East Africa]. It produces pearls, transparent stones, cotton muslins, and tortoiseshell.

Andhra Pradesh and Odisha Regions

About these places is the region of Masalia [present-day Masulipatam] stretching a great way along the coast before the inland country; a great quantity of cotton garments is made there. Beyond this region, sailing toward the east and crossing the adjacent bay, there is the region of Dêsarênê [present-day Odisha area], the habitat of an elephant known as Bôsarê. Beyond this, the course trending toward the north, there are many barbarous tribes, among whom are the Kirradai [Kirātas], a race of men with flattened noses, very savage; another tribe, the Bargysoi [Bhargas]; and the Horse-Face People, who are said to be cannibals.

Ganges Region, Malaysia, and Sumatra

After these, the course turns toward the east again, and sailing with the ocean to the right and the shore remaining beyond to the left, the Ganges region comes into view, and near it the very last land toward the east, Chrysê ["Golden"; indicating the Malaysian Peninsula and Sumatra]. There is a river near it called the Ganges, and it rises and falls in the same way as the Nile. On its bank is a market town that has the same name as the river, Ganges. Through this place are brought malabathrum and Gangetic nard and pearls, and muslin cotton garments of the finest sorts, which are called Gangetic. It is said that there are gold mines near these places, and there is a gold coin which is called *kaltis*. And just opposite this river there is an island in the ocean, the last part of the inhabited world toward the east, under the rising sun itself; it is called Chrysê [Sumatra], and it has the best tortoiseshell of all the places on the Erythraean Sea.

China

After this region under the very north, the sea outside ending in a land called Thina ["Qin"; i.e., China], there is a very great inland city called Thina, from which silk floss, yarn, and cloth are brought on foot through Bactria to Barygaza and are also exported to Limyrikê by way of the river Ganges. But this Thina is not easy of access; few men come from there, and seldom. The country lies under the constellation Ursa Minor and is said to border on the farthest parts of the Pontus [Black Sea] and the Caspian Sea, next to which lies Lake Maeotis [Sea of Azov]; all of which empty into the northern ocean.

Every year on the borders of the land of Thina there comes together a tribe of men with short bodies and broad, flat faces, and by nature peaceable; they are called Sêsatai, and are almost entirely uncivilized. They come with their wives and children, carrying great packs and plaited baskets of what looks like green grape-leaves. They meet in a place between their own country and the land of Thina [the eastern Himalayan foothills]. There they hold a feast for several days, spreading out the baskets under themselves as mats, and then return to their own places in the interior. And then the natives watching them come into that place and gather up their mats, and they pick out from the braids the fibers which they call *petroi*. They lay the leaves closely together in several layers and make them into balls, which they pierce with the fibers from the mats. And there are three sorts; those made of the largest leaves are called the large-ball malabathrum [aromatic leaves of *Cinnamomum tamala*]; those of the smaller, the medium-ball; and those of the smallest, the small-ball. Thus, there exist three sorts of malabathrum, and it is brought into India by those who prepare it.

The regions beyond these places are either difficult of access because of their excessive winters and great cold or else cannot be sought out because of some divine influence of the gods.[1]

* *

INTRODUCTION TO THE *VOYAGE TO HUANGZHI*

The *Voyage to Huangzhi* is the title I have given to an itinerary starting from coastal South China and voyaging to India and Sri Lanka, written by the Han historian Ban Gu 班固 (32–92 CE) and included in the "Treatise on Administrative Geography" (Dili zhi 地理志) in his *History of the Han* (Han shu 漢書).

The outbound voyage began at the ports of Hepu (in Gulf of Tonkin) and Xuwen (in the nearby Hainan Strait), arriving after five months at a country called Duyuan, which was probably on the eastern coast of the

Malaysian peninsula (see map 2). Another four months voyage brought
the boats to a place called Yilumo, likely in present-day Myanmar, at
the mouth of the Irrawaddy. The itinerary then takes one upriver on the
Irrawaddy for 20 days then briefly overland to arrive at a kingdom called
Fugan-Dulu, probably Pagan in central Myanmar. After sailing back
down the Irrawaddy, the itinerary then goes back out to sea and crosses
the Bay of Bengal to southeastern India. Though scholars disagree on the
identification of several of the intermediate points on this journey, there
is consensus that the final destination of "Huangzhi" is a transliteration
of Kanchi, the great holy city in southeastern India. While in the Kanchi
area, the Chinese trade embassies probably saw Greco-Roman merchants
for the first time. According to the *Navigating* text, these merchants
were actively trading in this same area and had established a merchant
diaspora community at Podukê (Arikamedu site, Puducherry), not far
from Kanchi. This also explains why the *Navigating* text uses a Greek
transliteration of a Chinese term for the large Malay ships that plied these
water (Greek: *kolandiophônta*; Chinese: *kunlun bo* 崑崙舶, "SE Asian
ships"). The return voyage started from a country south of Huangzhi
called Sichengbu, probably Sri Lanka, then traveled through the Straits of
Malacca, stopping off at Pizong (Pisang Island) before the voyage north
to Han territory in central Vietnam.

Though the timing and direction of the monsoon winds in East Asia
and South Asia are slightly different from those in the western Indian
Ocean, it still should have been possible to make this round-trip voyage
in a single year. However, the *Voyage to Huangzhi* says that some trade
envoys could take several years to return (if they managed not to die).
The first leg of the voyage from China was undoubtedly carried out
on large Chinese oceangoing merchant ships. Han nautical technology
was quite advanced, and pottery tomb models of ships show that the
Chinese had already begun to employ a sternpost rudder by that time,
centuries before such advantageous devices were used in the West.
However, the text then mentions that the Chinese trade envoys were
transferred to "barbarian merchant ships" toward their final destination.

This transshipment probably took place in eastern Malaysia or Sumatra, because the Chinese sailors were less familiar with the dangerous waters of the Straits of Malacca or the Bay of Bengal.

Roman trade in the Indian Ocean was facilitated by the state to a certain degree, for it provided garrisoned waypoints across the Eastern Desert in Egypt, the first short leg of the journey, and possibly some garrisons in ports in the Red Sea and India. The risk and reward for the entire operation was mostly the responsibility of the merchants themselves, financed by wealthy families. The Roman state was more than happy to tax the imports at the conclusion, at a fixed rate of 25 percent. The Chinese trading missions described by Ban Gu seem entirely state sponsored, staffed by eunuch officials who were responsible for decorating the Han palaces and outfitting the harem. But given that some of these Indian imports also ended up in other wealthy Han tombs, there must have been private trading ventures as well. The Chinese trade envoys brought gold bullion and plain silk and exchanged these for large Indian pearls and gemstones, as well as semi-opaque glass which had been imported from the Roman Empire. This was one avenue by which Roman glass entered China. Certain amounts also arrived overland on the traditional Silk Road, via Kashmir.

The text claims that these Indian and Southeast Asian kingdoms had been sending tribute missions to the Chinese court since the late second century BCE, but there is no record of these early missions in the imperial annals. There is confirmation in those annals of a live rhinoceros that was presented in tribute by the king of Huangzhi in 2 CE. This was a political stunt by the regent and soon-to-be usurper Wang Mang (45 BCE–23 CE), who sent envoys to ply the king of Huangzhi with presents, prompting him to send the magnificent beast. Chinese tradition said that when a sage ruler occupies the throne, barbarian tribes would send tribute of auspicious animals. Wang Mang used the Indian gift of the rhinoceros as an omen that supported his eventual usurpation of the Han throne. The same political ploy would be attempted again by the Yongle Emperor of the Ming dynasty, during the voyages of Zheng He in the 15th century (chapter 9), when he interpreted the gift of a giraffe by the king of Malindi as the auspicious Chinese *qilin* beast, another omen of sagely rule.

PRIMARY SOURCE AND ALTERNATE ENGLISH TRANSLATION

Ban Gu 班固, *Han shu* 漢書. Zhonghua shuju, 1962.

Wang Gungwu, "A Passage to Huang-chih," in *The Nanhai Trade: The Early History of Chinese Trade in the South China Sea*. Times Academic Press, 1998.

* * * * * * * * * * * * * * * BRIEF EXCERPT * * * * * * * * * * * * * * * *

Voyage to Huangzhi, by Ban Gu

Starting from the frontier barriers in Rinan [Vietnam], that is from Xuwen [present-day Zhanjiang] or Hepu [on Gulf of Tonkin], voyaging by boat for around five months, one arrives at the kingdom of Duyuan [Kuala Dungun, in eastern Malaysia]. Again, traveling by boat for around four months, one arrives at the kingdom of Yilumo [probably Bago region of Myanmar]; Again, voyaging by boat for approximately 20 days or more, one arrives at the kingdom of Shenli [in Myanmar, upstream on the Irrawaddy]. From there, one travels on foot for about ten days or more and will arrive at the kingdom of Fugan-Dulu [Pagan, central Myanmar]. From Fugan-Dulu, going by boat for two months or more, one arrives at the kingdom of Huangzhi [Kanchi in southeastern India].

The customs of the people [in these countries] are of a similar type to those of Zhuyai Commandery [Hainan Island]. These countries are quite expansive, their populations large, and many strange products are found there. Ever since the reign of Emperor Wu [r. 140–87 BCE], they have all come to court to present tribute.

There are chief interpreters subordinate to the director of eunuch attendants who go to sea with those men who answer their muster call. They buy bright pearls, *biliuli* [Sanskrit: *vaiḍūrya* "beryl," referring to semi-opaque Roman glass], rare gemstones, and strange beasts, being entrusted with gold and various plain silks to offer in exchange. They receive food rations in the countries where they arrive and are also provided with female companionship. They are transferred to merchant ships of the "southern barbarians" and delivered to their final destination. The barbarians consider trade profitable but also engage in piracy and murder.

The Chinese travelers suffer from the hardship of encountering high winds and waves, along with the danger of drowning. Even those who don't perish can still take several years to travel there and return. But the largest pearls can be just under two Chinese inches [approx. 4.75 cm] in circumference.

In the middle of the Yuanshi reign period of Emperor Ping of the Han [r. 1–5 CE], Wang Mang acted as regent for the ruler. He wanted to dazzle the court with his majestic inner virtue, so he sent generous gifts to the king of Huangzhi and instructed him to send envoys bearing a live rhinoceros as tribute. Traveling from Huangzhi by boat for about eight months, they arrived at Pizong [Pisang Island in the Straits of Malacca]. And after another voyage of about two months, they arrived back in Rinan [Vietnam, in 2 CE], within the jurisdiction of Xianglin County.

South of the kingdom of Huangzhi, there is the country of Sichengbu [Sanskrit: Siṃhaladvīpa, Sri Lanka]. The interpreters and envoys of the Han court returned from here.[2]

* *

NOTES

1. Translation by Wilfred H. Schoff, *The Periplus of the Erythraean Sea* (Longmans, 1912), 22, 37–49, with corrections, updated spellings, or alternate readings from Lionel Casson, *The Periplus Maris Erythraei* (Princeton University Press, 1989). Headings inserted by Anthony J. Barbieri-Low.
2. Ban Gu 班固, *Han shu* 漢書 (Zhonghua shuju, 1962), 28C.1671. Translation by the author.

CHAPTER 4

HAZY NOTIONS OF FAR-OFF EMPIRES (77–265 CE)

Although Greco-Roman merchant sailors operating from Red Sea ports and official Chinese trading envoys sailing from the Gulf of Tonkin had each discovered how to employ the seasonally shifting monsoon winds to reach the shores of India by the first century BCE, Roman knowledge of the Far East was still very vague at that time, and the Chinese had never even heard of the Roman Republic or early Empire. Chinese first-hand knowledge of the Far West ended with Bactria, Parthian Iran, and India, and even their vague second-hand information stopped at Mesopotamia and Ptolemaic Egypt (chapter 2). With the continued vigorous expansion of the Chinese and Roman empires, geographic knowledge of distant worlds expanded commensurately.

Neither of the paired authors in this chapter, the Roman naturalist Pliny the Elder or the Chinese historian Yu Huan, actually traveled to the far-off lands that they described, but neither were their accounts fantastic imaginings, for they had access to the reports of travelers and emissaries related to their respective expanding empires. Their accounts are a curious mixture of accurate information concerning trade products, ports, and routes of contact combined with legend, hearsay, gross misunderstandings, and utopian projections.

Roman Knowledge of the Chinese World

In Greek and Latin texts, domains at the far eastern edge of the world were called by two different names, both identified by modern scholars as referring to China, or areas under Chinese influence. Whether these realms were known through the central overland Silk Road or via the Maritime Silk Road through the Indian Ocean largely determined which name was employed.

Overland, these people were known as the Seres, or "Silk People," and their land was called Serice, both ultimately deriving from the Chinese word for silk, si 絲. This is the older of the two Roman names for the Far East, dating to before the flourishing of the Indian Ocean trade. Over the maritime route, the realm of the Far East was called either Thina (such as in the *Navigating Around the Red Sea* text, chapter 3) or Sinae (as in the work of Claudius Ptolemy), with both of these names deriving from the dynastic name of Qin 秦.

The Roman author Pliny the Elder wrote at some length about the Seres and their products and customs (excerpted below). He drew some of his information from a royal envoy named Rachias, who arrived in Rome from Sri Lanka and who talked about the very tall "flaxen haired and blue-eyed" Silk People who lived to the north of India beyond the Himalayas. Some scholars think that these blue-eyed "Chinese" were Indo-European speaking groups who lived in the Tarim Basin.

The geographer Claudius Ptolemy (fl. ca. 100 CE), in his *Guide to Drawing a Map of the World* (Geōgraphikē hyphēgēsis), extended Greco-Roman conceptions of the inhabited world well eastward, past India and Sri Lanka, to include both of these East Asian realms, the Seres ("Silk People") to the north and the Sinae ("Qin people") to their south. For the section of his book focusing on Asia, Ptolemy drew heavily on the work of his predecessor, Marinus of Tyre (fl. ca. 100–110 CE), who in turn gathered his information from the accounts of sailors, like a merchant named Alexander who sailed from Cape Comorin (at the southern tip of India) to the port of Kattigara (Mekong Delta), as well as from accounts of overland travelers such as the Macedonian Maes Titianus, who sent agents to northwest China to trade for silk around 100 CE.[1] This might be the same "embassy" recorded in Chinese sources as coming from the country of Mengqidoule 蒙奇兜勒 ("Macedonia") in 100 CE.[2] Ptolemy states that knowledge of China first came to Rome because of commercial interests (i.e. silk trade).

Ptolemy gives precise coordinates for the capitals of the Seres and the Sinae, which belies how hazy his knowledge truly was. His description of Serice (the land of the Seres), surrounded by high mountains with a river running through it, should be identified with the Tarim Basin and not the Yellow River Valley.[3] It seems that when Greco-Roman traders or their agents traveled east overland to trade in silk, they seem to have gone no farther than the oasis cities of Xinjiang. The Sinae to their south were known to Ptolemy from the Indian Ocean route, which ended at a port called Kattigara, probably located in Vietnam's Mekong Delta.[4] So when Roman-era sources speak of the customs and appearance of the people of the Seres and Thina/Sinae, they are actually referring to persons in the oasis cities of Xinjiang and the peripheral Yue people of far southern China and Vietnam, respectively, and not to the traditional Chinese culture along the Yellow River and Yangzi.

CHINESE KNOWLEDGE OF THE ROMAN WORLD

The earliest descriptions of the wealthy and powerful kingdom of Da Qin 大秦 ("Greater Qin"), which most modern scholars have identified as the Roman Empire, are found in sources compiled during the first to fifth centuries CE in China. The geographical reference to Da Qin is rather fluid, for sometimes it seems to indicate the city of Rome in Italy, while other times it might refer to the empire as a whole or just to its eastern provinces in Egypt or Syria. Occasionally, it seems to vaguely refer to a fantastic realm at the edge of the world. The most detailed account of Da Qin in an early Chinese text is found in Yu Huan's 魚豢 *Brief History of the Wei* (Weilüe 魏略, ca. 239–265 CE), which is translated in this chapter.

Subsequently, the historian Fan Ye 范曄 (398–446) compiled from earlier sources his "Account of the Western Regions" (Xiyu zhuan 西域傳) for his *History of the Latter Han* (Hou Hanshu 後漢書). Fan details Chinese interaction with the city-states and kingdoms of Central Asia, northwest India, and lands farther west like Parthia and Rome. Fan copied some information from Sima Qian's report of Zhang Qian's journey (see chapter 2) and from Ban Gu's *History of the Han* (Hanshu 漢書), but updated information was provided from the report of Ban Gu's nephew Ban Yong 班勇 from 125 CE, who served as chief clerk of the Western Regions. This report probably also incorporated the findings of an envoy named Gan Ying 甘英, who had been sent by the general Ban Chao 班超 (32–102 CE), Ban Yong's father, to reach Da Qin.

As protector general of the Western Regions, Ban Chao had reestablished Chinese control of the Tarim Basin by force and renewed contacts with Central Asia between 73 and 95 CE, after a lapse of more than six decades. Having heard from other groups about the great kingdom of Da Qin, he sent his officer Gan Ying on an epic journey in 97 CE to attempt to reach Da Qin. Traveling overland from Kucha, through Gandhara and Kandahar, Gan Ying arrived in Characene, at the head of the Persian Gulf. Fan Ye records:

> In the ninth year of the Yongyuan reign period of Emperor He [97 CE], the protector general of the Western Regions, Ban Chao, sent Gan Ying as an envoy to Da Qin [Roman Empire]. He arrived at Tiaozhi [Characene, at the head of Persian Gulf], and gazed out over the Great Sea, desiring to cross it. However, the sailors on the western frontier of Anxi [Parthia] said to Ying:

> The ocean is vast, and even those travelers who meet with favorable winds are only able to cross it in three months. Those who

encounter delaying winds can take as long as two years. That is why those who set sail across the ocean always take three years of provisions. There is something about this sea that is apt to make men feel homesick and sentimental, and many of them have perished.

When Gan Ying heard this, he halted his trip.[5]

So when Gan Ying inquired about ocean passage to Roman Egypt through the Arabian and Red Seas, the sailors on the Parthian frontier exaggerated the difficulties of the voyage, which convinced him to turn back. It is believed that the Parthians wanted to prevent direct contact between Han China and Rome so they could maintain their monopoly on the overland silk trade. Gan Ying apparently headed home through Parthian territory, then up through Bactria, Kashgar, and back to Kucha along the northern Silk Road.

In a subsequent and even more remarkable entry, Fan Ye reports that envoys who claimed to have been sent by the Da Qin ruler Andun 安敦 [identified as Marcus Aurelius Antoninus; r. 161–180], arrived off the coast of Vietnam between October 12 and November 10, 166 CE, and were taken to the Latter Han court at Luoyang. [6] This is recognized in the original text as the first direct diplomatic contact with Da Qin, but modern scholars have long questioned whether these men were really official Roman envoys, since no record of this mission appears in Roman texts. The consensus is that they were Syrian or Egyptian Roman merchants who booked passage on Southeast Asian ships in India to make the second leg of the long sea journey from the Red Sea or Persian Gulf to China.

The Chinese court was eager to portray this trade mission as a submission of tribute from a distant kingdom, since this lent prestige to the reigning emperor, but the Chinese historian was suspicious that these men did not really come from Da Qin, because their tribute items of ivory, rhino horn, and tortoiseshell were not that remarkable and could be obtained locally in Vietnam. Where were all the famous Da Qin products like glass, jewels, and coral? The timing of the voyage is also noteworthy. Rome and Parthia had been at war since 161 CE. Roman armies under Avidius Cassius had just sacked the Parthian capital of Seleucia-Ctesiphon in late 165 CE, and the ongoing Antonine Plague ravaged the territories that normally facilitated trade between East and West. This might have made it imperative to find a direct route from the Red Sea to China, avoiding the now unstable overland route.

INTRODUCTION TO PLINY THE ELDER

Gaius Plinius Secundus (Pliny the Elder, 23–79 CE) was a Roman soldier, naval commander, lawyer, and high imperial official who authored one of the longest works to survive from antiquity, the encyclopedic *Naturalis Historia* (Natural History) in 37 books, which synthesized knowledge about the human-centered natural world, including geography, ethnography, botany, geology, astronomy, art history, and zoology.

He notoriously transmitted the accounts of the "monstrous races" of India found in authors like Ctesias (see chapter 1) but also incorporated new genuine information, drawn from merchant accounts, military ventures, and diplomatic envoys. His Seres ("Silk People") are encountered overland through Central Asia and are said to be isolationist, mild mannered, and culturally primitive. He also mentions contact with the Seres from those traveling north from India, based on the account of an envoy named Rachias (probably the generic Indian title raja, "noble prince"), who was sent from Sri Lanka to Rome. These Seres are quite tall, with "flaxen hair and blue eyes," and because they are unable to communicate with civilized merchants, they engage in a silent trade. Pliny was concerned that the demand for Chinese silk corrupted the morals of Romans, who wore transparent gauze in public. He was also worried that the trade in luxuries with Arabia, India, and China drained the Roman economy of millions of coins per year. His greatest blunder was in stating that silk was derived from a floss combed from trees.

PRIMARY SOURCE AND ALTERNATE ENGLISH TRANSLATION

Rackham, H., W. H. S. Jones, and D. E. Eichholz, trans. *Pliny, Natural History*. 10 vols. Harvard University Press, 1938-63.

Yule, Henry. *Cathay and the Way Thither: Being a Collection of Medieval Notices of China*. Rev. ed. by Henri Cordier. 4 vols. Hakluyt Society, 1916.

EXCERPTS FROM *NATURAL HISTORY*, BY PLINY THE ELDER

* * * * * * * * * * * * * * * * **BRIEF EXCERPT** * * * * * * * * * * * * * * * *

The Seres, or Silk People

After leaving the Caspian Sea and the Scythian Ocean, our course takes a bend towards the Eastern Sea as the coast turns to face eastward. . . . The first human occupants are the people called the Seres [the Chinese], who are famous for the woolen substance obtained from their forests; after a soaking in water, they comb off the white down of the leaves, and so supply our women with the double task of unravelling the threads and weaving them together again; so manifold is the labor employed, and so distant is the region of the globe drawn upon, to enable the Roman matron to flaunt transparent raiment in public. The Seres, though mild in character, yet resemble wild animals, in that they also shun the company of the remainder of mankind and wait for trade to come to them. . . .

We, however, have obtained more accurate information during the principate of Claudius [41–54 CE], when an embassy actually came to Rome from the island of Taprobanê [Sri Lanka]. The circumstances were as follows: Annius Plocamus had obtained a contract from the Treasury to collect the taxes from the Red Sea; a freedman of his while sailing round Arabia was carried by gales from the north beyond the coast of Carmania [present-day Kerman Province, Iran], and after a fortnight made the harbor of Hippuri [present-day Kudiramalai Point, northwest Sri Lanka], where he was entertained with kindly hospitality by the king, and in a period of six months acquired a thorough knowledge of the language. Afterward, in reply to the king's enquiries, he gave him an account of the Romans and their emperor.

The king among all that he heard was remarkably struck with admiration for Roman honesty, on the ground that among the money found on the captive the *denarii* [a Roman silver coin] were all equal in weight, although the various figures on them showed that they had been coined by several emperors. This strongly attracted his friendship, and he sent four envoys, the chief of whom was named Rachias [raja, "noble prince"]. From them we learnt the following facts [about Taprobanê]: it contains five hundred towns, and a harbor facing south, adjacent to the town of Palaesimundus, which is the most famous of all the places in the island and a royal residence, with a population of 200,000. . . .

They also told us that the side of their island facing toward India is 1,250 miles long and lies southeast of India; that beyond the mountains of Emodus [Hindu Kush and Himalayas], they also face toward the country of the Seres, who are known to them by intercourse in trade as well, the father of Rachias having traveled there; and that when they arrived there the Seres always hastened down to the shore to meet them.

That people themselves, they told us, are of more than normal height and have flaxen hair and blue eyes, and they speak in harsh tones and use no language in dealing with travelers. The remainder of the envoy's account agreed with the reports of our traders—that commodities were deposited on the opposite bank of a river by the side of the goods offered for sale by the natives, and they took them away if satisfied by the barter. . . .

But the title "happy" belongs still more to the Arabian Sea, for from it come the pearls that that country sends us. And by the lowest reckoning, India, the Seres, and the Arabian Peninsula take from our empire 100 million *sesterces* [a Roman brass coin, about 26 g.] every year—that is the sum that our luxuries and our women cost us, for what fraction of these imports, I ask you, now goes to the gods or to the powers of the lower world?[7]

* *

INTRODUCTION TO YU HUAN

Yu Huan 魚豢 was a Chinese historian who held the post of court gentleman at the court of Cao Rui 曹叡 (r. 226-239), Emperor Ming of the Wei dynasty (220–265 CE). He wrote the unofficial *Brief History of the Wei* (Weilüe 魏略), a book in 50 chapters, now lost except for a few parts concerning foreign lands that are preserved as quotations in Pei Songzhi's 裴松之 (372–451 CE) commentary to the *Record of the Three Kingdoms* (Sanguozhi 三國志) of Chen Shou 陳壽 (233–297 CE).

Yu Huan never left North China to visit the far-off places he talks about. His account of the kingdom of Da Qin (Roman Empire), translated here, is indeed second-hand information and full of hearsay, but it is not complete fantasy, for it contains some very accurate information about how to get to Da Qin via overland and sea routes, exhaustive lists of natural products for export (many of which are corroborated in the *Navigating* text), gold-to-silver exchange rates, enumerations of vassal states, as well as institutions like postal relay stations. Some of this concrete information may have come from the official report of the thwarted Chinese envoy to Da Qin, Gan Ying, or been collected from the supposed Roman embassy from Marcus Aurelius, which arrived in China in 166 CE, or the testimony of the Roman merchant Qin Lun 秦論, who arrived in Vietnam by sea in 226 CE and stayed for over a decade before returning home.[8] Since Yu Huan's account still refers to the Parthian Empire (Anxi 安息) in Iran and does not mention Sassanian Persia (Bosi 波斯), his information probably predates the fall of the Arsacid dynasty of Parthia in 224 CE.

Yu Huan's account contain strange and ironic misconceptions, such as touring Roman emperors collecting petitions in a leather pouch or the citizenry peacefully replacing their emperor with the worthiest man during crises. The notice of lions and tigers endangering travelers would seem ridiculous for Italy but makes more sense if this refers to the eastern provinces. As was the case with Roman interpretations of the Chinese world, our authors have taken the situation on the periphery as descriptive of the imperial core.

There is also an attempt to familiarize the Romans (and to explain the origin of the name Da Qin, "Greater Qin"), by divulging that the Romans themselves declared that they were originally a branch of the Chinese people. They were said to look quite like Chinese people in appearance except for their clothing. Consequently, they are projected to be "tall, upright, and just."

PRIMARY SOURCE, ALTERNATE ENGLISH TRANSLATIONS, AND SELECTED STUDIES

Chavannes, Édouard. "Les Pays d'occident d'après Le *Wei Lio*: Avant-Propos." *T'oung Pao* 6, no. 5 (1905): 519–571.

Hill, John E. "The Peoples of the West, from the *Weilue* 魏略 by Yu Huan 魚豢: A Third Century Chinese Account Composed between 239 and 265 CE Quoted in *zhuan* 30 of the *Sanguozhi*, published in 429 CE." Silk Road Seattle Project, Walter Chapin Simpson Center for the Humanities, University of Washington, September, 2004. https://depts.washington.edu/silkroad/texts/weilue/weilue.html.

Hirth, Friedrich. *China and the Roman Orient.* Kelly and Walsh, 1885.

Leslie, D. D., and K. H. J. Gardiner, *The Roman Empire in Chinese Sources.* Bardi Editore, 1996.

Yu Huan 魚豢, *Wei lüe* 魏略, in Chen Shou 陳壽, *Sanguo zhi* 三國志 (Zhonghua shuju, 1984).

EXCERPT FROM THE *BRIEF HISTORY OF THE WEI*, BY YU HUAN

The Country of Da Qin

The country of Da Qin ["Greater Qin," i.e., the Roman Empire] is referred to in early texts as Lijian [Alexandria]. It is located west of Anxi [Parthia] and Tiaozhi [Characene, in southern Mesopotamia] and west of the Great Sea [Arabian Sea].

Routes to Da Qin

From the city of Angu [possibly Gerrha, in the western Persian Gulf] on the frontier of Anxi, one boards a ship and travels directly to Haixi [Egypt]. If one encounters favorable winds, he will arrive in two months. If the winds are delayed, it might take an entire year, and if there is no wind at all, it could take three years.

This country [Roman Egypt] is located west of the [Red] Sea; therefore it is commonly called Haixi ["West of the Sea"]. There is a river [the Nile] that emerges from this country, and to its west, there is another Great Sea [the Mediterranean]. Haixi has a city called Wuchisan [Alexandria]. From below this country, going directly north, one arrives at the city of Wudan [possibly Tanis in the Nile delta]. Going southwest, you cross another river [a branch of the Nile]; one can cross it by boarding a boat and traveling for one day. Going southwest again, you cross another river [or Nile branch], which takes a day to ford. In all, there are three great metropolises.

However, if you take the overland route from the city of Angu [Gerrha], you travel directly north to the area of Haibei ["North of the Sea," upper Mesopotamia and Syria], then travel directly westwards to reach Haixi [Egypt], finally traveling due south to arrive at the city of Wuchisan [Alexandria]. You have to ford one river, but traveling by boat, you can cross this in one day. You then circle around the coast [probably to the port of Appolonia in Cyrene, Libya], then, in general, it should take about six days to cross the Great Sea and arrive at this country [of Rome].

The People, Customs, and Ruler of Da Qin

The country of Da Qin has more than 400 smaller cities and towns and extends for several thousand Chinese miles in every cardinal direction. The seat of the king's government is located beside a river estuary [Tiber River]. The outer and inner city walls are made of stone.

The land produces pine, cypress, sophora, and catalpa trees, bamboo and common reeds, as well as willow and parasol trees and a variety of other flora. The custom of the common people is to cultivate the five grains. They raise horses, mules, donkeys, and camels as beasts of burden and cultivate silkworms.

As is their custom, they have many extraordinary magicians. Some can spit fire from their mouths, tie themselves up and break free without assistance, and juggle as many as twelve balls in the air with ingenious skill.

Their country has no permanent ruler, for when there is some calamity or untoward natural occurrence in the land, they immediately install a worthy man to be the new king and dismiss from service the former king, who dares not even feel resentment over this treatment.

Their ordinary people are tall, upright, and just. They look like the people of the Middle Kingdom but wear foreign-style clothing. They themselves say that they were originally one branch of the Chinese people. They have always wanted to send emissaries through to China, but the Anxi [Parthians] plotted to maintain their own profits [from the silk trade], so the Da Qin ambassadors were unable to get through.

Even vulgar persons are still able to write in the barbarian script.

As for their institutions, their public and private buildings are constructed in multiple stories. They have set up banners and flags, signal drums, small chariots with white canopies, and postal relay stations, just like those in the Middle Kingdom.

From the realm of Anxi [Parthia], you can also circuit around the sea towards the north [Black Sea and Mediterranean] to arrive at this country.

The common people are well connected to each other. Every 10 Chinese miles [approx. 5 km] along the roads, there is a courier station, and every 30 miles [approx. 15 km], there is a relay-station hostel. Along the whole route, there are no robbers or bandits, but there are fierce tigers and lions that can cause harm to travelers. If one does not travel the roads in a caravan, he would be unable to get through safely.

The country has installed petty vassal kings, numbering in the several dozens. The city where the king governs from [Rome] is more than 100 Chinese miles [50 km] in circuit and has bureaucratic offices and secretaries for official documents.

The king possesses five palaces, with each palace 10 miles [5 km] apart from the next. Each morning, the king goes to one of the palaces to handle affairs of state, and at sunset, he resides there for the night. On the next day, he travels to another palace, completing a round in five days.

The king appoints thirty-six generals [maybe referring to the *consilium principis*]. Each time there is a discussion of state matters, if just one of these generals does not show up, there will be no discussion.

When the king goes out on tour, he always has an attendant follow him holding a leather pouch. If there are people who want to make an oral petition, the attendant takes down their statement in writing and tosses it into the pouch. Once the king returns to the palace, he inspects the petitions and passes judgment on the matters.

They use rock crystal [transparent glass] to construct the columns of the palace as well as dining vessels and other items. They manufacture bows and arrows. . . .

The country produces fine linen cloth. They make gold and silver coins. One gold coin is equivalent to ten silver coins.

They have a particularly fine woven cloth, which is said to employ the down of the "water-sheep" [i.e., "sea silk," from the attachment fibers of the *Pinna nobilis* mollusk]. It is called Haixi [Egyptian] cloth. The six domestic animals of this country all emerge from the water. Some say that they do not only use the wool from sheep but also employ tree bark or the fibers of wild silkworm cocoons to produce thread, and weave this into things like rugs, Taftan blankets or carpets, and felt curtains, all of them very fine. Their colors are more vivid than those manufactured in the various countries east of the sea [in Persia].

They often make a profit by obtaining Chinese silk and then unraveling it to make foreign silk damasks. Thus, they frequently engage in maritime trade with the various countries in the Parthian Empire. The sea water is bitter and unfit to drink, so travelers rarely can reach this country [of Da Qin].

* *

Other Products of Da Qin

The mountains produce nine different colors of second-quality gemstones [probably fluorite], namely azure, blood-red, yellow, white, black, bright green, purple, rose-red, and magenta. The nine-colored stones that come from the mountains of Yiwu [near present-day Hami, Xinjiang] are also of this type. . . .

Da Qin has much gold, silver, copper, iron, lead, tin, tortoises good for divination, white horses with vermillion manes, *haiji* rhino horn, hawksbill tortoiseshell, black bears, blood-red hornless dragons [probably the resin of the Socotra dragon tree], venom-evading rats [mongoose], conch shells, giant clams producing mother-of-pearl, carnelian, "southern gold," kingfisher feathers and quills, elephant tusks, veined jade, bright-moon pearls, night-luminescent jewels [diamonds], genuine white pearls, amber, regular coral, opaque glass in ten colors, namely blood-red, white, black, bright green, yellow, azure, magenta, pale-blue, rose-red, and purple, exquisite jade, red coral, rock crystal [transparent glass], rose-colored garnets, realgar, orpiment, blue-green jade [jasper], multicolored jade, ten varieties of wool rugs, including yellow, white, black, bright green, purple, blood red, scarlet, magenta, golden, and pale blue tinged with yellow, polychrome Taftan carpets, polychrome or nine-colored Taftan carpets of lesser quality, gold-threaded embroideries, various colors of damask twill cloth, gold-painted cloth, reddish-purple *chi* cloth, *falu* cloth, reddish-purple *chiqu* cloth, "cloth that can be washed in fire" [asbestos], delicate cloth from Luode [Rhodes?], *baze* cloth, *dudai* cloth, *wensu* cloth, polychrome *tao* cloth, curtains woven from gold thread on a scarlet ground, polychrome bed canopy curtains, *yiwei* aromatic, myrrh, storax, *didi* [possibly Greek: *staktē*, "myrrh oil"], rosemary, *douna* [Sanskrit: *dhūna*, resin of Shala tree], monkshood, frankincense, turmeric and saffron, rue resin, and loosestrife, altogether twelve kinds of aromatics from herbs and trees. . . .

Yu Huan's Final Assessment

It is commonly considered that a fish in a small pond cannot comprehend the vastness of the Yangzi or the ocean and that a creature such as the mayfly knows nothing of the different climate of the four seasons. Why is that? It is because the place where the former resides is tiny, and the lifespan of the latter is short. When I now read extensively about the outer barbarian tribes and the various kingdoms like Da Qin, the subject is still so vast that it dumfounds and confuses me! How much more confounding then are the theories put forth by Zou Yan [who envisioned a world of nine continents surrounded by a great ocean, third c. BCE] or the speculative cosmology put forth in the *Classic of Supreme Mystery* [of Yang Xiong, 53 BCE–18 CE]?

Alas, I am limited to traveling on foot within the puddle left by the hoofprint of an ox, nor do I possess the longevity of a Pengzu [the Chinese Methuselah]. I am in no position to take advantage of favorable winds to sail swiftly away or to mount the mythical steed Yaoniao and look out over distant vistas. I merely strive to project my vision to gaze down upon the earth as if from the sun, moon, and stars, and to enable my thoughts to take flight toward the remotest regions of the world, and nothing more.[9]

NOTES

1. See Ptolemy, *Guide to Drawing a Map of the World* (book 1, chapter 11). Translation in Henry Yule, *Cathay and the Way Thither: Being a Collection of Medieval Notices of China.* Revised edition by Henri Cordier (Hakluyt Society, 1916), 1:187–189.
2. Fan Ye 范曄, *Hou Hanshu* 後漢書 (Zhonghua shuju, 1965), 4.188, 88.2910.
3. See Ptolemy, *Guide* (book 6, chapter 16). Translation from Yule, Cathay, 1:194–195.
4. See Ptolemy, *Guide* (book 7, chapter 3). Translations in Yule, Cathay, 1:195–196.
5. Fan Ye, *Hou Hanshu*, 88.2918. Translation by the author.
6. Fan Ye, *Hou Hanshu*, 88.2919-20. Translation by the author.
7. Pliny, *Natural History,* translated by H. Rackham (Harvard University Press, 1938-1963), 2:377–79, 2:401–5, 4:53, with slight modification and updated spelling.
8. See Yao Silian 姚思廉, *Liang shu* 梁書 (Zhonghua shuju, 1973), 54.798.
9. Yu Huan 魚豢, *Wei lüe* 魏略, in Chen Shou 陳壽, *Sanguo zhi* 三國志 (Zhonghua shuju, 1984), 30.86063. Translation and headings by the author.

CHAPTER 5

THE DECENTERING WORLDVIEWS OF UNIVERSAL RELIGIONS (400–650 CE)

During Late Antiquity and the early medieval period at both ends of Eurasia, classical worldviews and cosmologies were dislodged and decentered by the rise of the universal religions of Buddhism and Christianity, whose spread and influence were fostered by their symbiotic connection to major empires.

Buddhism rode the imperial favor of the Mauryan kings and the later Kushan monarchy to arrive at the doorway of China in the second century CE. It then benefitted from the patronage of the Northern Dynasties (386–581) and the Sui (581–618) and Tang (618–907) empires to establish its hold. While the Middle Kingdom of China remained at the center of the worldview in political discourse and secular literature, Chinese Buddhist monks embraced a different cosmology that placed India at the center of a spiritualized landscape, with China sitting unaccustomedly at the periphery.

Christianity became inextricably tied to the Roman Empire in its later centuries, following its tendrils into central and northern Europe and riding the waves of its trade diaspora into India. Christian presence in India later increased with missionaries and settlers from Persia, bringing with them their particular "Church of the East" sect of Christianity, with its unorthodox Christology and Mariology (later labeled Nestorianism). Moreover, the increasing orthodoxy of fifth- and sixth-century Christianity rejected pagan literature and science, and the reliance on revealed scripture almost completely dislodged the well-established cosmology of Claudius Ptolemy, with his mathematical proofs of a spherical earth surrounded by a rotating spherical heavens. Scriptural fundamentalists like the former merchant Cosmas Indicopleustus, excerpted in this chapter, argued for the primacy of an Old Testament–based cosmology, comprised of a flat rectangular earth with Jerusalem at the center, covered by a barrel-vaulted heaven.

This chapter excerpts the complementary accounts of two Chinese Buddhist monks who ventured to India on spiritual pilgrimage in search of scriptures and relics in the fifth and seventh centuries and pairs those with the account of a Christian merchant from Alexandria who traveled throughout the Indian Ocean during the first half of the sixth century.

INTRODUCTION TO FAXIAN

While the diplomat-general Zhang Qian first informed Emperor Wu of the Han dynasty about the northern portions of the Indian subcontinent and its ongoing trade with the southwestern frontier of the Han empire (chapter 2), he never personally traveled to India, only reaching northern Afghanistan. According to the Han historian Ban Gu, the Han empire under Emperor Wu also carried out oceangoing trade with eastern India for luxuries like pearls (chapter 3). Subsequently, Latter Han envoys like Gan Ying (chapter 4) would travel as far as the Persian Gulf but would bypass the Indian subcontinent, as it wasn't a priority in East-West relations.

All that would change with the introduction of Buddhism to China from India (via Central Asia) during the Latter Han period (25–220 CE). The religion was brought to China by Sogdian traders, Kushan and Parthian missionaries, and other foreigners along the Silk Road but eventually attracted the native elites, who converted to this new religion of salvation.

Indian Buddhist texts were subsequently imported and translated into Chinese, but their quality was uneven, and their coverage of the whole Buddhist Canon was scattershot and incomplete. Monks like Faxian 法顯 (ca. 340–ca. 422) were particularly distressed about the lack of authentic *vinaya* texts, those codes that regulated the conduct of monks in monastic orders. Faxian made a vow to travel to India to obtain such texts and visit pilgrimage sites.

Faxian left Chang'an in the spring of 399 CE, already 60 years old. He crossed the "River of Sand" desert in the Tarim Basin, visited Khotan on the southern Silk Road (401 CE), then traversed the Pamir Mountains and the Karakoram Pass into what is now Pakistan (see map 3). He arrived in Central India by 404 CE and visited all the sites associated with the historical Buddha. He sojourned in the great city of Pāṭaliputra (modern Patna) for three years (405–407), where he studied Sanskrit and copied texts. He moved on to Tamrālipti (present-day Tamluk) for a couple years (408–409) before traveling to Sri Lanka, where he remained for more than two years copying texts (409–411) before he departed for China on a merchant vessel (August 411). After two harrowing sea voyages, he arrived back in China in September 412. He completed a draft of his *Record of Buddhist Kingdoms* (Foguo ji 佛國記) in 414 (revised and supplemented in 416) then spent his remaining years translating the texts he brought back, assisted by the great translator from Kashmir, Buddhabhadra.

Map 3. Faxian's and Xuanzang's itineraries.

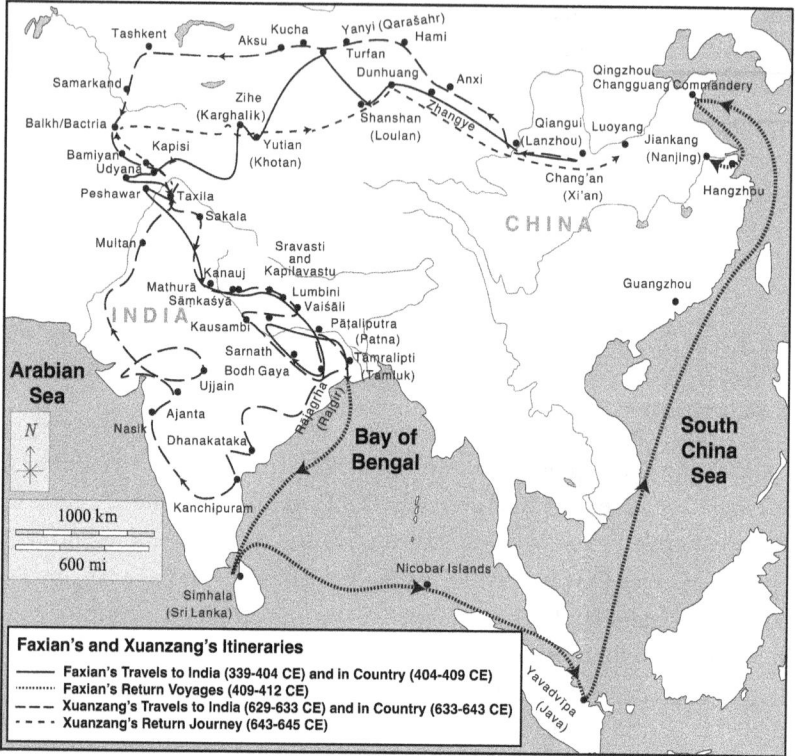

Source: Map created by Zhang Yexu and the author.

Faxian's remarkable *Record of Buddhist Kingdoms* has been a valuable eyewitness source for historians of Buddhism, scholars of Indian history, and those who study long-distance trade. For example, in the final section of his account, we learn that oceangoing trade between India and China was carried out aboard large vessels that carried over two hundred merchants, taking advantage of the monsoon winds and relying solely on the sun and stars for navigation. Just as in later periods (chapter 6), merchant vessels went from India or Sri Lanka to Java and Sumatra, then on to Canton in China.

Monks or missionaries who journey to distant lands for pilgrimage or to proselytize have different concerns than diplomatic envoys like Megasthenes or Zhang Qian (chapter 2). Faxian (and Xuanzang after him) were compelled to brave the hazardous journey to India, because they sought direct experience of the original Holy Land of Buddhism, where Buddha walked and preached and where more authentic scriptures could be found to guide Chinese believers. Faxian was obsessed with visiting physical relics of the Buddha (his tooth, his skull bone, etc.) and places where he had lived or preached. In a revealing passage, he laments being born in the "peripheral, barbarian land" of China, so far from the epicenter of his faith.

As a true believer, Faxian paints a rosy, almost utopian picture of India. This provides a not-so-subtle implied contrast with China in the early fifth century, which was torn apart by constant warfare and suffered under oppressive regimes. When Faxian refers to Central India along the Ganges, he uses the term "Zhongguo" 中國 (Middle Kingdom) to translate the Sanskrit term "Madhyadeśa" (Middle region). This is the same title that had been used to refer to Chinese regimes on the Central Plains since the Warring States period. So when Faxian talks about wonderful conditions in the "Middle Kingdom" of India, such as having no household registration, or harsh laws, or serfs bound to the land, the bitter irony of this contrast with contemporary China would be only too evident to his readers.

PRIMARY SOURCE, ALTERNATE EUROPEAN TRANSLATIONS, AND SELECTED STUDIES

Deeg, Max. *Das Gaoseng-Faxian-Zhuan als religionsgeschichtliche Quelle.* Harrassowitz Verlag, 2005.

Faxian 法顯. *Mémoire sur les pays Bouddhiques.* Translated by Jean Pierre Drège. Les Belles Lettres, 2013.

Giles, H. A. *The Travels of Fa-hsien (399–414 A.D.), or Record of the Bud-dhistic Kingdoms.* Cambridge University Press, 1923.

Li Rongxi. "The Journey of the Eminent Monk Faxian." In *Lives of Great Monks and Nuns.* Numata Center for Buddhist Translation and Research, 2002.

Sen, Tansen. "The Travel Records of Chinese Pilgrims Faxian, Xuanzang, and Yijing: Sources for Cross-Cultural Encounters Between Ancient China and Ancient India." *Education about Asia* 11, no. 3 (Winter 2006): 24–33.

Zhang Xun 章巽. *Faxian zhuan jiaozhu* 法顯傳校注. Shanghai guji chubanshe, 1985.

EXCERPTS FROM *RECORD OF BUDDHIST KINGDOMS*, BY FAXIAN (414 CE)

Faxian's Departure from China

In the past, when I [Faxian] was in Chang'an [present-day Xi'an], I despaired at the fragmentary state of the *vinaya* [Buddhist monastic rules] in China. For this reason, in the first year of the Hongshi reign period of the Latter Qin kingdom [of Yao Xing, r. 394–416], which was the cyclical year *jihai* [399 CE], I made a commitment, along with Huijing, Daozheng, Huiying, Huiwei, and others, to travel to Tianzhu [India] in search of books on the precepts for moral conduct and the monastic rules.

First, we set out from Chang'an, traversed the Long Mountains [in southeastern Gansu, western Shaanxi], and arrived at the kingdom of Qiangui [Qifu Qiangui 乞伏乾歸; r. 388–400, 409–412, of the Western Qin, whose capital during 399 was near present-day Lanzhou], where we passed our summer retreat [an annual time for Buddhist study and meditation]. When our summer retreat was finished, we traveled onward to the kingdom of Noutan [Tufa Noutan 禿髮耨檀, regent, 399–401, r. 402–414 CE, of the Southern Liang, capital near present-day Xining, Qinghai]. Passing over the Yanglou Mountains [Yangnü Mountains, north of Xining], we then arrived at the seat of Zhangye Commandery [in central Gansu]. At that time, Zhangye was experiencing great unrest [being under a months-long siege] and all the roads were cut off. The ruler of Zhangye, Duan Ye [r. 398–401], eagerly and attentively received us, bidding us to remain and serving as our *dānapati* [almsgiving patron]. Consequently, we met the monks Zhiyan, Huijian, Sengshao, Baoyun, Sengjing, and others. Delighted to learn that we shared the same goal, we passed the summer retreat together there [May–August, 400 CE]. When summer retreat was over, we resumed our journey, arriving at the oasis of Dunhuang [northwestern Gansu]. There are defensive barriers there that stretch for about 80 Chinese miles east to west [approx. 35 km], and about 40 miles north to south [approx. 17.5 km]. Altogether, we remained there for a month and a few days. My four original companions and I set off westward in the suite of an envoy, after bidding farewell to Baoyun and the others. Li Hao, the governor of Dunhuang, gave us provisions for our crossing of the River of Sand desert [the Taklamakan].

Within the River of Sand desert there are numerous evil demons and blazing-hot windstorms. If a party encounters them, all will perish, with no one being spared. In the desert, there are no birds flying above nor any beasts roaming down below. One can look all around, as far as the eye can see, trying to determine his location, but nobody is aware of any reference point. One can only rely on the bleached skeletons of dead men to mark the way.

Buddhist Kingdoms in Central Asia and India

In general, in the countries west of the River of Sand desert [in Central Asia], and in all the kingdoms of India, the monarchs all sincerely believe in the teachings of the Buddha. When they make offerings to the community of Buddhist monks, they would take off their heavenly crown and together with their royal clansmen and corps of ministers would distribute the food with their own hands. After the monks had finished dining, a carpet was spread out on the ground, and the king would sit on it, facing the most senior monks. In front of the assembly of monks, the king would never dare to sit on his own raised bed platform. Thus, the ritual that kings employed to make such offerings during the time when the Buddha was still in this world have been transmitted down to today.

People, Government, and Customs of Central India

To the south of this [across the Indus], the territory is called the Middle Kingdom [Sanskrit: Madhyadeśa, i.e., Central India along the Ganges under the Gupta Empire]. In this Middle Kingdom, cold and heat are harmoniously balanced, and there is neither frost nor snow. The people are prosperous and happy. They have no household registration system or official laws. Those who cultivate royal land are required only to submit a portion of the land's produce but are free to come or go as they please. The king's administration does not employ a web of laws and corporal punishment to ensnare the population. Those who are guilty are given only cash fines according to the severity of their crime. Even for those who repeatedly commit heinous crimes or plot rebellion, the punishment does not exceed the severing of the right hand, and that is all. The king's guards, attendants, and ministers are all provided with regular provisions and salaries.

The population of the entire country does not kill living beings, nor do they drink alcohol or eat onions and garlic, with the exception of the *caṇḍāla* [untouchable outcastes]. The name *caṇḍāla* means "bad person," and they live apart from others. When they enter a city or a marketplace, they are supposed to strike a wooden board to differentiate themselves so that people can recognize and avoid them and won't suddenly bump into them. The people in this country do not raise pigs or chickens, nor do they sell captured slaves. The marketplaces have no butchers or wine sellers. For commercial transactions, they use cowrie shells or animal teeth. Meat is only sold by outcaste *caṇḍāla* hunters or fishermen.

Heartbroken Outsiders

When Daozheng and I arrived at the Jetavana Monastery [near present-day Shravasti, Uttar Pradesh] and thought about how the World Honored

One [Shakyamuni Buddha] had resided there for 25 years, we were heartbroken to have been born in a peripheral barbarian land [China]. Along with our colleagues, we had traveled through a string of kingdoms. Some of these men have returned home [to China], while others have died [namely, the monks Huijing and Huiying]. Today, when we saw the vacant seat once occupied by the Buddha, we were greatly moved, and our hearts were heavy with grief.

The monks emerged from the monastery and asked me and my companions, "Which country did you come from?"

We replied, "We came from the land of the Han [China]."

The group of monks exclaimed, "How extraordinary! Men from a peripheral land were able to come here to seek the teachings of the Buddha!"

They said to each other, "None of the teachers and abbots in our line of succession has ever seen a Han Chinese monk arrive here."

Pāṭaliputra

Of all the countries in Central India, the capital city of this one [Pāṭaliputra, present-day Patna] is certainly the largest. The people are rich and prosperous, and they vie with one another in performing benevolent and righteous acts. . . .

The elders and householders of this kingdom each establish welfare houses for good deeds, hospitals, and pharmacies in the city. The destitute, orphaned, disabled, and sick people in the country can all go to these establishments and have their needs provided for. The doctors observe the patient's condition and prescribe, at their discretion, the appropriate food, drink, or medicine, so they can rest and recover. Those who are healed depart on their own.

King Ashoka's Hell on Earth

Long ago, when King Ashoka [Mauryan emperor, r. ca. 268–ca. 232 BCE] was a small child, he was playing in the street when Shakyamuni Buddha happened to be passing by, begging for food. The little boy was delighted and immediately picked up a clod of earth and gave it to the Buddha. The Buddha returned with it and later plastered it over the ground where he performed his walking meditation. As karmic retribution for this good deed, Ashoka was later made an Iron Wheeled King and ruled over Jambudvīpa [the continent south of Mt. Meru, i.e., "Greater India"].

Once, when he was making an inspection tour of Greater India in his iron-wheeled chariot, King Ashoka saw between the ranges of the Greater and Lesser Enclosing Mountains of Iron [that encircled the world] a hell for punishing the wicked. He asked his entourage of ministers, "What kind of place is this?"

They replied, "This is where Yama, King of Demons, punishes sinners."

The king considered to himself, "If the King of Demons is able to construct a hell to punish sinners, then why shouldn't I, as the ruler of men, also construct a hell to punish criminals and sinners?"

So he asked his ministers, "Who could make such a hell for me and preside over the punishment of sinners?"

The ministers replied, "Only the wickedest of men could do that."

Thereupon, the king sent his courtiers everywhere to search for such an evil man. They came upon a man on the bank of a pond, tall, strong, and swarthy of complexion, with blonde hair and blue-green eyes. He could catch fish with his feet. He would call out to birds and beasts, who promptly came to him, whereupon he would shoot and kill them. None were able to escape. Having acquired this person, the courtiers returned and presented him to the king.

The king secretly ordered him, saying, "I want you to build a square enclosure with towering walls, and inside plant all sorts of flowers and fruit trees. Fashion a fine bathing pool as well. Decorate the place with ornaments and make it so stately and magnificent, that it will make people desire to enter. I want the gates constructed very sturdily. If a person happens to enter, immediately seize him and subject him to all types of torture for his sins. No one should be able to escape. Even if I enter the gates, punish me also for my sins and don't release me. I now appoint you as the Lord who Constructed Hell."

Once, a lowly monk accidently entered the gates of this hell when he was begging door to door for food. The warden saw him and immediately wanted to punish him. The monk was terrified but requested a brief stay to allow him to finish his noon meal. Meanwhile, another person entered the hell. The warden put that man into a large mortar and pounded him with the pestle until red blood gushed out. When the monk witnessed this, he thought upon how this corporeal body is indeed impermanent, prone to suffering, and essentially empty, just like a bubble or froth. Suddenly, he became an *arhat* [a perfected person who has achieved Nirvana].

Then the warden seized the monk and put him into a cauldron of boiling liquid, but his mind and countenance remained serene. The fire suddenly went out, and the liquid cooled. A lotus flower grew from the cauldron, and the monk was seated atop it.

The warden immediately went to report to the king, "Something very strange has occurred in the hell. I would like Your Majesty to come and see it."

The king stated, "I had previously made a commitment, so now I dare not go."

The warden replied, "This is not a trivial matter. Your Majesty should go there in all haste. We can alter our previous agreement."

The king followed the warden back and entered the hell. The monk expounded the teachings of the Buddha for the benefit of Ashoka, and the king was able to have faith in the doctrine and interpret it as well. He then destroyed the hell he had built and repented for all the numerous evil deeds he had committed before.

The Bodhi Tree

From this point forward, King Ashoka had great faith in the Three Jewels [the Buddha, his teachings, and the monastic order], and often went under the Bodhi tree [where Shakyamuni Buddha had achieved enlightenment], where he repented his sins and reproached himself, observing the Eight Precepts [not to kill, steal, have sex, lie, drink liquor, sleep in a high bed, wear makeup, sing and dance, or eat after midday].

The queen consort asked the courtiers, "Where does the king always go on tour?"

The courtiers replied, "He is often below the Bodhi tree."

The queen waited until Ashoka was not present and sent men to cut down the tree. When the king came and saw what had happened, he fainted and fell to the ground. The various ministers sprinkled water on his face, and after a long while, he regained consciousness. The king built a wall of bricks around the four sides of the stump and irrigated the roots with one hundred pitchers of cow's milk. He prostrated himself on the ground, his four limbs splayed out, and made a vow, "If this tree does not live again, I will never rise from the ground." When the vow was complete, the tree began to sprout branches again from the stump and continues to grow until today, where it is now nearly 10 Chinese staves high [24.5 m].

Homesickness

It had been many years since I had departed China, and everyone whom I regularly met and interacted with were foreigners. The hills, rivers, grass, and trees before my eyes were all unfamiliar. Moreover, I had become separated from those who had walked alongside me, either because they had remained behind or had passed away. When I turned to look at the shadows, I saw only my own. My heart harbored a persistent feeling of sadness. Suddenly, next to this jade statue of the Buddha [at the Mount Fearless Monastery in Sri Lanka], I spotted a merchant making an offering of a white silk fan from Jin dynasty China. I couldn't help but feel great despair, and tears filled my eyes.

* * * * * * * * * * * * * * * **BRIEF EXCERPT** * * * * * * * * * * * * * * * *

The Return Voyage

Upon acquiring these Sanskrit Buddhist scriptures [from Sri Lanka], I booked passage [mid-August 411 CE] on a large merchant vessel that could carry more than two hundred people. Since ocean travel was very difficult and dangerous, a smaller vessel was towed at the stern, just in case the larger ship became damaged or destroyed.

Once the trade winds that were more favorable to navigation appeared, the merchant ship that I had boarded started sailing eastward, but after traveling for about two days, it suddenly encountered a typhoon, and the vessel sprang a leak and began to take on water. The merchants on the larger ship wanted to rush toward the smaller vessel, but those men on the auxiliary ship feared that too many people would come aboard, so they severed the tow rope. The merchants were terrified that their lives were in imminent danger, for they feared that the boat would become completely swamped. So they threw all their bulky cargo into the sea. I also discarded my purified drinking water bottle, my washbasin for ritual hand ablutions, and other miscellaneous items into the ocean, but I was afraid that the merchants would throw my Buddhist sūtras and statues overboard as well.

The only thing I could do in this situation was to invoke the name of the Boddhisatva Avalokiteśvara [Guanyin] wholeheartedly and place my life into the hands of all the monks back in China, praying, "I have traveled so far to seek the teachings of the Buddha. May your awesome spiritual power carry me on the currents to my destination."

The typhoon lasted for 13 days and 13 nights, until we finally arrived on the shore of an island [possibly one of the Nicobar Islands]. And when the tide receded, we discovered where the ship had been leaking and patched it up. Then, we resumed our course.

The sea is swarming with pirates, and if a ship encounters some, none can escape with their lives. The boundless ocean stretches as far as the eye can see, and one can't recognize what is east or what is west. One can navigate only by observation of the sun, moon, and stars. When it became overcast and rainy, the ship was simply pushed forward by the tailwinds, without any standard bearing. As dusk descended into the darkness of night, one saw only great billowing waves crashing into one another, with the occasional fiery flash of lightning, revealing sea monsters like giant soft-shelled turtles or Chinese alligators. The merchants were panic stricken and did not know which direction we were heading. The ocean was unfathomably deep, and there was nowhere we could drop anchor. When the sky cleared, we could finally get our bearings, resume the proper course, and make progress. If we had hit a submerged reef, there is no chance we would have survived.

It was like this for around 19 days, when we finally arrived at a kingdom called Yavadvīpa [Java]. In this country, heretical teachings and Brahminism flourished, while Buddhism was insignificant. I remained in this country for five months.

Then I embarked on another large merchant ship that also held more than 200 persons. Furnished with 50 days of provisions, we set sail on the 16th day of the fourth month [May 12, 412 CE]. I observed my summer retreat while onboard the boat. We headed directly northeast, bound for Canton.

A little over a month later [June 412 CE], at the time of the second night watch, we encountered a black cyclone with torrential rain. The traders and merchant passengers were all terrified, but at that moment, I again focused my mind and invoked Avalokiteśvara and the whole monastic community in China. Thanks to the protection of their awesome spiritual power, I was able to make it until daybreak.

When the weather had cleared, the various Brahmin merchants gathered together and discussed, "Because we are transporting this Buddhist monk, it has caused us to have bad luck and to encounter this great suffering. Let's maroon this monk on the shore of some island. We can't let a single person place all of us in such danger."

My original benefactor [who sponsored Faxian's passage] spoke up, declaring, "If you put this monk ashore, then you had better land me as well! Otherwise, you should just kill me! Because if you maroon this monk, then when we arrive in Chinese territory, I will report you to the king of the country. The king of China is a believer in Buddhism and esteems monks and the Buddhist community." The merchants hesitated and dared not put me ashore.

As the sky was continually cloudy, there were significant mistakes and deviations in the observations the pilot employed to set the course, and as a result, more than 70 days had passed without sighting land, and provisions of food and water were almost exhausted. So we had to take salt water from the sea for cooking. We rationed and distributed the fresh water, allowing each person only two Chinese pints [approx. 409 ml], but soon even that would be depleted. The merchants considered the matter together, saying, "Ordinarily, it should only take fifty days to reach Canton. We have now already exceeded that by many days. Haven't we gone off course?" So, they immediately turned the ship towards the northwest, seeking the coastline.

Following 12 days and nights, they arrived at the south-facing shoreline below Mt. Lao, in the jurisdiction of Changguang Commandery [present-day Laoshan, Qingdao, on the southern Shandong Peninsula, 2,000 km from Canton], and soon acquired fresh water and vegetables.

* *

But because their passage had been dangerous and difficult, and they had been filled with worry and anxiety for so many days, they found it hard to believe. But when they suddenly reached this shore and saw goosefoot plants like they remembered, they knew this was Chinese territory. But they didn't see any residents or the tracks of persons, so they still didn't know where exactly this place was. Some said that they had not yet reached Canton, while others said that they had gone far past it. No one knew for certain.

Then, some men rowed into a river inlet in a small dinghy to look for people, desiring to ask where this place was. They found two hunters whom they brought back with them to the ship and ordered me to translate when they interrogated the men.

I first comforted them, then calmly asked, "Who are you?"

They replied, "We are disciples of the Buddha."

I inquired again, "What are you seeking by entering these mountains?"

They falsely claimed, "Tomorrow is the fifteenth day of the seventh month [September 6, 412 CE]. We want to obtain peaches to offer to the Buddha to mark the end of the summer retreat [and for the Festival of All Souls]."

I again asked, "What country is this?"

They replied, "This is Qingzhou Province, in the jurisdiction of Changguang Commandery, which is governed by the ruling house of the Eastern Jin."

When the questioning was over, the merchants were delighted. They gave the hunters some of their goods from the hold as gifts and asked them to bring them to the administrative seat of Changguang Commandery.

Faxian's Reception and Translation Work in China

The governor of Changguang, Li Yi, honored and revered the teachings of the Buddha, so when he heard that a monk carrying Buddhist sūtras and statues had arrived after sailing across the ocean in a ship, he took attendants with him and went to the seashore to welcome and receive the sūtras and statues of the Buddha with due honor and return with them to the commandery seat.

The merchants then went on to the Yangzhou region [centered on the Eastern Jin capital of Jiankang, present-day Nanjing], while the regional inspector of Yanzhou and Qingzhou, Liu Daolian, invited me to remain with him throughout the winter and into the summer.

When my summer retreat was over, I wanted to hasten immediately to Chang'an, having been away from my fellow monks for such a long time, but because of the importance of the affairs I had undertaken, I instead went south to the Eastern Jin capital of Jiankang, where with the collaboration of a meditation master [the renowned Buddhabhadra, who came from India around 406-8 CE], I published translations of some of the Buddhist scriptures and monastic regulations that I had brought back.

I, Faxian, set out from Chang'an [399 CE] and after six years arrived in Central India [404 CE]. I sojourned there for a further six years [404–409 CE] and returned over the course of the following three years [409–412 CE], arriving in Qingzhou Province. I traveled through nearly thirty countries. From west of the River of Sand desert all the way to India, the excellence of both the dignified deportment and the transformative teachings of all the Buddhist monks is impossible to describe in detail. I thought only of how the various masters back in China were not able to become thoroughly informed in these matters, so I disregarded my insignificant life and voyaged across the sea to return, facing every sort of danger and difficulty. Fortunately, I received the awesome spiritual aid of the Three Venerables [the Buddha, his teachings, and the monastic order], so that even when I was in danger, I was able to make it through safely. Therefore, I wrote down on bamboo and silk an account of what

I had experienced, in the hope that it would enable worthy men to share in what I had heard and seen. This was in the *jiayin* year [414 CE].[1]

Introduction to Xuanzang

The brilliant Buddhist monk Xuanzang 玄奘 (born Chen Hui 陳褘, 602–664), was the most famous pilgrim to trek from China to India in search of authentic scriptures and the most influential on later Chinese culture. Xuanzang entered the Buddhist clergy as an adolescent and quickly distinguished himself for his piety and erudition. He became dissatisfied with the incomplete condition of Chinese translations of Buddhist texts and developed a goal to travel to the original home of Buddhism to obtain authentic scriptures, study with Indian masters, and visit sites associated with the Buddha.

He departed Chang'an in 629 CE, despite a recent imperial order banning travel abroad (at least for laymen). To get to India, he followed roughly the same land route that Faxian had traveled two centuries prior (see map 3). First traversing the Gansu Corridor, he hazarded a nearly fatal crossing of the Gobi Desert before reaching the Turfan Oasis and then followed the northern route of the Tarim Basin through Kucha, over the Tianshan Mountains to Tashkent and Samarkand, then across the Amu Darya River to Bactria, over the Hindu Kush into Bamiyan in Afghanistan (where he saw the famed cliff statues), then finally into the subcontinent through Gandhara (modern Peshawar), Kashmir, and along the Ganges. He sojourned in India for 13 years, residing at the courts of kings, making pilgrimage to dozens of sites, and studying for five years at the Nalanda Academy, where he mastered Sanskrit and Indian rhetoric. He later toured the south and west of India and returned to China by the land route, taking the southern route through Khotan and Dunhuang on his way back to a hero's welcome in the Tang capital, eager to begin translating the hundreds of Buddhist texts he brought back.

After Xuanzang returned to China from his odyssey, he had an audience with Emperor Taizong (r. 626–649) in Luoyang (spring 645 CE). The ruler was amazed at the wealth of detailed information that the monk recalled about the countries he visited, more accurate than earlier Chinese accounts of India, so he urged Xuanzang to write it down in a report. This would have also aided the state in future diplomatic or military encounters with these regions. Xuanzang's book appeared a little over a year later (autumn 646 CE) and was compiled from the monk's detailed notes, aided by his assistant Bianji 辯機. It was called *Record of the Western Regions of the Great Tang Dynasty* (Da Tang Xiyu ji 大唐西域記). The *Record* describes 128 major and minor polities in present-day Xinjiang, Uzbekistan, Afghanistan, Pakistan, and India, noting their location, size, customs, products, and the status of Buddhism there. It is mostly a dry, impersonal account, occasionally livened up with Xuanzang's retelling of local legends of bygone kings or tales from the lives of the Buddha.

Xuanzang's interests were primarily those of a devout pilgrim, for he was mostly interested in the physical relics, statues, and stupas of the historical Buddha and all the sites associated with his journey towards enlightenment. The account in *Record of the Western Regions* is complemented by a biography of Xuanzang by his disciple Huili 慧立 that relates the more dramatic and miraculous parts of the journey, such as his privations crossing the Gobi, his trek over the Tianshan and Hindu Kush Mountains, and his numerous escapes from bandits, culminating in his triumphal return to China.

The section of Xuanzang's book translated here in slightly abridged form is his "General Overview of India" (Yindu zongshu 印度綜述), organized under 17 different headings, including housing, clothing, food, religion, death rituals, writing, education, law, and the caste system. This interrupts his chronological and geographically ordered description of the countries he visited, inserted at the point when Xuanzang approached the Indus from the northwest.

His account is a masterful ethnography of the Indian subcontinent, highlighting both shared traits and regional diversity. It's an invaluable tool for those studying Indian history and society of the seventh century. This type of ethnography of foreign lands was pioneered in East Asia by the Han historian Sima Qian (ca. 145–ca. 86 BCE) and was decidedly a product of an imperial mindset, for only a subject of a universalizing empire could produce such an encyclopedic and analytical account of the "others" on their frontiers. As such, it contains the expected descriptions of barbaric oddities to shock the civilized Chinese audience, such as the naked Jain ascetics, aged men who drowned themselves following a party given by friends, and a taboo against eating scallions and garlic. But as in other such accounts, like Sima Qian's ethnography of the Xiongnu or Herodotus's account of the Persians, there are criticisms of the author's home culture implicit in the comparison. Xuanzang's praise of the simplicity and fairness of the Indian legal system and taxation is a veiled complaint against their draconian and onerous counterparts in Tang China.

Legends surrounding Xuanzang's trip to India had already begun to circulate during his lifetime. By the Southern Song period, these had been combined with popular folk tales, including some with a cudgel-wielding monkey as Xuanzang's protector. These tales culminated in the famous 16th-century Ming novel *Journey to the West* (Xiyou ji 西遊記).

PRIMARY SOURCE, ALTERNATE ENGLISH TRANSLATIONS, AND SELECTED STUDIES

Beal, Samuel. *Si-yu-ki: Buddhist Records of the Western World.* 2 Vols. Trübner, 1884.

Huili 慧立. *A Biography of the Tripiṭaka Master of the Great Ci'en Monastery of the Great Tang Dynasty.* Translated by Li Rongxi. Numata Center for Buddhist Translation and Research, 1995.

Waley, Arthur. *The Real Tripitika.* Allen and Unwin, 1931.

Xuanzang 玄奘. *Da Tang xiyu ji jiaozhu* 大唐西域記校注. Edited and annotated by Ji Xianlin 季羨林. 2 vols. Zhonghua shuju, 2000.

Xuanzang, 玄奘. *The Great Tang Dynasty Record of the Western Regions.* Translated by Li Rongxi. Numata Center for Buddhist Translation and Research, 1996.

EXCERPTS FROM THE "GENERAL OVERVIEW OF INDIA" IN *RECORD OF THE WESTERN REGIONS OF THE GREAT TANG DYNASTY,* BY XUANZANG (646 CE)

Explaining the Name of India

Examining in detail the name Tianzhu 天竺 [the name for India since the Latter Han], there is considerable difference of opinion, leading to an utterly chaotic situation. In the olden days [during the Former Han], it was called Shendu 身毒, while later, some called it Xiandou 賢豆. Nowadays, according to the correct pronunciation, it should be pronounced Yindu 印度 [India]. The Indians themselves call their country by different names, according to the locality where they live, and each region has different customs. But distant lands [like Persia and China] have adopted their own names for the general area, calling that region that they consider the most beautiful as India. . . .

There are many castes in India, which divides its numerous people into groups according to racial type. Among these, the Brahmins are considered to be especially pure and noble. India received another more elegant name from this group, which has been handed down as a customary appellation. Ignoring regional differences and the boundaries between local states, the land is generally called by them the Brahmin Country.

Territory

Let us now talk about the extent of the territory of India. The total perimeter of the five traditional regions of India [center, north, south, east, and west] is more than 90,000 Chinese miles [approx. 45,000 km], bordering the Indian Ocean on three sides and backing onto the Great Snowy Mountains [Himalayas and Hindu Kush] to the north. The country is broad in the north and narrower toward the south, so its shape is like that of the half-moon. The overall territory is divided into more than 70 individual countries. The climate is particularly hot, and the land is well watered and damp. Mountains are quite numerous in the north, and the infertile soil of the foothills is saline and alkaline. The open countryside of the eastern region is crossed by many rivers, and the soil is rich and moist. Fertile, embanked farm fields abound. The south is covered with luxuriant vegetation, but the soil of the western region is mostly hard and barren. . . .

Measurements and Their Conversion

As for the names of their units of measurement, there is one they call the *yojana*. The *yojana* was derived from the distance the ancient sage king's army could march in a single day. According to one old traditional reckoning, a *yojana* would be the equivalent of 40 Chinese miles [20 km], but according to the national custom in India today, it equals only 30 miles [15 km]. The sacred teachings of the Buddha record that it only corresponds to 16 miles [8 km].

Dividing this distance further, one *yojana* is split up into eight *krośa*. The distance of a *krośa* derives from how far away one can hear the lowing of a large ox. A *krośa* is divided up into 500 "bows" [Sanskrit: *dhanu*], and a *dhanu* is split up into four "forearms" [Sanskrit: *hasta*]. The length of one *hasta* is divided into 24 "fingers" [Sanskrit: *aṅgula*]. One *aṅgula* is made from the length of seven "barleycorns" [Sanskrit: *yava*]. Continuing on to smaller sizes, there are seven further subdivisions, including a louse, a nit, a grain of dirt that can fall through a crack and float in a sunbeam, an ox hair, a filament of lamb's fleece, rabbit down, a metal particle or water vapor, down to a fine speck of dust, which can be further divided into seven pieces of infinitesimally fine dust, until you can divide it no further without arriving at nothingness. Thus, this is called the "infinitely small" [Sanskrit: *paramāṇu*; i.e., an atom].

Time and the Seasons

When it comes to the revolution of yin [the dark, female principle] and yang [the bright, masculine principle] and the successive stations of the sun and moon, although what Indians call them may be different, the intervals of time and the seasons are the same. The position of the moon in the night sky relative to certain constellations is used as the standard by which to mark out the months.

The briefest duration of time is called a *kṣaṇa* ["instant"]; 120 *kṣaṇa* make a *tatkṣaṇa,* [1.6 seconds], 60 *tatkṣaṇa* make a *lava* [1 minute, 36 seconds], 30 lava make a *mahūrta,* [48 minutes], 5 *mahūrta* make one *kala* ["long hour"], 6 of these *kala* constitute a day and night, [according to Buddhist practice], but common custom says that the entire day and night is divided into 8 *kala*. . . .

The year [according to common custom] can also be divided into six seasons. From the 16th day of the first lunar month until the 15th day of the third lunar month is the season of "gradually increasing heat." From the 16th day of the third month until the 15th day of the fifth month is the season of "fullness of heat." From the 16th day of the fifth month until the 15th day of the seventh month is the "rainy season." From the 16th day of the seventh month until the 15th day of the ninth month is the season of "flourishing growth." From the 16th day of the ninth month until the 15th day of the 11th month is the season of "gradual cooling." From the 16th day of the 11th month until the 15th day of the first lunar month is the season of "fullness of cold.". . .

Monks in India follow the Buddha's holy teachings to live in retreat and meditate during the rainy season, either during the first three months of the year or during the last three months. . . .

Houses and Their Furnishings

Regarding towns, villages, and wards where commoners live, their square perimeter walls are broad and lofty, while both the larger avenues and thoroughfares, as well as smaller streets and lanes, are full of twists and turns. Shopkeeper's stalls open right onto the street, sometimes blocking the middle of the road, and a flag-topped marketplace tower straddles the avenue. The residences of butchers, fishermen, prostitutes, actors, executioners, and dung collectors are marked with special banners and are relegated to wards outside the town itself. If persons from these groups come and go from the town to conduct their business, they must keep to the left side of the road.

As for house construction and the building of enclosure-walls or town walls, because the terrain is so low and damp, most city walls are built up from layered bricks, while other various residential walls might be constructed by binding together staves of bamboo or wood. Houses and towers are built from wooden planks and have flat roofs. The walls are smeared with lime plaster and capped with ceramic tiles or unfired bricks. The construction techniques for exceptionally tall buildings [like pagodas] are the same as those seen in China. Roofs are thatched or sometimes topped with tiles or wooden planks, and the walls are plastered as decoration. The floor is smeared with cow dung, which is thought to purify it, and fresh flowers are scattered on top. This is very different [from the customs of China].

The Buddhist monasteries and convents are built in a very extraordinary fashion. Four towers soar up from the corners, and the main pavilions rise to three stories high. The rafters, eaves, ridgepoles, and roofbeams are decorated with fantastic carvings and openwork, and all the doors, windows, and walls are painted with polychrome images.

The interiors of ordinary commoners' dwellings are lavishly outfitted, while the exteriors remain plain and simple. The height and breadth of interior rooms and main halls display some variation, while the architectural form of towers and pavilions does not stick to any defined style. Doors open to the east, just as the king's throne faces east [unlike in China, where it faces south].

As for the furniture for sitting and resting, everyone employs rope-bound, wooden folding chairs. The seats of the royal family, high officials, scholars, common people, and great families are distinct in terms of their decoration but are constructed according to the same standards. The king's throne is taller and more spacious, inlaid with pearls and gems. It is called the Lion Throne. It is covered in fine cotton cloth, and a jewel-encrusted footstool is placed in front of it. All the ordinary officials carve their seats with different decorations and encrust them with jewels and other rarities, each according to their taste.

Clothing and Adornment

None of their upper or lower garments, outerwear, or underwear are cut and tailored. They value brilliant white color and disparage polychrome dyed fabrics. A man wears a piece of cloth wrapped around his waist, winds it round up to his armpits, then drapes it like a stole across his upper body, leaving his right shoulder bare. Women wear an apron that hangs down to the ground and an upper garment that completely covers both shoulders. Women's hair is gathered into a small bun atop their head, with the rest of it hanging loose. Some men shave off their whiskers or engage in other unusual customs. Some people wear floral headdresses or drape pearl and jade pendants over their bodies. Many of their garments are made of something called *kauśeya*, or from fine cotton cloth. *Kauśeya* is silk from the wild silkworm. They also have garments made of *kṣauma* cloth, which is a sort of linen, *kambala* cloth, which is woven from fine fleece, and *harali* cloth, which is woven from the fine fur of a wild animal. This animal's fur is so soft and fine that it can be spun and woven. Hence, the material is highly treasured and can serve as clothing.

The climate in northern India is bitterly cold, so they tailor their clothing to be short and tight-fitting, very similar to the dress of the barbarians north of China. The apparel of the heretics [Hindus and Jains] exhibits a confusing variety of styles and tailoring. Some [like devotees of Krishna] wear peacock tailfeathers, some [like the Kapālamālin devotees of Shiva] sport necklaces of human skulls, some [like the Nirgrantha Jain ascetics] go totally naked, while others cover their bodies with straw or bark garments. Some pull out all their head hair and shave off their mustache, while others sport tousled hair all about their temples but gather the hair on top into a neat, mallet-shaped bun. Their lower and upper garments have no set style and may be dyed either red or white.

Buddhist monks wear only the traditional "three garments" [patch-robe, upper garment, and underwear], the saṅkakṣikā upper garment, or a waist-skirt as vestments. The tailoring of the three garments is different according to the tenets of each sect or school. The trim on the hem of the garments might be wide or narrow, and the pleats may be large or small. The saṅkakṣikā upper garment covers the left shoulder and conceals the two armpits, being gathered on the right and opening to the left. It is tailored to extend just past the waist. The waist skirt is not fastened with a regular belt or cord. When wearing one, one gathers the fabric up to form numerous pleats then ties it off with a silk ribbon. The shape of the pleats is different among the various sects, and the colors might be yellow or red.

The costume of members of the Kṣatriya and Brahmin castes is pure and simple. They live a clean, frugal lifestyle. The clothing and ornaments worn by kings and high ministers are quite different. They adorn their heads with garlands of flowers or bejeweled crowns and bedeck their bodies with gold bracelets and necklaces of precious gems. Rich merchants and great traders tend to just wear bracelets. Most people in India go around barefoot and few persons wear shoes. People intentionally stain their teeth, sometimes red, sometimes black. They neatly trim their hair to even it up and pierce their earlobes. Indians have long noses and deep-set eyes. Such is their general appearance.

Hygiene Around Eating and Drinking

As for maintaining their personal hygiene, they are very particular, and nothing can alter their resolve in this regard. When Indians are about to eat or drink, they always wash their hands first. No one eats leftovers from the previous evening's meal, and persons don't pass around the same serving vessels. Ceramic and wooden serving containers are always discarded as soon as they are through being used, but gold, silver, bronze, and iron vessels are rubbed clean and given a bright polish. When they are done eating, they chew on a willow twig to cleanse their mouths. Until one has finished bathing and rinsing his mouth, he will not engage in close contact with anyone. They also pour water over their hands each time after urinating. Indians apply various aromatics to their bodies, such as sandalwood or turmeric perfume. When the king is about to bathe, they strike the drums and serenade him with hymns, accompanied by stringed instruments. When making sacrifices or praying at temples, everyone must first bathe and cleanse their hands.

Writing

Examining their written characters in detail, they were fashioned by the creator god Brahmā, who originally handed down the forms of 47 alphabetic letters. The letters are combined to form words that represent different things and are inflected according to changing circumstances. The writing system has evolved considerably and branched into numerous variants, and its usage has broadened gradually over time. There have been some slight alterations due to variations in locality and peoples, but generally, these dialects are not that different from the original source. Of course, the language of Central India is the most precise and correct form. With harmonious diction and elegant tone, it is identical to the voice of the gods. Its spirit is clear and resounding, and it serves as a standard for the people. Neighboring borderlands and foreign countries follow an erroneous pronunciation and vie with each other in hastening the vulgarization of the language. No one adheres to the pure style [of Central India]. As for writing down speeches or recording events, each kingdom has someone who manages that. Historical annals and royal proclamations are called *nīlapiṭa* ["dark-green collection"]. Both good and bad events are written down, and disasters and auspicious events are thoroughly recorded.

Education

To commence elementary education of the young and lead them forward, one first guides them with the spelling primer called the *Book in Twelve Chapters*. After seven years of age, one instructs them in gradual stages with great works from the five fields of knowledge. The first field contains treatises on the elucidation of sounds, which glosses ancient words and elucidates their meanings, divided into annotated and itemized lists. The second field contains works on elucidation of technical arts, which teach technical arts [like mathematics, astronomy, and music], mechanical skills, the movements of yin and yang, and calendrical calculation. The third field contains works on elucidation of the medical arts, including exorcising pathogenic agents, guarding against malevolent influences, pharmacology, acupuncture, and moxibustion. The fourth contains works on elucidating causation [formal logic], which teaches how to determine the orthodox from the perverse and to thoroughly investigate the difference between true and false. The fifth contains works on elucidating one's inner being [philosophy and religion], which fully demonstrate cause and effect and the subtle principles of the Five Vehicles [modes of religious practice that convey karmic reward].

Among the Brahmins, they study the doctrines of the four *veda-śāstra*. The first is called "longevity," which discusses nurturing the life force and cultivating one's inner nature. The second is called "sacrifice" and discusses the guidelines for sacrifices and prayers. The third is called "peaceful regulation" and discusses ritual decorum and ceremonial rites, divination, military strategy, and troop formations. The fourth is called "techniques," and it relates to extraordinary abilities, technical methods, exorcistic incantations, and medical arts.

Teachers must have extensively studied the profound and subtle principles in these works themselves and completely penetrated their most abstruse mysteries. Only then are they able to explain their general sense to students, gradually guiding them toward understanding more profound statements. Teachers seek to stimulate students and properly guide them along the path of learning. They try to carve rotten wood [unpromising students] into something useful and encourage those of poor intelligence to make some progress. If a scholar has great intellectual capabilities and a nimble mind and he harbors a determination to withdraw from the world and live in seclusion, he will lock himself away and concentrate on study until the project has been completed successfully.

When one has reached the age of 30, his ambitions are well established, and his studies have been completed. After he has garnered an official position and salary, the first thing a former student will do is to repay his teacher's kindness. Some scholars are broadly knowledgeable about antiquity and have a fondness for refined pursuits. They live contentedly in seclusion, holding fast to their principles and untouched by the vicissitudes of the material world. They move about unconstrained, beyond the tumult of events, and remain indifferent to either favor or humiliation. They enjoy a widespread reputation, but even though lords and kings may greatly honor and appreciate the reclusive scholar, no sovereign is able to make him demean himself into accepting an official appointment. Even so, the state respects the wise and farsighted, and the common people esteem those who are brilliant, so they are offered both lofty praise and important appointments. Thus, people develop great ambitions to become scholars, enjoying their pursuit of the liberal arts with no sense of fatigue. In their pursuit of the Way and acting in accordance with the moral principle of humaneness, they will not consider a thousand miles too far to travel. Though his family may be very wealthy, a man may aim to travel to far-off lands in search of knowledge, begging for food to get by. Those who value knowledge of the true Way feel no shame in poverty.

Indulging in casual amusements, idleness, gourmet delicacies, or extravagant clothing is not only considered immoral but is also inconsistent with contemporary customs. Doing so brings shame and a bad reputation.

Buddhism

Different people have divergent understandings of the true meaning of the Buddha's teachings, according to differences in their abilities and character. Since the time when the Buddha lived was so long ago, the true dharma [teachings of the Buddha] has been diluted to a greater or lesser degree, so each follower must rely on his own mental abilities to achieve enlightenment.

The different Buddhist sects stand in stark confrontation, and debates rage like surging ocean waves. Although the various schools and their specialists travel along different roads, their final destination is the same. There are a total of 18 Lesser Vehicle [Hīnayāna] Buddhist sects, each wielding its own sharp rhetoric. The adherents of Greater Vehicle [Mahāyāna] and Lesser Vehicle Buddhism dwell in complete segregation.

Some devote themselves to entering a quiet state of deep, seated meditation, while others engage in walking or standing forms of meditation. The practice of meditation and the acquisition of wisdom are separate endeavors, just as noisy clamoring in debate or quietly meditating are distinctly different practices. Accordingly, each fraternity of monks has instituted its own rules and prohibitions.

Regardless of whether we are talking about *vinaya* [monastic rules], *abhidharma* [treatises], or sūtra [threaded books], these all belong to the corpus of Buddhist scriptures. One who can fully expound on one of these categories of works is exempted from manual labor assigned by the director of duties at the monastery. If he can expound on two categories of texts, he receives superior accommodations and furnishings. He who can explain three categories is assigned servants to deferentially attend upon him. One who can expound upon four categories of scriptures is allotted some "pure men" [Buddhist laymen] as attendants. If he can explain five categories, he is allowed to ride around in a howdah on an elephant. He who can expound on six categories of works is also allowed an entourage and guards.

When one has achieved lofty moral distinction, his achievements will also be honored in an exceptional way. Occasionally, there will be a gathering where monks will discuss and evaluate who is superior and who is inferior, clearly differentiating the good from the bad and demoting the benighted ones while promoting the intelligent. If there is one who can deliberate using subtle words, who can praise or censure the most profound theories to arrive at the correct one, who has elegant and exquisite diction, and whose mind is accommodating and nimble during debates, then on account of this, he is allowed to sit high on a bejeweled elephant, surrounded by an entourage of attendants as thick as a forest. If, on the contrary, a member of one of the Buddhist sects makes arguments that are devoid of substance, the sharp edge of his statements is blunted by his opponents, his arguments lack reason though they spill out in profusion, and he promotes heterodox principles though his words are pleasant to hear, then they will smear his face with red and white clay, sprinkle dust all over his body, and abandon him in the wilderness or toss him into some ditch. Thereby, they commend the good and censure the bad, as well as demonstrate who is wise and who is stupid.

Accordingly, people delight in cultivating the Buddhist way and diligently study while they live at home. Following his own preference, a man may eventually leave his family to become of monk or may return to lay life. If one falls into error and violates one of the monastic rules, he is punished in front of the other monks. If it is a minor violation or a first offense, he will just be publicly berated. On a second offense, no one will speak to him for a while. For a more serious offense, the group will no longer live with the offender. If no one will live with him, that means he has been ostracized and despised by the other monks. Once one has been rejected like this, he tries to find a place to live. If he cannot find a place to settle and suffers hardship in a land far from home, he may have no choice but to don his original clothes and return to lay life.

Indian Castes

Regarding the differentiation into endogamous castes, there are four separate groups within the system. The first are called the Brahmins, who are defined by their purity of moral conduct. They hold fast to their principles and abide by the correct way. Their personal integrity is absolutely spotless. The second are called the Kṣatriya. This is the royal caste, for generation after generation, a member of this group has ruled as monarch. They take benevolence and tolerance as their moral standard. The third is called the Vaiśya, the merchant caste. They trade goods and engage in commerce, pursuing profit both near and far. The fourth is called Śūdra, the peasant caste. They devote all their strength to working the land and toil away at sowing and reaping. These four castes are differentiated by their level of purity, and one can marry only a member of the same caste, for the high born and the lowly must travel different paths. Both paternal and maternal relatives are not allowed to marry each other either. Once a woman has taken a husband, she can never remarry for the rest of her life. Outside of these four castes, there are many other types of men, who each dwell together according to their kind. It would be difficult to record all of them in detail.

The Military

Only members of the Kṣatriya caste can rightfully form a hereditary dynasty of kings. Occasionally, regicides or usurpations of the kingship have occurred, when men from a different caste have declared themselves ruler.

A country's warriors are selected from among the fiercest and most valiant men. The military occupation is passed down from father to son, and consequently, they can achieve thorough mastery of martial skills. During times of peace, they serve as palace guards, but during military campaigns, they are mobilized as the vanguard of the army.

There are four divisions of the military: infantry, cavalry, chariots, and elephants. They cover the elephants with sturdy armor and outfit their tusks with sharp metal barbs. A single general is settled atop an elephant, transmitting orders and directing the troops from up there. There are two soldiers stationed on either side who control the elephant for him. A war chariot is drawn by four horses driven by a commanding officer and assisted by arrayed ranks of foot soldiers pressed up against the wheels on each side. The cavalry is often dispersed to fend off incursions or to pursue retreating enemy soldiers. The infantry are equipped with only light weaponry and are selected from among daring and courageous men. They carry a great shield and wield a pike or a long sword, advancing to the front in tight formations. In general, all their weapons are pointed and sharp. They consist of spears, shields, bows and arrows, daggers, swords, battle axes and hatchets, halberds, wooden spikes, javelins, iron-bladed throwing discs, and the like. These have been in use for generations.

Law and the Administration of Justice
Regarding the prevailing mood and customs of India, although the people are impatient by nature, their intentions are honest and simple. They don't seek to obtain wealth by improper means and can be exceedingly self-effacing and yielding in etiquette. They are terrified of being condemned in the netherworld but consider the undertakings of the mortal realm to be of scant importance. They don't practice scheming and treachery and are faithful to their oaths and pledges. Their political teachings value unpretentious simplicity, and their manner is gentle and amiable. On the rare occasion that bands of brutal, rebellious rabble violate the country's laws or plot to endanger the sovereign, should the evidence be quite clear, the offenders are usually just confined to prison and not executed, for such matters of life and death are allowed to run their course through the action of karma, but these incarcerated persons are no longer considered part of the human community. If they violated the norms of propriety and righteousness or went against the principles of loyalty and filial piety, they would be punished by having their noses, ears, hands, or feet

severed. Some would be driven out of the country and others banished to a remote wilderness. Outside of these crimes, some faults and infractions can be redeemed with a payment of money. In trying cases and taking oral depositions, they do not employ whippings or canings. When the person being interrogated responds to the questions truthfully, the judge passes sentence fairly, based on the facts of the situation. Some people adamantly deny that they are guilty, for they are ashamed to admit their own shortcomings and want to gloss over their faults. When the judge still wants to get to the truth of the situation and manage the case properly, there are four methods of ordeal: water, fire, weighing, and poison.

In the water ordeal, the criminal and a large rock are placed in two separate sacks that are connected by a cord and then submerged in a deep river, to assess whether he has spoken truthfully or falsely. If the person sinks but the stone floats, then he is guilty; if the person floats while the stone sinks, then he has not concealed anything. In the ordeal by fire, they heat an iron plate and make the defendant squat upon it. Then they make him tread on it, force him to press his palms into it, and order him to lick it with his tongue. If there is nothing to the allegations, he will suffer no harm, but if the allegations are true, he will be injured. For those who are too cowardly to endure the scorching hot iron plate, they are allowed to take a flower bud and toss it toward the flames. If they are innocent, the flower will open, but if they are guilty, the flower will be scorched. The ordeal by weighing works like this: the person and a stone are placed on each side of a balanced scale to obtain proof based on their relative weight. If the charges are baseless, the person will descend and the stone will rise; if the charges are true, then the stone will be heavier and the person lighter. In the ordeal by poison, they take a black ram and make an incision in its right thigh and then insert a portion of food eaten by the accused, along with a mixture of poisons. If the person is guilty, the poison will spread and the ram will die; if the person is innocent, the poison will fade away and the ram will revive. By using these four methods, they can guard against a multitude of mistaken judgements.

Paying Respect

As for the forms of paying respect, they consist of nine different levels of formality: (1) expressing regards with comforting words, (2) bowing the head to show respect, (3) raising both hands in the air and bowing, (4) folding the hands together at chest level, (5) genuflecting, (6) kneeling with thighs and body erect, (7) prostrating on hands and knees, (8) prostrating with knees, elbows, and head touching the ground, (9) stretching out fully prone with knees, elbows, and head on the ground. Of these nine variations, the most extreme form is really no more than a single Chinese ritual prostration [the kowtow]. Kneeling on the ground and praising a person's virtues is called the highest form of respectful tribute. When a man is far away from the person being honored, he will clasp his hands and knock his forehead on the ground. When the person is nearby, he will kiss the soles of the man's feet and caress his heels. When one expresses thanks for a commission received from a superior, he must gather up the skirts of his robe and kneel.

When a venerable or worthy person has received the reverence of others, he must at least say a few reassuring words. Sometimes, he will touch the head of the inferior or pat him on the back, earnestly speaking words of guidance, all to express intimacy and generosity.

After a Buddhist monk has been respectfully addressed by another, he simply replies with good wishes, but he won't stop others from prostrating themselves before him. Indians will also circumambulate around a revered person [as they would a statue of a deity or a stupa], sometimes just once, sometimes repeating three times. If one harbors a long-cherished wish in his heart, the number of circumambulations will correspond to the nature of the wish.

Illness and Death

In general, when someone in India takes seriously ill, he will fast for seven days. Within this time period, many will fully recover, but if the illness is not completely resolved, only then will they begin to take medicine. The types and names of their medicines are different from ours, and their doctors' healing techniques and their ability to diagnose and predict the course of an illness exhibit some key differences.

When a loved one dies and a funeral is held, the mourners wail loudly and shed tears together. They rend their clothes, pull out their hair, strike their foreheads, and pound their chests. Indians have no mourning-dress regulations [as in China, based on proximity to the deceased], and the length of appropriate mourning periods is not fixed. There are three different types of ritual for a funeral and disposal of the corpse: the first is called "fire burial," wherein firewood is heaped into a pyre and the corpse is burned. The second is called "water burial," whereby the corpse is cast into the water and is swept away with the river current. The third is called "wilderness burial," where the corpse is discarded in the forest to be devoured by wild beasts.

When a king nears death, he preemptively abdicates in favor of his presumptive heir, who will take charge of the funeral proceedings to determine the proper order of precedence within the state. While the king is still alive, he is given an honorific title that reflects his virtue, but after he is dead, he is not labeled with a posthumous name that passes judgment upon his reign [as in Chinese dynasties].

In a household where a death has occurred, everyone will refrain from eating, but after the funeral is over, they all return to their normal activities with no prohibitions. Those who have accompanied the body of the deceased to the funeral are considered temporarily unclean. They must ritually wash themselves outside the city walls before they can again enter the town.

When persons have grown extremely old and their fated day of death is approaching, or if they are suffering from a terminal illness, they fear the limit of their lifespan has reached its maximum extent, are wearied of the mundane world and want to cast aside this human realm, scorn matters of life and death, and wish to distance themselves from the ways of the world. Thereupon, relatives and close friends will throw him a banquet with music, and then embark on a boat, paddling him out to the middle of the Ganges where he will drown himself. They call this "obtaining a rebirth in heaven." One in ten persons will do this, but I have not seen it with my own eyes.

The large group of men who have left their families to become Buddhist monks are not allowed by monastic rules to wail for the dead, so when their parents die, they simply chant a sūtra to repay their kindness, though they do pay careful attention to the proper rituals and piously carry out sacrifices to ensure their future happiness and endow them with resources for the underworld.

Taxation

Since governance and moral education in India are tolerant and generous, the key operations of the state are also simplified. Households are not registered with the state, and people are not liable for labor service or capitation taxes. The agricultural fields owned by the king are divided into four portions. The first provides for the needs of the state, while also provisioning sacrificial vessels filled with grain for sacrifices to the gods. The second portion is given out as fiefs to the prime minister and other senior officials. The third part is given out as rewards to brilliant and far-sighted scholars of great erudition and talent. The fourth portion is given to monastic and temple organizations of the various religions to plant blessed fields for karmic reward. As a result, the tax burden on the people is very light, and the demands for labor service are greatly reduced. Each family is provided with a division of inheritable, taxable land based on the number of people in the household. If one's family is cultivating royal fields, the land is taxed at one-sixth of the harvest. In their pursuit of profit, merchants come and go freely, transporting their goods for sale. At customs checkpoints located near major river fords or strategic passes, they pay only a minor tax and then are allowed to pass through. For major state construction projects, the people's labor is not exploited without compensation, for persons will be fairly paid according to the value of the work they have accomplished. Soldiers are recruited according to an assessment of the mission, whether that is to garrison the frontier, go on a punitive campaign, or to defend the palace. Military officials post a recruitment notice, with the financial reward indicated, and await those persons willing to enlist. The governors of prefectures and counties, ministers who assist the ruler, regular officials, and their aides are each given an allotment of land and support themselves with the production of that fief.

* *

Products

Just as the climate and soil conditions of each area differ, so too do the products of the land vary. There are various types of flowers, plants, fruits, and trees, each with a different name. There are the so-called mango, tamarind, butter-tree fruit, Indian jujube, wood apple, gooseberry, Indian persimmon, cluster fig, banana, coconut, and jackfruit. All the varieties are difficult to exhaustively record, but the ones most valued by the society's inhabitants have been briefly spoken about here.

When it comes to the Chinese jujube, chestnut, astringent persimmon, or common persimmon, they are unheard of in India. As for various fruits like pears, crabapples, peaches, apricots, and grapes, they had been imported from Kashmir and are often cultivated in different areas. Every country in India plants pomegranate and orange trees.

In working the land for agriculture, they sow and reap, plow and weed, planting each variety of seed according to the proper season. Each person is left to choose when to labor or when to be idle. Of those crops for which the land is suitable to produce, rice and wheat are the most plentiful. Vegetables include ginger, mustard, melon, calabash, and elephant foot yam. Scallions and garlic are rare, and those willing to eat them even scarcer, for if anyone in a family eats them, he will be driven out of town.

When it comes to milk, yogurt, animal fat, butterfat, cane sugar, toffee candy, mustard oil, and various types of wheat biscuit, they are frequently eaten or used in food preparation. Indians from time to time will serve sumptuous dishes like fish, lamb, roebuck, and deer. But hairy animals like cattle, donkey, elephant, horse, pig, dog, fox, wolf, lion, monkey, and ape are not eaten as a rule. Those who eat them will be shamed and despised by everyone as sordid and filthy. They must reside in segregated areas outside the city walls and will no longer be seen in human society.

There are also differences in the consumption of wine and other beverages according to tastes and the particular caste. Wine from grape juice or sugar cane is the drink of the Kṣatriya caste. Strong, undiluted grain alcohol is the beverage of choice among the Vaiśya caste, among others. Buddhist monks and Brahmins only drink grape juice or sugar cane syrup, so these can't really be called wine. Those outside the caste system or of base status are not associated with a particular beverage.

Utensils that Indians regularly rely upon vary considerably in usefulness and quality, but vessels for daily use are never in short supply. Although they do use vessels like our Chinese woks and cauldrons for cooking, they do not know about steamers. Most vessels are fashioned of unfired earthenware, and they rarely use copper. They eat their food from a single vessel, and the various flavors mingle. They use their hands to ladle out beverages and eat with their fingers, for generally they do not have scoops or chopsticks. Only the sick might employ a bronze spoon.

Gold, silver, native copper, white jade, and rock crystal are natural products of the country. The coastal areas export large quantities of precious rarities and gems of various colors, with different types and names, to exchange for needed commodities. As for the currency used to facilitate exchange, they have gold and silver coinage as well as large and small pearls.[2]

INTRODUCTION TO COSMAS INDICOPLEUSTUS

The *Christian Topography* (Χριστιανικὴ Τοπογραφία) is an illustrated Greek treatise written in stages between 543 and 550 CE, attributed by 11th-century commentators to a man called Cosmas Indicopleustus ("Mr. World Order, the Indian Navigator"). We do not know the author's real name, but he tells us in his book that he was a former merchant from Alexandria who had traded throughout the Indian Ocean maritime networks for decades. Though devoutly Christian, there is no evidence that he was an ordained monk. He names his religious teacher as Patrikios, the Greek name of Mar Aba I (d. 552 CE), who eventually became the Christian archbishop of Persia. This meant that Cosmas was an adherent of the Church of the East, later labelled as Nestorian Christianity, a sect declared heretical at councils in 431 and 451 CE but which flourished in Syria and Persia and later made its way to China.

The original core of *Christian Topography* (books 1-5) was composed shortly after 543 CE as a polemic against the cosmology of pagan Greek geographers like Claudius Ptolemy, who envisioned an eternal universe with a stationary spherical earth, surrounded by one or more rotating celestial spheres. Cosmas sought to overthrow this cosmology by using Old Testament scripture. According to Cosmas's exegesis, the earth was a flat rectangle, twice as broad from east to west, surrounded by an ocean. Beyond this ocean was a band of inaccessible land, the eastern portion of which housed the original paradise of the Garden of Eden. This terrestrial world was covered by a barrel-vaulted heaven, with the barrier of the "firmament" separating the mortal world from the Kingdom of God. The northwest was dominated by a lofty mountain, behind which the sun disappeared at night. Four great navigable gulfs encroached into this landmass, the Roman Gulf (i.e., Mediterranean), Caspian Sea, Red Sea, and the Persian Gulf (with the Indian Ocean).

Though his world geography shared similarities with the pre-Pythagorean Greek view of the shape of the inhabited world, Cosmas claimed inspiration from the design of the Tabernacle built in the wilderness by Moses to house the Ark of the Covenant. After God had revealed all knowledge, including the shape of the universe, to Moses on the mountain, the prophet built the Tabernacle shrine as a microcosm of that universe. Cosmas argued that the rectangular shape of the altar table in the outer chamber symbolized the shape of the inhabited world, while the wavy fringe of the tablecloth represented the surrounding ocean. Drawing inspiration from revealed scripture rather than observation and mathematics, Christians like Cosmas reoriented the worldview of the early Middle Ages. This would continue to have an influence on later medieval world maps that placed Jerusalem at the center of a flat world, surrounded by an ocean.

Cosmas made later additions to his *Christian Topography*, expanding his ideas and responding to critics, and parts of a different book by Cosmas, a descriptive geography of the entire world, were appended by later editors. This is where his account of India and Sri Lanka (excerpted below) probably came from originally.

Cosmas provides another account of Indian elephant taming, a subject that perpetually fascinated Western readers (see chapter 2). Like earlier classical authors, Cosmas identifies China as the country of origin of silk but refers to this realm by a new name, Tzinitza, derived from the Persian placename, Chinistan, but still based on the title of the long-defunct Qin dynasty. Just as with the earlier Roman accounts of the East, the Tzinitza of Cosmas does not seem to refer to the heartland of Chinese civilization along the Yellow River but rather to Indochina or the Malaysian peninsula, for he describes China as being "surrounded on the left by the ocean." But he accurately states that "beyond Tzinitza [to the east] there is neither navigation nor any land to inhabit." Interestingly, Cosmas considers the overland Silk Road to be shorter and more efficient in transporting large quantities of silk from China to Persia than the

Indian Ocean route, contradicting modern economic historians who emphasize the greater efficiency of ocean transport.

Cosmas blends both the merchant and religious viewpoints seen in some of our other travel accounts. So while Cosmas's cosmology and geography are directly shaped by his Old Testament orthodoxy, his worldly observations still focus on ports, trade routes, merchandise, and coinage. Cosmas describes an Indian Ocean trading world very similar to that seen in the anonymous first century, *Navigating Around the Red Sea* (chapter 3). From his home port in Alexandria, he had personally plied these waters as a merchant for decades, traveling through the Red Sea, into Ethiopia, and on to the Persian Gulf, western India, and Sri Lanka, though some of his information about the subcontinent may have been secondhand.

However, several aspects of this Indian Ocean trading world had shifted since the first century, and the entire system was rapidly collapsing at the time Cosmas was writing. He mentions many of the same Eastern trade goods (silk, pepper, aromatic woods, cotton, gemstones) and trading ports as the *Navigating* text, but many of the political actors had changed. The Sassanians had replaced the Parthians as trade intermediaries, and the Christian religion had made some major inroads into Persia and some smaller beachheads in India. However, the picture that Cosmas paints of this world really dates to two or three decades earlier, when he was still an active merchant. The bubonic plague had entered Egypt on Indian Ocean trading ships in 541 CE, transferred from northeast India and Chinese Central Asia as a consequence of the trade in silk. This so-called Plague of Justinian would decimate the urban populations of the Mediterranean, killing perhaps 40 percent of Constantinople's citizens and one-fourth of the population of the Mediterranean, destroying the urban consumer economy that drove the trade in Indian luxuries. The earlier Antonine Plague and Third Century Crisis in the Roman Empire had caused downturns in the trade with the East before, but this new cataclysm rang its death knell. The epidemic, along with momentous political shifts in Ethiopia and southern Arabia, opened the door for Sassanian and later Arab control of the entire Indian Ocean trading networks in subsequent centuries (see chapter 6).

SOURCE TRANSLATION AND SELECTED STUDIES

Bowersock, Glen. *Throne of Adulis: Red Sea Wars on the Eve of Islam.* Oxford University Press, 2013.

McCrindle, J. W. *The Christian Topography of Cosmas, an Egyptian Monk.* Hakluyt Society, 1897.

Kominko, Maja. *The World of Kosmas: Illustrated Byzantine Codices of the Christian Topography.* Cambridge University Press, 2013.

Wolska-Conus, Wanda. *La topographie chrétienne de Cosmas Indicopleustes: Théologie et sciences au VIe siècle.* Presses universitaires de France, 1962.

EXCERPTS FROM *CHRISTIAN TOPOGRAPHY,* BY COSMAS INDICOPLEUSTUS (543–550 CE)

The Country of China

Yet if Paradise did exist in this earth of ours, many a man among those who are keen to know and enquire into all kinds of subjects would think he could not be too quick in getting there; for if there be some who, to procure silk for the miserable gains of commerce, hesitate not to travel to the uttermost ends of the earth, how should they hesitate to go where they would gain a sight of paradise itself?

Now this country of silk is situated in the remotest of all the Indies and lies to the left of those who enter the Indian sea, far beyond the Persian Gulf, and the island called by the Indians Sielediba [Sanskrit: Siṃhaladvīpa, i.e., Sri Lanka] and by the Greeks, Taprobanê. It is called Tzinitza [Persian: Chinistan; i.e., China] and is surrounded on the left by the ocean, just as Barbaria [northeast Africa] is surrounded by it on the right. The Indian philosophers, called the Brachmans [Brahmins], say that if you stretch a cord from Tzinitza to pass through Persia, onward to the Roman dominions, the middle of the earth would be quite correctly traced, and they are perhaps right. For the country in question deflects considerably to the left, so that the loads of silk passing by land through one nation after another reach Persia in a comparatively short time, while the route by sea to Persia is vastly greater. For just as great a distance as the Persian Gulf runs up into Persia, so great a distance and even a greater has one to run, who, being bound for Tzinitza, sails eastward from Taprobanê, while besides, the distances from the mouth of the Persian Gulf to Taprobanê and the parts beyond through the whole width of the Indian sea are very considerable. He then who comes by land from Tzinitza to Persia shortens very considerably the length of the journey. This is why there is always to be found a great quantity of silk in Persia. Beyond Tzinitza there is neither navigation nor any land to inhabit. . . .

The Spread of Christianity

For the Christians who were at one time persecuted by the Greeks and Jews have conquered and drawn their persecutors over to their own side. In like manner, we see that the church has never been destroyed but that its adherents have been greatly multiplied and that similarly, the whole earth has been filled with the doctrine of the Lord Christ and is still being filled and that the Gospel is preached throughout all the world. This I avouch to be the veritable fact, from what I have seen and heard in the many places which I have visited.

Even in Taprobanê [Sri Lanka], an island in Further India, where the Indian sea is, there is a church of Christians, with clergy and a body of believers, but I know not whether there be any Christians in the parts beyond it. In the country called Male [Malabar Coast], where the pepper grows, there is also a church, and at another place called Calliana [present-day Kalyān, near Mumbai], there is moreover a bishop, who is appointed from Persia. On the island called Dioscoridês [Socotra, off the Horn of Africa], which is situated in the same Indian sea and where the inhabitants speak Greek, having been originally colonists sent thither by the Ptolemies who succeeded Alexander the Macedonian, there are clergy who receive their ordination in Persia, and are sent on to the island, and there is also a multitude of Christians. I sailed along the coast of this island but did not land upon it. I met, however, with some of its Greek-speaking people who had come over into Ethiopia. And so likewise among the Bactrians and Huns and Persians, and the rest of the Indians, Persarmenians [Sassanian Armenians], and Medes and Elamites, and throughout the whole land of Persia, there is no limit to the number of churches with bishops and very large communities of Christian people, as well as many martyrs, and monks also living as hermits. . . .

Description of the Island of Taprobanê

This is a large oceanic island lying in the Indian sea. By the Indians, it is called Sielediba [Sri Lanka], but by the Greeks Taprobanê, and therein is found the hyacinth stone [a reddish-brown variety of zircon]. It lies on the other side from the pepper country [Malabar Coast]. Around it are numerous small islands [Laccadive Islands] all having fresh water and coconut trees. They nearly all have deep water close to their shores. The great island, as the natives report, has a length of 300 *gaudia* [an hour's walking distance], that is, of 900 miles, and it is of the like extent in breadth. There are two kings on the island, and they are at feud the one with the other. The one has the hyacinth country, and the other the rest of the country where the harbor is and the center of trade.

* * * * * * * * * * * * * * **BRIEF EXCERPT** * * * * * * * * * * * * * * * *

Taprobanê [Sri Lanka] is a great mart for the people in those parts. The island has also a church of Persian Christians who have settled there, and a Presbyter who is appointed from Persia, and a Deacon and a complete ecclesiastical ritual. But the natives and their kings are heathens. In this island they have many temples, and on one, which stands on an eminence, there is a hyacinth as large as a great pinecone, fiery red, and when seen flashing from a distance, especially if the sun's rays are playing round it, a matchless sight. The island being in a central position, is much frequented by ships from all parts of India and from Persia and Ethiopia, and it likewise sends out many of its own. And from the remotest countries, I mean Tzinista [China] and other trading places, it receives silk, aloeswood, cloves, sandalwood, and other products, and these again are passed on to marts on this side, such as Male, where pepper grows, and to Calliana, which exports copper and sissoo-logs, and cloth for making dresses, for it also is a great place of business. And to Sindu [present-day Sindh, Pakistan, near mouth of Indus], also where musk and castor is procured and spikenard, and to Persia and the Himyarite Kingdom [in Yemen], and to Adulis [a port on Red Sea]. And the island receives imports from all these marts which we have mentioned and passes them on to the remoter port while at the same time exporting its own produce in both directions.

Sindu is on the frontier of India, for the river Indus, that is, the Phison [one of the rivers flowing out of Eden in Genesis], which discharges into the Persian Gulf, forms the boundary between Persia and India. The most notable places of trade in India are these: Sindu, Orrhotha [on west coast of Kathiawar Peninsula], Calliana, Sibor [present-day Chaul], and then the five marts of Male that export pepper: Parti, Mangarouth [Mangalore], Salopatana, Nalopatana, Poudopatana ["New Town," situated somewhere between Mangalore and Calicut]. Then out in the ocean, at the distance of about five days and nights from the continent, lies Sielediba, that is, Taprobanê. And then again on the continent is Marallo, a mart exporting chank shells, then Caber [near Tharangambadi, at mouth of Kaveri River], which exports *alabandenum* [garnet], and then farther away is the clove country, then Tzinista, which produces the silk. Beyond this, there is no other country, for the ocean surrounds it on the east. This same Sielediba then, placed as one may say, in the center of the Indies and possessing the hyacinth receives imports from all the seats of commerce and in turn exports to them, and is thus itself a great seat of commerce.

Sopatrus and the Contest of Coins

Now I must here relate what happened to one of our countrymen, a merchant called Sopatrus, who used to go there on business but who to our knowledge has now been dead these five and thirty years past. Once upon a time, he came to this island of Taprobanê on business, and as it chanced a vessel from Persia put into port at the same time with himself. So the men from Adulis with whom Sopatrus was went ashore, as did likewise the people of Persia, with whom came a person of venerable age and appearance. Then, as the way there was, the chief men of the place and the custom-house officers received them and brought them to the king. The king, having admitted them to an audience and received their salutations, requested them to be seated. Then he asked them, "In what state are your countries, and how go things with them?" To this they replied, "They go well." Afterward, as the conversation proceeded, the king inquired, "Which of your kings is the greater and the more powerful?"

The elderly Persian snatching the word answered, "Our king is both the more powerful and the great and richer, and indeed is king of kings, and whatsoever he desires, that he is able to do.

Sopatrus on the other hand sat mute. So the king asked, "Have you, Roman, nothing to say?"

"What have I to say," he rejoined, "when he there has said such things? But if you wish to learn the truth, you have the two kings here present. Examine each, and you will see which of them is the grander and the more powerful."

The king, on hearing this, was amazed at his words and asked, "How say you that I have both the kings here?"

"You have," replied Sopatrus, "the money of both—the *nomisma* [the gold *solidus* of Byzantium] of the one, and the *drachm* [silver Sassanian coin, about 4 g.], that is, the *miliaresion* [a Byzantine name for silver coinage] of the other. Examine the image of each and you will see the truth."

The king thought well of the suggestion, and nodding his consent, ordered both the coins to be produced. Now the Roman coin had a right good ring, was of bright metal and finely shaped, for pieces of this kind are picked for export to the island. But the *miliaresion*, to say it in one word, was of silver, and not to be compared with the gold coin. So the king after he had turned them this way and that, and had attentively examined both, highly commended the *nomisma*, saying that the Romans were certainly a splendid, powerful, and sagacious people. So he ordered great honor to be paid to Sopatrus, causing him to be mounted on an elephant and conducted round the city with drums beating and high state. These circumstances were told us by Sopatrus himself and his companions, who had accompanied him to that island from Adulis; and as they told the story, the Persian was deeply chagrined at what had occurred. . . .

* *

Indian Elephants

The kings of various places in India keep elephants, such as the king of Orrhotha, and the king of Calliana, and the kings of Sindu, Sibor, and Male. They may have each 600 or 500, some more, some fewer.

Now the king of Sielediba gives a good price both for the elephants and for the horses that he has. The elephants he pays for by cubit [approx. 47 cm] measurement. For the height is measured from the ground, and the price is reckoned at so many *nomismata* [Byzantine gold coins] for each cubit, 50 it may be, or a hundred, or even more. Horses they bring to him from Persia, and he buys them, exempting the importers of them from paying custom. The kings of the continent tame their elephants, which are caught wild, and employ them in war. They often set elephants to fight with each other for a spectacle to the king. They keep the two combatants apart by means of a great crossbeam of wood fastened to two upright beams and reaching up to their chests. Several men are stationed on this and that side to prevent the animals meeting at close quarters, but at the same time to instigate them to fight one another. Then the beasts thrash each other with their trunks till one of them gives in. The Indian elephants are not provided with large tusks, but should they have such, the Indians saw them off, that their weight may not encumber them in action. The Ethiopians do not understand the art of taming elephants, but should the king wish to have one or two for show, they capture them when young and subject them to training.

Now the country abounds with them, and they have large tusks that are exported by sea from Ethiopia even into India and Persia and the Himyarite Kingdom and the Roman dominion. These particulars I have derived from what I have heard.[3]

NOTES

1. Excerpted from Zhang Xun 章巽, *Faxian zhuan jiaozhu* 法顯傳校注 (Shanghai guji chubanshe, 1985), 2–7, 54, 72, 103, 123–124, 151, 167–178. Translation and headings by the author.
2. Xuanzang 玄奘, *Da Tang xiyu ji jiaozhu* 大唐西域記校注 (Zhonghua shuju, 2000), 1:161–218. Translation by the author. Headings in the original.
3. Translation from J. W. McCrindle, *The Christian Topography of Cosmas, an Egyptian Monk* (Hakluyt Society, 1897), 47–49, 118–120, 363–374, with updates to modernize and Americanize the spelling. Headings added by the author.

Chapter 6

Collisions and Connections Between New Empires (750–850 CE)

The interactions between the Roman and Han Empires, the Eurasian mega-polities of antiquity, involved significant indirect trade in luxuries but very little personal contact or experience living within the other realm. As a result, Chinese knowledge of the Roman Empire was often hazy and distorted, and Roman information about China was equally misguided and inaccurate (chapters 3 and 4).

With the rise of two new great empires during the medieval period, the Arab empire in the West (632–1258 CE) and the Tang dynasty (618–907 CE) in the East, commercial, military, and diplomatic exchanges became much more direct and intense. By 850 CE, thousands of Arabic merchants were now resident in the Chinese emporium of Canton, overseen in their trading enclave by their own Muslim judge. Arabic rulers sent periodic "tribute" missions to the Chinese court, and numerous Chinese artisans, doctors, and soldiers resided in the Abbasid capitals. Consequently, for this period, we are fortunate to have two parallel first-hand accounts of the religion, law, economy, produce, and customs of these two great empires, written by merchants and sojourners from opposite ends of Eurasia.

Introduction to Du Huan

Du Huan's 杜環 *Record of a Journey* (Jingxing ji 經行記; ca. 762) is the earliest first-person account of Islam and its practices to reach China. It also provides a detailed portrait of the early Abbasid Caliphate (750–1258 CE) of the Arabs, with its initial capital at al-Kufah, just before the move to Baghdad in 762 CE.

Du Huan had been serving under the Chinese general of Korean descent, Gao Xianzhi 高仙芝 (d. Jan., 756 CE) at the Battle of the Talas River (751 CE) when he was captured along with 20,000 other Chinese prisoners, including papermakers and other artisans. He was brought to Iraq, where he lived under some sort of supervision for the next decade, though he seems to have been allowed to take side trips to places like Damascus (see map 4). In 761, he was finally allowed to leave, returning to China aboard an Arabic merchant ship headed for Canton. He wrote his *Record of a Journey* about what he had seen.

Du Huan's complete work is lost, but portions of it survive in quotations in his more famous relative Du You's 杜佑 (735-812) work, *Encyclopedic History of Institutions* (Tongdian 通典, 801 CE). We can determine from those fragments that the original work was structured as an itinerary that started in Central Asia, where Du Huan was taken prisoner, and continued into the heart of the Arabic and Byzantine worlds, returning to China by sea, going through Sri Lanka to Canton. Surviving passages describe Ferghana, Tashkent, Samarkand, Merv, Damascus, the Abbasid Caliphate, Persia, Egypt, the Byzantine Empire, and Sri Lanka. Du Huan had personally visited many of these places, but some entries were based upon second-hand knowledge or earlier accounts (see map 4).

For the most part, Du Huan's account is factual and straightforward, but it also included the usual shocking cultural comparisons. For example, Du was surprised that Arabic women had to veil their faces when they left home and that eating pork and drinking alcohol were prohibited. He accurately describes Muslim prayer and beliefs but notes with some surprise (from a Buddhist perspective) that Muslims could eat meat after sunset during the ritual fasting of Ramadan or that taking an animal's life could be seen as a pious act. From a Confucian perspective, he was astonished that people did not prostrate themselves before their ruler or their own parents, only before Allāh. He also seems to make an indirect criticism of the harsh Chinese legal system when he mentions that Islamic law does not implicate people for the crimes of their close relatives.

We learn from Du Huan the names of several other Chinese persons who were living in the Abbasid capital working as painters or silk weavers. Some of these men may have originally been prisoners of war like himself, but many probably emigrated voluntarily to set up shop in the new Arab capital.

Du Huan also provides the first details in any Chinese text of the Eastern Roman Empire, called Fulin 拂菻 (Arabic: Rūm, "Rome"). He equates Fulin with the name Da Qin ("Greater Qin"), which had been used to refer to the Roman Empire, or at least its eastern provinces, since the Han dynasty (chapter 4). He provides some particulars about the extent of Constantinople's walls, the size of its army, the quality of glass and other handicrafts produced there, the dress and eating habits of its residents, and their incessant hostilities with the Arabs. However, it is unlikely that Du Huan ever visited Constantinople.

Later, Du tells his readers of the Turks who lived north of Ukraine "whose feet resemble the hooves of oxen and who delight in eating human flesh." These are typical monstrous races thought by people from both East and West to live at the margins of the civilized world. The Kingdom of Women, reported by Du Huan to be located somewhere in the Mediterranean area west of Constantinople, constitutes another standard trope, a gender-inverted world where women ruled. A century earlier, the great Buddhist pilgrim Xuanzang had already mentioned this Kingdom of Women, located southwest of Constantinople on an island in the sea. This story is not necessarily a retelling of the Greek myth of the Amazons, for a Kingdom of Women was mentioned in the wilds of the far west as early as the *Guideways Through Mountains and Seas* (chapter 1).

The curious "ghost market" (*guishi* 鬼市), which Du Huan situates west of Constantinople, does not belong to the realm of imagination like the Country of Women, for similar trading arrangements that facilitated transactions between peoples who could not communicate have been recorded throughout history. Herodotus (4.196) wrote that the Carthaginians traded for gold with the peoples on the west coast of Africa in this way. Pliny had mentioned a silent trade for silk in second-hand accounts of the Silk People of Chinese Central Asia (chapter 4). Modern ethnographers have recorded the "silent trade" in a variety of places, including Africa, Siberia, and Southeast Asia.

Map 4. Route of Sulaymān the Merchant and reconstructed itinerary of Du Huan.

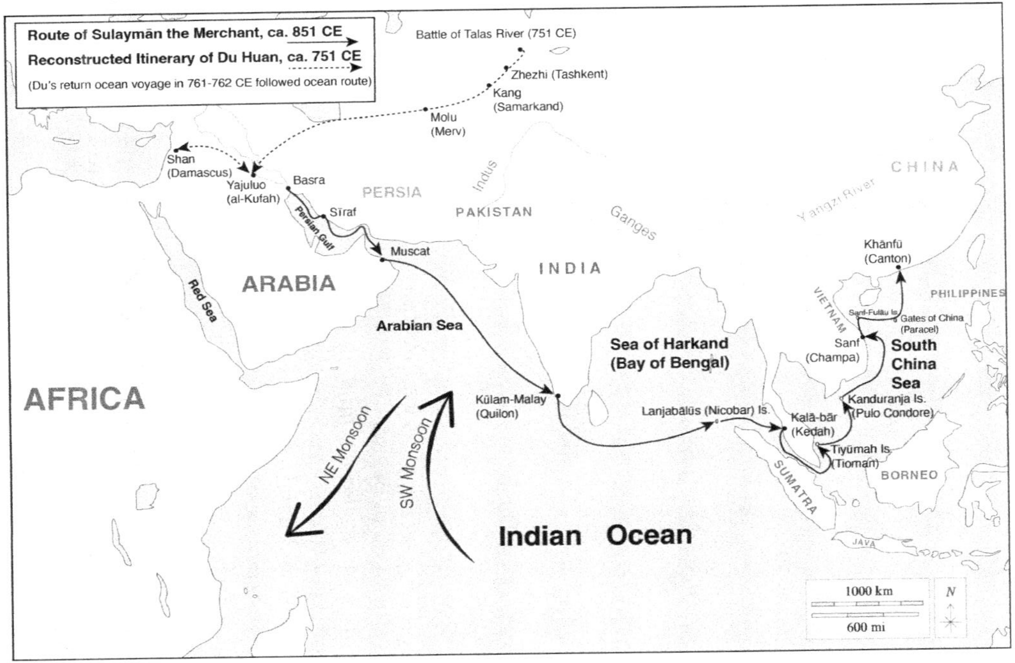

Source: Map created by the author.

Source Text, Alternate English Translation, and Selected Studies

Akin, Alexander. "Notes on the West by a Chinese Prisoner of War." *Harvard Middle Eastern and Islamic Review* 5 (1999–2000): 77–102.

Du Huan 杜環. *Jingxing ji jianzhu* 經行記箋注. Zhonghua shuju, 2000.

Kotyk, Jeffrey. *Sino-Iranian and Sino-Arabian Relations in Late Antiquity.* Brill, 2024.

Park, Hyunhee. *Mapping the Chinese and Islamic Worlds: Cross-Cultural Exchange in Pre-Modern Asia.* Cambridge University Press, 2012.

Excerpts from *Record of a Journey*, by Du Huan (ca. 762 CE)

* * * * * * * * * * * * * * * **BRIEF EXCERPT** * * * * * * * * * * * * * * *

Country of Dashi

The country of the Dashi 大食 [Persian: Tāzīk, the Arabs] is also called Yajuluo [al-Kufa, capital of the Abbasid Caliphate, 749–761, 170 km south of Baghdad]. The Caliph of Dashi carries the title of *mumen* [*amīr al-mu'minīn,* "commander of the faithful"] and his capital is situated here. The men and womenfolk of this place are stalwart and handsome in appearance. The garments they wear are fresh and clean, and their bearing is refined and graceful. When a woman leaves the gates of her house, she must conceal her face.

No matter whether a person is noble or humble, five times a day he prays to Heaven. They eat meat during ritual fasting [after nightfall during Ramadan] and consider killing a living creature to be a pious deed.

They wear silver belts with pendant silver daggers. They have prohibited the drinking of alcohol and banned music. A quarrel between persons never comes to blows.

There is also a hall for prayer that holds tens of thousands of people [Great Mosque of Kufa]. On every seventh day, the Caliph comes out to attend prayers. He mounts a high dais and expounds on Islamic law for the assembled masses, saying:

The life of man is quite difficult. This is the way of Heaven and cannot be changed. Treacherous or wicked behavior like kidnapping and robbery, speaking falsehoods about even trivial matters, placing others in danger while securing one's own safety, deceiving the poor and oppressing the humble; if one's crime is among these, nothing is considered more heinous. If there is a battle and one is killed by the enemy, he will certainly be reborn in paradise. If he slays the enemy, he shall obtain limitless blessings here on earth.

The Caliph controls a great territory and receives the benefit of all its resources. His followers increase like a rapidly flowing river. Their laws are lenient, and their funeral customs are frugal.

Inside the walls of their cities and within the gates of their villages, every product of the earth is available. It is like the hub of a wheel, where all the spokes converge. Ten thousand types of merchandise are found here in abundance and at cheap prices. Brocaded and embroidered cloth, pearls, and seashells fill their market stalls. Camels, horses, donkeys, and mules crowd the lanes and alleyways. They make residences out of densely laid carved stones and have something like our imperial carriage from the Middle Kingdom. On each festival day, the nobility are presented with glass vessels and brass bowls in innumerable quantities.

They have fine polished rice as well as wheat flour, which are no different from those found in China. Their fruits include almonds and dates. The bodies of their turnips are spherical and as large as a two-liter measure, with a wonderful flavor. Their other vegetables are identical to those found in the various other countries. The largest varieties of grapes are as big as chicken eggs. Among their fragrant oils, two are particularly valuable: one is called *yesaiman* [jasmine] and the other is called *moniushi*. The most valuable aromatic herbs include one called *chasaibeng* and another called *qialuba* [chinaroot greenbrier].

They have Chinese-style looms for making figured damask silks, as well as goldsmiths, silversmiths and painters, and we observed that Han Chinese craftsmen who originated the production of paintings there included Fan Shu and Liu Ci from the capital of Chang'an. Master silk weavers in residence there include Yue Huan and Lü Li from the Hedong area [present-day Taiyuan, Shanxi].

They even harness camels to pull carriages. As for their Arabian steeds, popular legend in China says that they were produced from the union of a dragon and a horse on the shores of the Western Sea. Their abdomen has a smaller girth, but their forelegs are longer than that of Chinese horses, and the finest among them can run a thousand Chinese miles. Their camels are relatively small and sturdy, and their backs have only a single hump. The good ones can travel a thousand Chinese miles in a day as well. They also have the "camel bird" [ostrich], which grows taller than four Chinese feet [approx. 1.44 m]. Its feet are like the hooves of a camel, and its neck is strong enough to support a man riding it, traveling for five or six miles. They lay eggs with a capacity as great as three Chinese pints [approx. 600 ml]. They also have the *jidun* tree [Arabic: *zaytūn*, "olive"], whose fruit resembles summer jujubes. They can be pressed to make an oil that is eaten to banish the effects of pestilential miasmas. The climate is warm, and the land has no ice or snow. Many people suffer from malaria and dysentery. Within one year, as many as five out of ten people might die from these.

Recently, the Arabs have swallowed up 40 or 50 countries and made them their vassal subjects. Many of their soldiers are sent to garrison these areas. Their border extends all the way to the Western Sea.

* *

Three Religious Laws

In all the various countries that I passed through on my overland journey, it turns out that there is only one race of foreigner, but there are several types of religious teaching. There is the religion of the Arabs [Islam], there is the religion of Greater Qin [Roman and Nestorian Christianity], and there is the religion of the Xunxun [Arabic: Zamzama; Zoroastrians]. It is among the Zoroastrians that the practice of incest is the most widespread. They do not speak when taking meals. Under the religious law of the Arabs, if one's brother, child, or other relative is judged guilty under the law, even though a person may have been slightly at fault with them, he will not be implicated through mutual responsibility.

They do not eat the flesh of pigs, dogs, donkeys, or horses. They do not prostrate themselves as a show of reverence before the king of the country nor before their own parents. They do not believe in ghosts and spirits, but only sacrifice to Heaven [Allāh]. Their custom is that every seventh day is a holiday, when they also do not buy and sell or come and go from places. They will only drink wine and make merry at the end of the day. . . .

Country of Shan

The country of Shan [Arabic: al-Shām; Damascus, Syria] lies on the western border of the country of the Arabs (see map 4). It is several thousand Chinese miles in circumference. They construct houses by stacking courses of stones to make the walls and cover the roof with ceramic tiles. Grain prices there are particularly cheap. There is a great river [Euphrates] which flows east from here and enters al-Kufa. Merchants buy grain here for sale elsewhere. They come and go continuously. Most of the people are tall and sturdily built. Their robes are broad and loose fitting, sort of resembling Confucian robes. The country of Shan is divided into five administrative zones. It boasts more than 10,000 infantry and cavalry soldiers. To the north, it borders on the Khazar Turks, and

north of the Khazars [northern Ukraine or southern Russia] there are
the Turks whose feet resemble the hooves of oxen and who delight in
eating human flesh.

Country of Fulin

The country of Fulin [Parthian: Frwm ; Arabic: Rūm; Byzantine Empire]
lies thousands of Chinese miles west of the country of Shan, separated by
mountains. It is also called Greater Qin [Roman Empire]. The people there
have ruddy complexions. The menfolk all wear plain, undyed clothing,
whereas the womenfolk adorn themselves with pearls and silk brocades.
They like drinking wine and have a fondness for dry biscuits. Their
artisans are very ingenious and quite skilled at weaving. Sometimes, men
of Fulin are held captive in other countries, but they stubbornly cling to
their traditional customs until death, without changing. Nothing in the
world can compete with the marvelous glassware they produce.

The walls of their royal capital [Constantinople] measure 80 square
Chinese miles [20 km^2; archaeologically, 14 km^2], and the extent of their
territory in each direction is several thousand Chinese miles. They have
approximately 1,000,000 crack troops and are constantly trying to repulse
the Arab armies. To the west, they border on the Western Sea [probably
the Sea of Marmara], and to the south, they approach the Southern Sea
[Mediterranean]. To the north, they are in contact with the Khazar Turks.

In the middle of the Western Sea there is a marketplace where a tacit
agreement exists between buyers and sellers: "If I go, you leave, and when
you come back, I return home." First, the seller spreads out his goods on
the ground and departs. Afterward, the prospective buyer reciprocates
and matches these items, placing goods he considers to be of equal value
alongside the seller's goods. They wait until an acceptable exchange
value is reached, and only then they each collect their goods. They call
this the "ghost market."

I also heard that to the west of this place is the Kingdom of Women, where women become impregnated under the influence of water and give birth.[1]

INTRODUCTION TO *ACCOUNTS OF CHINA AND INDIA* (851 CE)

The anonymous compilation of Arabic merchant sailor reports, *Accounts of China and India* (Akhbār Al-Ṣīn wa'l-Hind), dated 851 CE, contains the earliest eyewitness, reasonably accurate accounts of the government, society, and economy of China to be found in any Middle Eastern or Western work, predating the revelatory accounts of William of Rubruck and Marco Polo (chapters 7 and 8) by four centuries. It forms an ideal parallel journey to Du Huan's account of the Arab world. The compilation shares many generic traits with the Greco-Roman *Navigating Around the Red Sea* text (chapter 3), especially its concern with itineraries, ports of call, and trade.

The authorship of *Accounts of China and India* is sometimes attributed to one Sulaymān al Tajir ("Sulaymān the Merchant"), because he is the only informant named in the compilation, but it is clear from the choppy structure and repetitions that this was a collection of oral and written eyewitness accounts gathered from those Arab and Persian merchants who had made the arduous journey to Canton to trade for the riches of the East. The text was probably originally compiled to entertain and edify an urban, cosmopolitan audience in the Abbasid Caliphate.

The sea route that went all the way from the Persian Gulf to China may have been in operation since the second century CE, when merchant "envoys" from the Roman world made their way to the Han capital (chapter 4), and Persian merchants had plied those waters in later centuries, but this text contains one of the first detailed descriptions of it. The *Accounts of China and India* mentions that these voyages were made on "Chinese ships" (*al-sufun al-Ṣīnīyah*) by which it is meant "ships of the China trade," not junks constructed or operated by Chinese merchants.

Taking advantage of the monsoon winds, the Arab dhows left the port of Sīraf in Iran, sailed past the rocky capes of Oman and out the mouth of the Persian Gulf, landing in southwestern India about a month later. They then crossed the Bay of Bengal, sometimes stopping in the Nicobar Islands, before landing a month later on the western Malaysian peninsula. They subsequently passed through the Straits of Malacca and then set sail for the Champa Kingdom in central Vietnam. The final destination was a place the Arabs called Khānfū (Canton), the emporium in the far south of the Tang empire. The outbound voyage took about four to six months (see map 4).

The *Accounts of China and India* text gives fascinating details about how trade was conducted and taxed in Canton and how the foreign settlement was overseen, with disputes adjudicated by Muslim judges. In his supplemental text appended to *Accounts of China and India*, the 10th-century geographer and traveler Abū Zayd al Sīrāfī tells us that in 877 CE (actually 879 CE), the brutal Chinese rebel Huang Chao 黃 巢 massacred Canton's inhabitants, including all the 120,000 Muslim, Christian, Jewish, and Zoroastrian merchants living there.

Those who related the accounts in this compilation were fascinated and shocked by Chinese social customs. They noted how Chinese women did not cover their hair and how Chinese men had sex with menstruating women and even with boy prostitutes at Buddhist temples. The Chinese also ate unclean animals and did not slaughter them properly according to Islamic law. Shockingly, Chinese people did not wash their backsides after defecating but wiped themselves with paper, a valuable commodity in the Arab world.

But beyond these notices meant to shock the reader, the Arabic merchant sailors also made the first notices of Chinese tea and porcelain. Chinese ceramics, alongside silk, formed the foundation of the Arabic trade with the Far East, as indicated by the contents of the famous Belitung shipwreck and the scattered finds of Tang ceramics throughout the Indian Ocean basin and Persian Gulf. The observant merchants also recorded detailed information about Tang government and society, including the long-running practice of stabilizing grain prices, tea and salt monopolies, government passports for traveling merchants, and contracts signed by tracing finger joints, a practice confirmed by signed wills from the "library cave" at Dunhuang. They even represented Chinese official titles in Arabic script faithfully enough that we can still identify them.

The Arabic merchant sailors' reports sometimes repeated hearsay and legend or veered into the realm of the fantastic. The account of the *al-darā* bell was a conflation of the genuine historical Chinese practice of the "drum of grievances," with an analogous Persian legend. But the story of the village of pygmies in Taihu County is just a fanciful tale.

Despite their condescension toward Chinese idolatry and impurity, the *Accounts* demonstrate a certain admiration for China and its people, at least in comparison to India. The Arabs felt that China was a more civilized and healthier place than India and that its people were white and handsome (like the Arabs viewed themselves) rather than dark and primitive, which is how they looked upon many people in the coastal and island areas of India.

SOURCE TEXT, ALTERNATE ENGLISH TRANSLATIONS, AND SELECTED STUDIES

Ahmad, S. Maqbul. *Arabic Classical Accounts of India and China*. Indian Institute of Advanced Study, 1989.

Chaffee, John W. "Merchants of an Imperial Trade." In *The Muslim Merchants of Premodern China: The History of a Maritime Asian Trade Diaspora, 750–1400*. Cambridge University Press, 2018.

Kotyk, Jeffrey. *Sino-Iranian and Sino-Arabian Relations in Late Antiquity*. Brill, 2024.

Mackintosh-Smith, Tim. *Two Arabic Travel Books*. Edited by James E. Montgomery. New York University Press, 2014.

Park, Hyunhee. *Mapping the Chinese and Islamic Worlds: Cross-Cultural Exchange in Pre-Modern Asia.* Cambridge University Press, 2012.

Sauvaget, Jean. *'Aḫbār aṣ-Ṣīn wa l-Hind. Relation de la Chine et de l'Inde rédigée en 851.* Belles Lettres, 1948.

Excerpts from *Accounts of China and India* (851 CE)

Trade and Settlement at Canton

All this is in their hands, and so their merchandise is rare. Among the reasons for the scarcity of goods is the fires that occur sometimes in Khānfū [Canton], which is a haven for ships and the meeting place for commerce between Arabs and Chinese. The fires that destroy the merchandise happen because the houses there are built of wood and split reeds [bamboo]. Another cause of this scarcity is that the ships that sail to and from China are sometimes wrecked or that those who sail on them are robbed or they are forced into long layovers, so that they sell their merchandise other places than within the Arab countries. It may also happen that the wind blows them off course and diverts them toward Yemen or elsewhere, and they sell their merchandise there. Sometimes, they also must remain for long periods to repair their vessels.

Sulaymān the Merchant mentions that at Canton, which is the gathering place for tradespeople, there is a Muslim whom the ruler of the Chinese invests with the power to settle disputes between the Muslims who head to that region, and this is due to a particular desire of the Chinese sovereign. During important festivals, it is he who directs the Muslims in prayer, delivers the *khutba* [sermon], and pronounces all the vows to supplicate for the sultān of the Muslims. The manner in which he exercises his office does not elicit any criticisms from the Iraqi merchants regarding his sentences, which all conform to the principles of justice, the stipulations in the Book of God Almighty, and with the laws of Islam.

The Ocean Trade Route to China

As for the places from which they come, it is reported that most of the ships bound for China are loaded at Sīraf [present-day Bandar-e Sīraf, Iran], the merchandise having first been transported to Sīraf from Basra, Oman, and elsewhere, and then they are loaded on the China boats at Sīraf [see map 4]. This is because of the abundant waves in the sea of this area and the shallowness of the water in certain places.

The distance between Basra and Sīraf by sea is 120 *farsakh* ["league"; approx. 600 km]. When the merchandise has been loaded at Sīraf, they take aboard fresh water from there, and then they *khaṭafū* ["snatch away"], which is a term used by seafarers that signifies "to set sail," for a place called Muscat, which is at the far end of the territory of Oman. The distance from Sīraf to this point is about 200 leagues [approx. 1,000 km]. In the eastern part of this sea, between Sīraf and Muscat, one finds, among other countries, the coastal areas of the Banī al-Ṣaffāq [an Arab tribe in western Iran] and the island of Ibn Kāwān [Qeshm Island, Iran]. In this sea are the mountainous reefs of Oman, between which is a place called al-Durdūr ["the whirlpool," near Cape Musandam, Oman], a narrow strait between two reefs through which small ships can pass but not the China ships. Among these reefs, there are two called Kusayr ["the exhausted one"] and ʿUwayr ["the one-eyed"], of which only a little bit appears above the waterline. After having passed these reefs, one arrives at a place called Ṣuḥār in Oman. One takes on fresh water at Muscat from a well one finds there. Here one finds a great number of sheep from Oman.

From there, the ships set sail for India, headed toward Kūlam-Malay [Quilon, India]. The journey from Muscat to Kūlam-Malay takes about a month with a moderate wind. Kūlam-Malay has a military post, subject to the country of Kūlam-Malay, which collects a duty on the China ships. One finds fresh water here, supplied from wells. The China ships are charged a duty of 1,000 *dirhams* [a silver coin of about three grams], while the other ships are charged from 10 to 20 *dīnār* [a small gold coin worth 20 *dirhams*]. The journey from Muscat to Kūlam-Malay and the

Sea of Harkand [Bay of Bengal] is around one month. One takes on fresh water at Kūlam-Malay.

Then the boats are "snatched away," that is to say they set sail for the Sea of Harkand. After crossing it, one arrives at a place called Lanjabālūs [Nicobar Islands], where the people do not understand Arabic or any of the languages that the merchants know. They are people who do not wear clothes, are white in skin color and beardless. Seafarers report that they have never seen their womenfolk. . . .

Then the ships set sail for a place called Kalā-bār [Kedah, Malaysia]. The word *bār* designates both the kingdom and the coastal area. This is part of the Kingdom of al-Zābaj [Srivijaya, centered on Sumatra], which is to the right of India. A king groups them all under his authority. They wear a waist towel, and the commoners and the highborn each only wear such a waist wrapper. One gets fresh water there from wells. They prefer well water to that from springs or rainfall. The distance between Kūlam-Malay and this place is not far; from Harkand [Bay of Bengal] to Kalā-bār is only a month's journey.

Then the ships go to a place called Tiyūmah [Tioman Island, off eastern Malaysia], where one finds fresh water if he wants it. To get there takes a voyage of 10 days. Then the ships set sail for a place called Kanduranja [Pulo Condore Island, off southern Vietnam]; 10 days. One finds fresh water there, if he needs some. This is similar with all the islands of the Indies; if one digs a well, he will find fresh water. There is a mountain there that dominates the landscape, which at times has served as a refuge for fugitive slaves and thieves.

The ships then sail to a place called Ṣanf [Champa, central Vietnam], a voyage of 10 days, and one finds fresh water there. This is where the aloeswood comes from called Ṣanfī aloeswood. It is ruled by a king. They are dark-skinned people, and each dresses in a sarong consisting of two pieces of cloth. After taking on fresh water there, one sails for a place called Ṣanf-Fulāu [Cù lao Chàm Island, off central Vietnam], which is an island in the sea of 10 days voyage, where one finds fresh water. Then one sets sail for a sea called Ṣankhā [Chinese: Zhanghai 長海, referring to South China Sea] and passes through the Gates of China [Paracel Islands], which are mountains in the sea separated from each other by a passage which is navigable for ships.

When God has brought them safe and sound to Ṣanf-Fulāu, the ships set sail for China, a voyage of one month, but the reefs through which they make passage are spread out over a voyage of seven days. When the ships have passed through the Gates of China and have entered the estuary [of the Pearl River], they proceed to take on fresh water in the locality of China where they will set anchor, that is to say, at Canton, which is a city.. . .

Chinese Dress, Produce, Trade Regulations, and Customs

The inhabitants of China, young and old, dress in silk during the summer as well as during the winter. The rulers themselves wear silk of the finest quality; the others wear what they can afford. During winter, men wear two layers of trousers, or three, four, or even five, if they can manage. What they are trying to do is to keep the lower part of their body warm, for they fear the abundance of dew on the ground. In summer, they wear a single silk shirt or some comparable garment. They do not wear turbans.

Rice constitutes their main food, but sometimes they make a stew to accompany it, which they pour over the rice and eat it. The rulers eat wheat bread and the meat from all sorts of animals, even pigs and others as well. For fruit, they have the apple, peach, citron, pomegranate, quince, pear, banana, sugarcane, melon, fig, grape, ribbed and glossy cucumber, hackberry [probably confused with the jujube], walnut, almond, hazelnut, pistachio, plum, apricot, the rowan, and the coconut. They do not have many date palms in the country, except for one here or there inside someone's house compound. They drink a fermented beverage made from rice, because they don't have grape wine in their country. None is exported to them either, so they don't know about it, and they don't drink it. It is also from rice that they make vinegar, wine, sweetmeats, and other similar products.

* * * * * * * * * * * * * * **BRIEF EXCERPT** * * * * * * * * * * * * * * *

They are an unhygienic people. They do not cleanse their backsides with water after they defecate but instead wipe themselves with Chinese paper. They eat carrion and other similar practices to those of the Magi [Zoroastrians]. Their religion, in fact, resembles that of the Magians.

Their women keep their heads uncovered and put hairpins in their hair. It sometimes happens that there can be as many as 20 combs of ivory or other material in a woman's coiffure. The menfolk cover their heads with something resembling our *al-qalānis* [bonnet].

Their custom regarding thieves is that they will execute them if they catch them.

. . .

It is said that the king of China has over 200 large cities, and each one has its ruler and a eunuch. Each of the cities has several other dependent towns under it. Among these is Canton, which is the port where the ships lay anchor, and twenty towns are under its jurisdiction. . . .

Their transactions are made through small copper coins. Although their treasuries are like the treasuries of other kings, no other king possesses such copper coins. They are the monetary currency of the country. They have gold, silver, pearls, brocades, and silk; all this is abundant among them, but these are all just commodities and property. It is the copper coins that constitute their currency. They import ivory, frankincense, copper ingots, and *al-dhabl* from the sea, which is the skin off the back of tortoises, as well as this *al-bishān* [Sanskrit: *viṣāṇā*, "horn"] that I have already described, which is the rhinoceros; they make belt ornaments from its horn.

They are rich in beasts of burden, though they don't have Arabian horses but other breeds instead. They have donkeys and plenty of camels with two humps. They possess a clay [*al-ghaḍār*, probably kaolin] of excellent quality, out of which they manufacture goblets that are as thin as glass bottles. One can even see the shimmer of the water right through them, although they are made of this clay.

When sailors enter their country coming from the sea, the Chinese seize all their goods and store them in warehouses, guaranteeing indemnity for six months, until the last of the sailors have arrived, after which they take a customs duty of three-tenths in kind and return the rest to the merchants. Whatever the government needs, it acquires at the highest price and pays for it immediately and does not wrong anyone in this regard. Among the things they buy is camphor at the rate of fifty *fakkūj* per *maund* [approx. 1.0 kg], each *fakkūj* being equal to 1,000 of their copper coins. If the government does not buy the camphor, it will fetch only half that price in other countries.

When a Chinese person dies, he is not buried until the following year, on the anniversary of his death. They place him in a coffin, which they leave in their house, and cover the body with quicklime that absorbs the fluids of the corpse and thereby preserves it. The body of a king, however, is treated with aloes essence and camphor. They mourn their dead for three years. Those who do not mourn are beaten with a wooden stave, whether he be man or woman. "Do you not grieve for your dead?" he is questioned. The deceased is buried in a grave, just like the Arabs make, but they do not withhold food from him, claiming that the deceased still eats and drinks. Indeed, they put food next to him at night, and when they find none of it left in the morning, they say that it was he who ate it. They continue to weep for the dead and to feed him as long as he is in the house. It sometimes happens that people will bankrupt themselves because of their care of the deceased, so that they have neither money nor land remaining which has not been spent on them. Formerly, they would bury with their king all that he possessed, including domestic items, clothing, and belts —their belts fetch a considerable price—but they had to abandon this custom, because the tombs of the dead were being violated and all that was interred with them was stolen.

Everyone among the Chinese, whether they are poor or rich, young or old, all learn to trace Chinese characters and to write.

* *

Chinese Government and Administration

The title of their rulers varies according to the degree of their prestige and the size of their cities. Thus, the ruler over a small town is called *ṭasūshī* [Chinese: *da cishi* 大刺史 "grand prefect"], which means "maintainer of the town." With a city like Canton, the title of its ruler is called *dayfū* [Chinese: *dufu* 都府 "area commander"], while its eunuch supervisor is called *ṭūquām* [Chinese: *dujian* 都監 "director-in-chief"]. The eunuchs are people of their own race who have been castrated. The supreme judge is called *luqshī sāmkūn shī* [Chinese: *lushi canjun shi* 錄事參軍事 "administrative supervisor"], and they have other titles the exact form of which I do not know.

None of them can occupy a position of authority at less than 40 years of age, for they say at that age a man is mature due to his life experiences. When a lesser ruler holds audience, he takes his place in his city upon a dais in a great hall, and before him is placed a bench. He is presented with papers on which legal disputes are written. Behind the administrator stands a man called a *lunjūn* [Chinese: *langzhong* 郎中 "court gentleman"]. If the ruler slips up or commits an error in one of the orders that he gives, the *lunjūn* returns it to him. They do not pay any attention to the words of a plaintiff unless his complaint is in writing. Before those who have petitions to present are allowed to enter into audience, someone stationed at the gate of the palace goes over their papers. If there is something wrong about them, he rejects it and returns it to him. Only those who are familiar with the laws and maxims can produce writings that are intended for the ruler. The scribe must write on his paper, "Written by so-and-so who is the son of so-and-so." Then, if it is found to be faulty, the scribe is blamed and beaten with a wooden stave. The governor does not hold audience to mete out justice before he has eaten and drank, so that he does not commit an error. The salary of each administrator is paid from the treasury of his town.

As for the great king [the Tang emperor], one does not see him except for once in every 10 months. "If the people see me too often," he says, "they would look down upon me. Authority is maintained only through kingly pride. The common people actually do not appreciate justice. Thus, one must act arrogantly and tyrannical towards them to be respected."

There is no land tax, but they are subject to a poll tax on the head of each male, according to his apparent wealth. If there is an Arab or other foreigner in China, he pays a tax only on his personal property in order for him to be able to retain the rest.

Whenever there is a rise in prices, the ruler takes food from his treasuries and sells it cheaper than the market price so that the inflation will not last. What goes into the treasury is only the poll taxes, and I think that 50,000 *dīnārs* go into the treasury of Canton every day, even though it is not one of their largest cities.

Important sources of revenue for the king's own use come from salt and an herb that they drink mixed with hot water, and which is sold in every city for considerable sums. They call it *al-sākh* [Chinese: *cha* 茶 "tea"]. It is leafier than alfalfa, and is a little more fragrant than that, but it is bitter. One boils water and then sprinkles the leaves over it. They consider it as an antidote for all ailments. The whole of what goes into the treasury comes from the poll tax, salt tax, and a tax on this herb.

In each town, there is an object called *al-dara*. This is a bell placed near the head of the ruler of the town and attached to a cord that extends as far as the public road. Between that spot and the ruler is a distance of about a league [approx. 5 km]. If the cord is shaken, the little bell will start ringing. So when anyone who has experienced injustice pulls this cord, the bell near the prince rings out, and the person is allowed to enter into audience to personally present his case and explain the wrong he has suffered. In each locality, one finds the same situation.

He who wishes to travel from one place to another in China receives two written documents, one from the king and one from a eunuch. The royal document is meant for the journey; it mentions the name of the man and that of those who are accompanying him, his age and that of his companions, and the tribe or clan to which he belongs, because all of those who are in China, be they Chinese, Arab, or others, must absolutely trace their genealogy to an identified group, whose name they bear. As for the document from the eunuch, it describes the money and the goods that the traveler has with him. This is because there are military posts along the road that examine the two documents, and when someone arrives there, they note down, "So-and-so, son of so-and-so, belonging to such-and-such tribe, came to us on such-and-such day, of such-and-such month, in such-and-such year, having with him such-and-such money and goods." This is done so that the man's money and goods might not become lost. If he loses something or happens to die, one can know how it was lost, and the remainder can be restored to him or his heirs.

Monetary Transactions and Debts

The Chinese act fairly in matters of monetary transactions and debts. When a man has a claim against another, he draws up a document binding him to his debt, and the indebted person also writes up a document and affixes his signature to it by means of tracing the joints of his middle and index fingers. The two notes are then joined together and folded, and something is written over the spot where they are joined. They are then separated, and the note acknowledging his debt is given to the debtor. Subsequently, if either party repudiates the other, he is told to produce the document that he holds. If the debtor asserts there is no claim against him or refutes a note that is written in his hand and bears his finger impression while that of the creditor has disappeared, the defaulting debtor is told, "Present a document declaring you have not contracted this debt; but if the creditor provides proof that contradicts your statement, regarding the debt that you have reneged on, you will receive twenty blows of the stick on your back and be fined 20,000 *fakkūj* of copper coins." One *fakkūj* is equal to 1,000 copper coins, that makes about 2,000 *dīnārs*. As for the 20 blows with the stick, that would probably result in his death. Therefore, no one in China agrees deliberately to make such a statement, for fear of losing his life and his fortune. I have never seen anyone who consented to do it. Thus, they bring fairness into their relationships, and no one loses a just claim, even though their transactions are carried out without resorting to witnesses or oaths. . . .

Miscellaneous Notices

They set up a stone stele 10 cubits high on which they inscribe a list of various illnesses and their remedies; for such-and-such an illness, there is such-and-such a cure. If a man is poor, he receives money from the state treasury to pay for his treatment.

They do not impose a tax on land. They only collect a poll tax from people that is proportional to the amount of money and goods that a man possesses. When a male child is born to one of them, one registers his name with the authorities, and when he is 18 years old, this poll tax is collected from him. When he reaches 80 years of age, this tax is no longer demanded from him, but on the contrary, a certain sum is paid to him as a pension from the state treasury. They say, "We received from him when he was a youth, so now we pay him back when he is old."

In every town, there is a school and a schoolmaster to instruct the poor and their children. These schoolmasters received their support from the government treasury.

Their womenfolk go about with their hair uncovered, but the men cover their heads.

In this country there is a village named Tāyū [probably Taihu 太 湖 County, in present-day Anhui] located in the mountains whose inhabitants are very small in stature. All of the short people in China originate from this village.

The people of China are handsome and tall. They are pure white in color with just a tinge of redness. There are no human beings who have blacker hair than the Chinese. Their women cut and style their hair.

. . .

In China, it so happens that when a prince under the authority of the emperor rebels, they cut his throat and eat him, because the Chinese eat the flesh of all those who have perished by the sword.

When the Indians and the Chinese wish to conclude a marriage, they exchange visits of congratulations, then gifts, then the marriage is celebrated publicly, accompanied by cymbals and drums. The gifts are usually in cash, according to their means. When one of them takes a wife and she commits the sin of adultery, she and her seducer are put to death throughout all the lands of India. If a man fornicates with a woman against her will, he alone is put to death, but if it was with her consent, they are both put to death.

. . .

The Chinese engage in sodomy with young slave boys, who are set up for this purpose to play the role of temple prostitutes for the Buddhists.

The walls of the Chinese are made of wood. The Indians build with stone, plaster, fired brick, and earth. This is sometimes the case in China as well. Neither the Chinese nor the Indians use carpets.

A Chinese or Indian man may marry as many women as he desires.

The food of the Indians consists of rice, while that of the Chinese consists of rice and wheat. The Indians don't eat wheat.

Neither the Indians nor the Chinese get circumcised.

The Chinese worship before the statues of idols, to which they pray and make entreaties. They also have sacred books.

The Indians grow long beards. I have seen the beards of some of them reach three cubits in length. They do not trim their mustaches. For the most part, the Chinese do not have beards, which is natural from birth for most of them. In India, when a man suffers a bereavement, he completely shaves his head and beard.

. . .

Both the Chinese and the Indians assert that their sacred statues speak to them, but it is only the priests who are doing the talking.

The Chinese and the Indians kill animals that they want to eat without slitting their throats and bleeding them. They hit them on the head until they are dead.

Neither the Indians nor the Chinese purify themselves by ablution when they are in a state of major pollution [e.g., after sexual intercourse]. The Chinese, after defecating, only wipe themselves with paper. The Indians perform an ablution every day before the midday meal, after which they take their food. The Indians will not have sex with a woman during her period; they even chase them out of the house to avoid being defiled by them. The Chinese, on the other hand, will have intercourse with menstruating women and won't drive them out of the house.

. . .

The Chinese have no native tradition of religious learning. Their religious practices originated in India. They claim that it was the Indians who brought them their Buddhas and taught them their religion. Both countries believe in reincarnation and differ only in the details of dogma and practice.

The study of medicine flourishes in India as does philosophy. The Chinese also study medicine, but their medical science mostly involves cauterization. They have knowledge of astronomy as well, but it is less developed than in India.

. . .

China is more pleasant and beautiful than India. Throughout most of India, there are no cities. China, on the other hand, has a considerable number of cities, all surrounded by ramparts.

China is a healthier place, and there are fewer diseases, for the air is more sanguine. One seldom finds blind or one-eyed men there nor anyone suffering from disease, but these are found in great numbers in India.

. . .

The Chinese are more handsome than the Indians and resemble the Arabs more in their manner of dress and in their domestic animals. The outfits that they wear during their ceremonies resemble those of the Arabs.[2]

NOTES

1. Du Huan 杜環, *Jingxing ji jianzhu* 經行記箋注 (Zhonghua shuju, 2000), 12–19, 21–23, 45–56. Translation by the author.
2. Translated from the Arabic by Sauvaget, *Aḥbār aṣ-Ṣīn wa l-Hind*. Translated from his French by the author, with some corrections suggested by Mackintosh-Smith, *Two Arabic Travel Books*. The headings have been added by the author.

Chapter 7

Spanning Eurasia During the Mongol Century (1253–1288)

During the early 13th century, in one of history's greatest ruptures, the shock cavalry of Genghis Khan and those of his sons and generals overran the entire Eurasian Steppe in two lightning campaigns of conquest and terror (1218–1225, 1235–1241), reaching the doorstep of Western Europe. Subsequently, Genghis's grandson Hulegu overthrew the Abbasid Caliphate in Baghdad in 1258. Beyond the horrific death toll and the panic they caused, these campaigns also broke the Muslim monopoly on trade and travel across Asia. Papal envoys, missionaries, and private merchants were the first to brave this new opening.

The Catholic popes and the crusader kings of Europe sought concrete information about the Mongols and their intentions toward the West and whether they might be open to an alliance against the Muslims in the Holy Land. The newly installed Mongol rulers of China and Persia were also interested in potential alliances and contact with European religious and secular authorities, leading to five decades of intense trans-Eurasian contact. The parallel travelers in this chapter both participated in this age of furious trans-continental diplomacy.

INTRODUCTION TO WILLIAM OF RUBRUCK

William of Rubruck (ca. 1215–ca. 1270) was a Flemish Franciscan monk who embarked on a two-year missionary odyssey (June 1253–June 1255) into Mongol-controlled lands that eventually took him to the court of the reigning Great Khan Möngke (r. 1251–1259), a grandson of Genghis Khan. Though he carried a letter of introduction from King Louis IX of France, William was not an official envoy from any king or pontiff, and his personal goal was to preach the Gospel and minister to Christians in Mongol domains. He was not the first European to visit the Mongol court and write an account, for others like the Franciscan John of Pian di Carpine had immediately preceded him (1245–1247).

Along with a fellow Franciscan Bartholomew of Cremona, William's journey began in the Holy Land, then up through Constantinople and the Black Sea, going overland from Crimea, crossing the Volga north of the Caspian Sea, and then over the south Russian steppe, ending at the Mongol capital of Karakorum, more than 5,000 kilometers away (see map 5). William being overweight exacerbated the challenges of the arduous journey.

Map 5. Travels of William of Rubruck and Rabban Sauma.

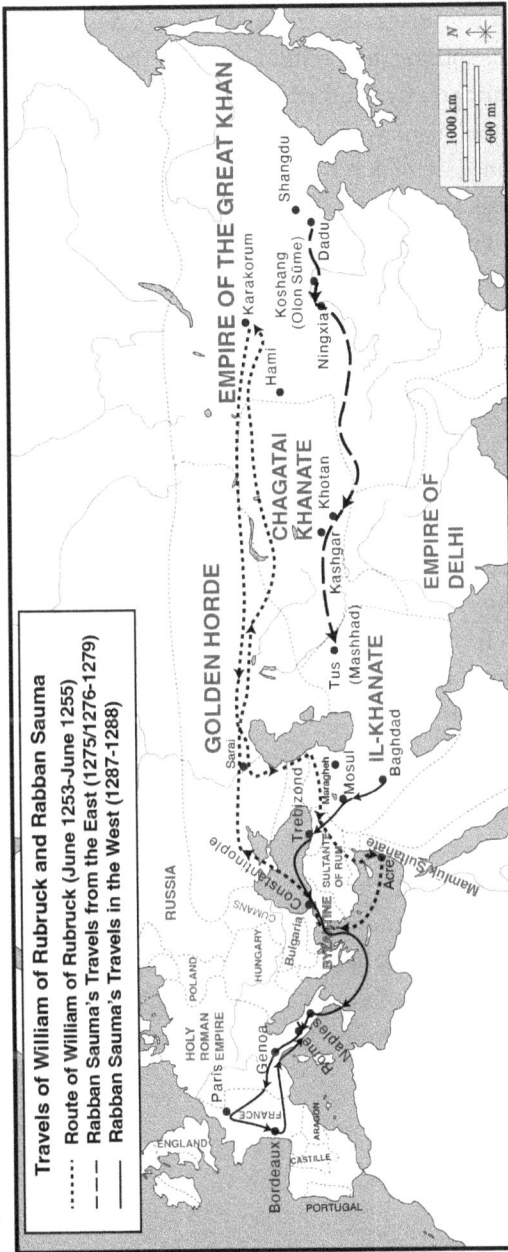

Source: Map created by Zhang Yexu and the author.

Though written eight centuries ago, Friar William's descriptions of Mongol housing, food, clothing, and religious beliefs, excerpted below, read like a modern ethnography. And while he makes several mistakes in geography and Mongol history, his account avoids the exaggerations of Marco Polo's account (see chapter 8). And while Polo hailed from a merchant family and was interested in commerce and valuable products, William was a devout Franciscan who had sworn off all attachment to material goods. Thus, William is much more attentive to issues of people and culture. While he inquired about the "monsters and human freaks" said to inhabit the East by Classical authors, he found no evidence of them, so began to doubt the veracity of those old tales.

William's revealing account provides the earliest description of East Asian Buddhist beliefs, rituals, and mantras. He was also the first to confirm that Cathay (the medieval European designation for North China, derived from the ethnonym Qidan 契丹 of the Liao Dynasty that had ruled North China, 916–1125) was the same as the land of the Seres ("Silk People"), known from Classical authors like Pliny (chapter 4). He was also the first Western author to accurately describe Chinese written characters and one of the earliest to mention paper money.

William was surprised to discover that he was not the only European at the Khan's court, for many others had been captured during Mongol campaigns in Hungary and Russia. Notably, there was William Buchier, a metalsmith from Paris who built for the Great Khan an enormous tree-shaped silver alcohol-serving fountain. William's sojourn at Möngke's court climaxed with a theological debate with his doctrinal opponents, the Buddhists and the Muslims. Rabban Sauma would participate in a similar rhetorical confrontation during his mission to Europe.

**PRIMARY SOURCE, ALTERNATE ENGLISH TRANSLATION, AND
SELECTED STUDIES**

Jackson, Peter. *The Mission of Friar William of Rubruck.* Hakluyt Society,
1990.

Olschki, Leonardo. *Marco Polo's Asia.* University of California Press,
1960.

Rachewiltz, Igor de. *Papal Envoys to the Great Khans.* Stanford University
Press, 1971.

Rockhill, William Woodville. *The Journey of William of Rubruck to The
Eastern Parts of the World, 1253–55.* Hakluyt Society, 1900.

**EXCERPTS FROM THE *TRAVEL ACCOUNT* (ITINERARIUM), BY
WILLIAM OF RUBRUCK (1253–1255 CE)**

* * * * * * * * * * * * * * * * **BRIEF EXCERPT** * * * * * * * * * * * * * * * * *

After having left Soldaia [Sudak in Crimea; June 1, 1253], we
came on the third day across the Tartars [a.k.a. "Tatar," another
ethnonym by which Mongols were called], and when I found
myself among them, it seemed to me of a truth that I had been
transported into another world. I will describe to you as well as
I can their mode of living and manners.

Mongol Yurts and Carts

Nowhere have they fixed dwelling places, nor do they know where
their next will be. They have divided among themselves Scythia
[northern Asia], which extends from the Danube to the rising of
the sun, and every captain, according as he has more or less men
under him, knows the limits of his pasture lands and where to
graze in winter and summer, spring and autumn. For in winter,
they go down to warmer regions in the south; in summer, they
go up to cooler ones toward the north. The pasture lands without

water they graze over in winter when there is snow there, for the snow serves them as water.

They set up the dwelling in which they sleep on a circular frame of interlaced sticks converging into a little round hoop on the top, from which projects above a collar as a chimney, and this framework they cover over with white felt. Frequently, they coat the felt with chalk, or white clay, or powdered bone, to make it appear whiter, and sometimes also they make the felt black. The felt around this collar on top they decorate with various pretty designs. Before the entry, they also suspend felt ornamented with various embroidered designs in color. For they embroider the felt, colored or otherwise, making vines and trees, birds and beasts.

And they make these houses so large that they are sometimes 30 feet in width. I myself once measured the width between the wheel tracks of a cart 20 feet, and when the house was on the cart, it projected beyond the wheels on either side five feet at least. I have myself counted to one cart 22 oxen drawing one house, 11 abreast across the width of the cart, and the other 11 before them. The axle of the cart was as large as the mast of a ship, and one man stood in the entry of the house on the cart driving the oxen.

Furthermore, they weave light twigs into squares of the size of a large chest, and over it from one end to the other they put a carapace also of twigs, and in the front end they make a little doorway and then they cover this coffer or little house with black felt coated with tallow or ewe's milk, so that the rain cannot penetrate it, and they decorate it likewise with embroidery work. And in such coffers they put all their bedding and valuables, and they tie them tightly on high carts drawn by camels, so that they can cross rivers without getting wet. Such coffers they never take off the cart.

When they set down their dwelling houses, they always turn the door to the south, and after that they place the carts with coffers on either side near the house at a half stone's throw, so that the dwelling stands between two rows of carts as between two walls.

The married women make for themselves the most beautiful luggage carts, which I would not know how to describe to you unless by a drawing, and I would depict them all to you if I knew how to paint. A single rich Mongol or Tartar has quite a hundred or two hundred such carts with coffers. Batu [a grandson of Genghis Khan, d. 1255 CE] has 26 wives, each of whom has a large dwelling, exclusive of the other little ones that they set up after the big one, and which are like closets, in which the sewing girls live, and to each of these large dwellings are attached quite 200 carts. And when they set up their houses, the first wife places her dwelling on the extreme west side, and after her the others according to their rank, so that the last wife will be in the extreme east; and there will be the distance of a stone's throw between the yurt of one wife and that of another. The *ordu* ["court camp"] of a rich Mongol seems like a large town, though there will be very few men in it.

One girl will lead 200 or 300 carts, for the country is flat, and they tie the ox or camel carts the one after the other, and a girl will sit on the front one driving the ox, and all the others follow after with the same gait. Should it happen that they come to some bad piece of road, they untie them and take them across one by one. So they go along slowly, as a sheep or an ox might walk.

* *

Rituals of the Home

When they have fixed their dwelling, the door turned to the south, they set up the couch of the master on the north side. The side for the women is always the east side, that is to say, to the left of the master of the house, he sitting on his couch with his face turned to the south. The side for the men is the west side, that is, on his right. Men coming into the house would never hang up their bows on the side of the women.

And over the head of the master is always an image of felt, like a doll or statuette, which they call the "brother of the master"; another similar one is above the head of the mistress, which they call the "brother of the mistress," and they are attached to the wall; and higher up between the two of them is a little thin one, who is, as it were, the guardian of the whole dwelling. The mistress places in her house on her right side, in a conspicuous place at the foot of her couch, a goat skin full of wool or other stuff, and beside it a very little statuette looking in the direction of the attendants and women. Beside the entry on the women's side is yet another image [Mongolian: *ongod*], with a cow's udder for the women, who milk the cows; for it is part of the duty of the women to milk the cows. On the other side of the entry, toward the men, is another statue with a mare's udder for the men who milk the mares.

And when they have come together to drink, they first sprinkle with liquor this image that is over the master's head, then the other images in order. Then an attendant goes out of the dwelling with a cup and liquor and sprinkles three times to the south, each time bending the knee, and that to do reverence to the fire; then to the east, and that to do reverence to the air; then to the west to do reverence to the water; to the north they sprinkle for the dead. When the master takes the cup in hand and is about to drink, he first pours a portion on the ground.

If he were to drink seated on a horse, he first before he drinks pours a little on the neck or the mane of the horse. Then when the attendant has sprinkled toward the four quarters of the world, he goes back into the house, where two attendants are ready with two cups and platters to carry drink to the master and the wife seated near him upon the couch. And when he has several wives, she with whom he has slept that night sits beside him in the day, and all the others must come to her dwelling that day to drink, and court is held there that day, and the gifts which are brought that day are placed in the treasury of that lady. A bench with a skin of milk, or some other drink, and with cups, stands in the entry. . . .

Food, Drink, and Wild Game

Of their food and victuals you must know that they eat all their dead animals without distinction, and with such flocks and herds it cannot be but that many animals die. Nevertheless, in summer, so long as lasts their *kumiss*, that is to say mare's milk, they care not for any other food. So then if it happens that an ox or a horse dies, they dry its flesh by cutting it into narrow strips and hanging it in the sun and the wind, where at once and without salt it becomes dry without any evil smell. With the intestines of horses they make sausages better than pork ones, and they eat them fresh. The rest of the flesh they keep for winter. With the hides of oxen, they make big jars, which they dry in admirable fashion in the smoke. With the hind part of the hide of horses they make the most beautiful shoes. . . .

This *kumiss*, which is mare's milk, is made in this manner. They stretch a long rope on the ground fixed to two stakes stuck in the ground, and to this rope they tie toward the third hour the colts of the mares they want to milk. Then the mothers stand near their foal and allow themselves to be quietly milked; and if one be too wild, then a man takes the colt and brings it to her, allowing it to suck a little; then he takes it away and the milker takes its place. When they have got together a great quantity of milk, which is as sweet as cow's as long as it is fresh, they pour it into a big skin or bottle, and they set to churning it with a stick prepared for that purpose and which is as big as a man's head at its lower extremity and hollowed out; and when they have beaten it sharply it begins to boil up like new wine and to sour or ferment, and they continue to churn it until they have extracted the butter. Then they taste it, and when it is mildly pungent, they drink it. It is pungent on the tongue like *rapé* wine [wine from unripe grapes] when drunk, and when a man has finished drinking, it leaves a taste of milk of almonds on the tongue, and it makes the inner man most joyful and also intoxicates weak heads and greatly provokes urine. They also make *caracosmos*, that is black or "great" *kumiss*, for the use of the great lords. . . . They churn then the milk until all the thicker parts go straight to the bottom, like the dregs of wine, and the pure part remains on top, and it is like whey or white must. The dregs are very white, and they are given to the slaves, and they provokes much to sleep. The clear [liquor] the lords drink, and it is assuredly a most agreeable drink and most efficacious.

Batu has 30 men around his camp at a day's distance, each of whom sends him every day such milk of a hundred mares, that is to say every day the milk of 3,000 mares, exclusive of the other white milk that is brought in by others. As in Syria, the peasants give a third of their produce, so it is these [Tartars] must bring to the *ordu* of their lords the milk of every third day.

As to cow's milk, they first extract the butter, then they boil it down perfectly dry, after which they put it away in sheep paunches that they keep for that purpose; and they put no salt in the butter, for on account of the great boiling down, it spoils not. And they keep this for the winter. What remains of the milk after the butter they let sour as much as can be, and they boil it, and it curdles in boiling, and the curd they dry in the sun, and it becomes as hard as iron slag, and they put it away in bags for the winter. In wintertime, when milk fails them, they put this sour curd, which they call *grut* [Turkish: *qurut* "cheese"], in a skin and pour water on it, and churn it vigorously till it dissolves in the water, which is made sour by it, and this water they drink instead of milk. They are most careful not to drink plain water.

The great lords have villages in the south, from which millet and flour are brought to them for the winter. The poor procure these things by trading sheep and pelts. The slaves fill their bellies with dirty water, and with this they are content. They catch also rats, of which many kinds abound here. Rats with long tails they eat not, but give them to their birds. They eat mice and all kinds of rats which have short tails. There are also many marmots, which are called *sogur* [Turkish: *soghur*], and which congregate in one hole in winter, 20 or 30 together, and sleep for six months; these they catch in great numbers. There are also *cuniculi* [possibly a jerboa or mongoose], with a long tail like a cat's, and on the end of the tail they have black and white hairs. They have also many other kinds of small animals good to eat, which they know very well how to distinguish.

I saw no deer there. I saw few hares, many gazelles. Wild asses I saw in great numbers, and these are like mules. I saw also another kind of animal which is called *arcali* [Mongolian: *arghali* "wild sheep"], which has quite the body of a sheep and horns bent like a ram's, but of such size that I could hardly lift the two horns with one hand, and they make of these horns big cups.

They have gyrfalcons in great numbers, which they all carry on their right hand. And they always put a little thong around the hawk's neck, which hangs down to the middle of its breast, by which, when they cast it at its prey, they pull down with the left hand the head and breast of the hawk, so that it be not struck by the wind and carried upward.

So it is that they procure a large part of their food by the chase. When they want to chase wild animals, they gather together in a great multitude and surround the district in which they know the game to be, and gradually they come closer to each other till they have shut up the game in among them as in an enclosure, and then they shoot them with their arrows.

Clothing and Customs

Of their clothing and customs, you must know, that from Cathay [North China], and other regions of the east, and also from Persia and other regions of the south, are brought to them silken and golden stuffs and cloth of cotton, which they wear in summer. From Russia, the Moxel tribe [Mokshas], and from Greater Bulgaria [middle Volga], and the Pascatur tribes [Bashkirs, living between the Volga and the Urals], which is greater Hungary, and the Kerkis [Kyrgyz, living on the upper Yenisei River], all of which are countries to the north and full of forests, and which obey them, are brought to them costly furs of many kinds, which I never saw in our parts, and which they wear in winter. And they always make in winter at least two fur gowns, one with the fur against the body, the other with the fur outside exposed to the wind and snow; these latter are usually of the skins of wolves or foxes or *papio* [probably lynx]; and while they sit in the dwelling, they have another lighter one. The poor make their outside gowns of dog and kid skins.

They make also breeches with furs. The rich furthermore wad their clothing with silk stuffing that is extraordinarily soft, light, and warm. The poor line their clothes with cotton cloth or with the fine wool which they are able to pick out of the coarser. With this coarser they make felt to cover their houses and coffers, and also for bedding. With wool and a third of horse hair mixed with it they make their ropes. They also make with felt covers, saddlecloths, and rain cloaks; so they use a great deal of wool. You [Louis IX] have seen the costume of the men. . . .

And the dress of the girls differs not from the costume of the men, except that it is somewhat longer. But on the day following her marriage, a woman shaves the front half of her head and puts on a tunic as wide as a nun's gown, but everyway larger and longer, open before, and tied on the right side. For in this the Tartars differ from the Turks; the Turks tie their gowns on the left, the Tartars always on the right. Furthermore, they have a headdress, which they call *bocca* [Mongolian: *boghtagh*], made of bark, or such other light material as they can find, and it is big and as much as two hands can span around, and is a cubit and more high, and square like the capital of a column. This *bocca* they cover with costly silk stuff, and it is hollow inside, and on top of the capital, or the square on it, they put a tuft of quills or light canes also a cubit or more in length. And this tuft they ornament at the top with peacock feathers, and round the edge [of the top] with feathers from the mallard's tail, and also with precious stones. The wealthy ladies wear such an ornament on their heads and fasten it down tightly with a fur hood, for which there is an opening in the top for that purpose, and inside they stuff their hair, gathering it together on the back of the tops of their heads in a kind of knot, and putting it in the *bocca*, which they afterward tie down tightly under the chin. So it is that when several ladies are riding together, and one sees them from afar, they look like soldiers, helmets on head and lances erect. For this *bocca* looks like a helmet, and the tuft above it is like a lance.

Gender Roles and Marriage Customs

And all the women sit astride their horses like men. And they tie their gowns with a piece of blue silk stuff at the waist and they wrap another band at the breasts and tie a piece of white stuff below the eyes which hangs down to the breast.

And the women there are wonderfully fat, and she who has the least nose is held the most beautiful. They disfigure themselves horribly by painting their faces. They never lie down in bed when bearing their children.

It is the duty of the women to drive the carts, get the dwellings on and off them, milk the cows, make butter and *grut* [cheese], and to dress and sew skins, which they do with a thread made of tendons. They divide the tendons into fine shreds and then twist them into one long thread. They also sew the boots, the socks, and the clothing. They never wash clothes, for they say that God would be angered by that and that it would thunder if they hung them up to dry. They will even beat those they find washing them. Thunder they fear extraordinarily; and when it thunders, they will turn out of their dwellings all strangers, wrap themselves in black felt, and thus hide themselves till it has passed away. Furthermore, they never wash their bowls, but when the meat is cooked, they rinse out the dish in which they are about to put it with some of the boiling broth from the kettle, which they pour back into it. They also make the felt and cover the houses.

The men make bows and arrows, manufacture stirrups and bits, make saddles, do the carpentering [on the framework of] their dwellings and the carts; they take care of the horses, milk the mares, churn the *kumiss* or mare's milk, make the skins in which it is put; they also look after the camels and load them. Both sexes look after the sheep and goats, sometimes the men, other times the women, milking them. They dress skins with a thick mixture of sour ewe's milk and salt. . . .

As to their marriages, you must know that no one among them has a wife unless he buys her; so it sometimes happens that girls are well past marriageable age before they marry, for their parents always keep them until they sell them. They observe the first and second degrees of consanguinity [i.e., one can't marry his biological mother or full sister], but no degree of affinity [i.e., no taboo against marrying one's step-mother, half-sister, or brother's widow]; thus [one person] will have at the same time or successively two sisters. Among them no widow marries, for the following reason: they believe that all who serve them in this life shall serve them in the next, so as regards a widow they believe that she will always return to her first husband after death. Hence, this shameful custom prevails among them, that sometimes a son takes to wife all his father's wives, except his own mother; for the residence of the father and mother always belongs to the youngest son, so it is he who must provide for all his father's wives who come to him with the paternal household, and if he wishes it he uses them as wives, for he does not consider himself injured if they return to his father after death. When then anyone has made a bargain with another to take his daughter, the father of the girl gives a feast, and the girl flees to her relatives and hides there. Then the father says, "Here, my daughter is yours: take her wheresoever you find her." Then he searches for her with his friends till he finds her, and he must take her by force and carry her off with a semblance of violence to his house.

Mongol Justice

As to their justice, you must know that when two men fight together, no one dares interfere, even a father dare not aid a son; but he who has the worst of it may appeal to the court of the lord, and if anyone touches him after the appeal, he is put to death. But action must be taken at once without any delay, and the injured one must lead him [who has offended] as a captive. They inflict capital punishment on no one unless he be taken in the act or confesses. When one is accused by a number of persons, they torture him so that he confesses. They punish homicide with capital punishment and also cohabiting with a woman not one's own. By not one's own, I mean not his wife or bondwoman, for with one's slaves one may do as one pleases. They also punish with death grand larceny, but as for petty thefts, such as that of a sheep, so long as one has not repeatedly been taken in the act, they beat him cruelly, and if they administer a hundred blows, they must use a hundred sticks. I speak of the case of those beaten under order of authority. In like manner false envoys, that is to say persons who pass themselves off as ambassadors but who are not, are put to death. Likewise, sorceresses, of whom I shall however tell you more, for such they consider to be witches.

Mourning, Funerary Customs, and Illness
When anyone dies, they lament with loud wailing, then they are exempt, for they pay no taxes for a year. And if anyone is present at the death of an adult, he may not enter the dwelling even of Möngke Khan for the year. If it be a child who dies, he may not enter it for a month.

Beside the tomb of the dead, they always leave a tent if he be one of the nobles, that is of the family of Genghis Khan, who was their first father and lord. Of him who is dead the burying place is not known. And always around these places where they bury their nobles there is a camp with men watching the tombs. I have not ascertained if they bury treasure with their dead. . . .

When anyone sickens, he lies on his couch and places a sign over his dwelling that there is a sick person therein and that no one shall enter. So no one visits a sick person, save him who serves him. And when anyone from the great *ordu* [court] is ill, they place guards all around the *ordu,* who permit no one to pass those bounds. For they fear lest an evil spirit or some wind should come with those who enter. They call, however, their priests, who are these same soothsayers. . . .

Northern China

There is also Great Cathay [North China], whose people were anciently I believe, called the Seres ["Silk People"]. From among them come the best silk stuffs, which are called *seric* [Mongolian: *sirkek*; Chinese: *si* 絲] by that people, and the people get the name of Seres from one of their cities. I was given to understand that in that region there is a city with walls of silver and towers of gold. In that land are many provinces, the greater number of which do not yet obey the Mongols, and between them and India there is a sea.

These Cathayans are small men, who in speaking aspirate strongly through the nose, and in common with all Orientals, have small openings for the eyes. They are most excellent artisans in all manners of crafts, and their doctors know full well the virtues of herbs and diagnose very skillfully the pulse; but they do not use diuretics, nor do they know anything about the urine; this I have seen myself. There are a great many of them at Karakorum, and it is their custom for all sons to follow the same trade as their fathers. . . .

Cathay is on the ocean. And master William [William Buchier, the goldsmith] told me that he had himself seen the envoys of a certain people called Caule [Chinese: Gaoli 高麗; Koryo dynasty Korea] and Manse [Persian: Mangi; Mongolian Nanggiya; Chinese: Nanjia 南家 "Southerners," i.e., Southern China], who live on islands the sea around which freezes in winter, so that at that time the Tartars can make raids there; and they [the Koreans] had offered [them] 32,000 *tumen* of *iascot* [pieces of silver] a year, if they would only leave them in peace. A *tumen* is a number containing 10,000.

The common money of Cathay is of paper, in length and breadth a palm, and on it they stamp lines like those on the seal of Möngke Khan. They write with a brush such as painters paint with, and they make in one figure the several letters containing a whole word. The Tibetans write as we do and have characters quite like ours. The Tanguts [former Xixia Kingdom] write from right to left like the Arabs, but they add lines upward from the bottom; the Uighurs, as previously said, downwards from the top.[1]

INTRODUCTION TO RABBAN SAUMA

Bar Sauma, mostly known by his honorific name Rabban Sauma ("Master Sauma"; ca. 1225–January, 1294) was an ethnically Uyghur or Önggüd Turkic Nestorian Christian, born in the Mongol capital, present-day Beijing. During the late 13th century, he and his disciple and friend Mark made a pilgrimage to the West, aiming to visit Jerusalem. Though frustrated in his original intention, Rabban Sauma was later sent on a special mission by the Mongol Il-Khan of Persia, Arghun (r. 1284–1291) to the rulers of the Western world to coordinate an attack on the Mamluks of Egypt, who controlled the Holy Land. This diplomatic overture received only lukewarm response, because political power in Western European states was fractured and the papacy was presently vacant, but Rabban Sauma was still able to fulfill his personal desire to visit churches like St. Peter's Basilica and to witness relics of Christ and the saints.

He became the first person from China to reach Western Europe and leave any record. He visited Constantinople, Rome, and France and had audiences with the pope and the kings of France and England. Rabban Sauma was in Western Europe a couple decades after William of Rubruck was in Karakorum, but his account still provides an excellent comparative foil. Rabban Sauma's original first-person travelogue has not been transmitted to us unaltered. His original account, now lost, was written in Persian but was later translated into Syriac then drastically shortened and edited by an unnamed compiler (possibly the Nestorian patriarch Mar Timothy II, 1318–1322) for incorporation into a larger history. Manuscripts of this history were discovered in 1883 in Kurdistan.

There is some evidence that suggests that Rabban Sauma and his companion Mark were officially sponsored and given credentials of safe passage by Kublai Khan for their pilgrimage to Jerusalem. They traveled a standard route to get from China to the Middle East (see map 5). They departed Beijing in 1275 or 1276 and first went to Mark's hometown of Koshang (Olon Süme, in present-day Inner Mongolia). They then traveled southwest along the Yellow River until they reached Tangut territory in present-day Ningxia. After an arduous two-month journey through the Gansu Corridor and along the southern Silk Road of the Tarim Basin, they arrived at the oasis of Khotan, where they remained for six months. Departing the Tarim through Kashgar, they passed northwest through Talas (Taraz, Kazakhstan), across the Syr Darya River in Uzbekistan, before traveling southwest into Khorasan (present-day Afghanistan), then part of the Il-Khanate of Persia (1279 CE).

Regional conflicts blocked them from reaching their ultimate goal of Jerusalem, so they would reside in the Il-Khanate for seven years, until King Arghun sent Rabban Sauma to the west in 1287. On this second journey, Sauma would travel up the Tigris through Mosul and to the Black Sea, where he boarded a crowded boat for Constantinople. Upon his return from Europe around September of 1288, he lived out his remaining years quietly in Maragheh, Persia, never to return to China.

One of the most fascinating passages in Rabban Sauma's account is his interrogation by the cardinals in Rome over the nature of Jesus Christ and the Holy Trinity. Nestorian Christians like Rabban Sauma believed that there was a divine Jesus, who was son of God, and a human Jesus, who was born from an entirely human Mary, and that the two entities coexisted without mixture or fusion in the figure of Jesus Christ. The Roman Catholic Church declared this a heresy since the Council of Ephesus in 431 CE. Regarding the Trinity, Nestorians believed that the Father was paramount, and that the Son and Holy Spirit derived from him, just as heat and light are caused by the sun. The Catholic Church considered the Father, Son, and Holy Spirit to be coequal and coeternal. Despite these doctrinal differences, the Catholic fathers were still amazed that the Eucharist of the mass was still similar between the two long-separated sects of Christianity.

It is instructive to compare Rabban Sauma's account of his pilgrimage to Christian holy places with that written by the Chinese Buddhist monk Faxian (chapter 5), who traveled to India. Like that earlier Buddhist pilgrim, Rabban Sauma had been born in China, a land cut off from the original home of his religion, and was driven by a desire to see the authentic places where his savior lived and preached and especially to witness and be blessed by physical relics. In Rabban Sauma's case, that meant pieces of the True Cross and the Crown of Thorns, as well as body parts of saints and martyrs.

PRIMARY SOURCE, ALTERNATE ENGLISH TRANSLATION, AND SELECTED STUDIES

Borbone, Pier Giorgio. *History of Mar Yahballaha and Rabban Sauma.* Verlag Tradition, 2021.

Budge, E. A. Wallis. *The Monks of Kûblai Khân, Emperor of China, or the History of the Life and Travels of Rabban Ṣawmâ, Envoy and Plenipotentiary of the Mongol Khans to the Kings of Europe, and Markôs Who as Mar Yahbhallâhâ III Became Patriarch of the Nestorian Church in Asia.* The Religious Tract Society, 1928.

Rossabi, Morris. *Voyager from Xanadu.* Kodansha, 1992.

EXCERPTS FROM *THE HISTORY OF MAR YAHBALLAHA AND RABBAN SAUMA*

Rabban Sauma and Rabban Mark Resolve to Travel to Jerusalem

One day they [Rabban Sauma and his friend Mark] came up with a plan: "If we left this land [of China] for the West, we would have a lot to gain in receiving the blessings of the shrines of the holy martyrs and the fathers of the Church; then if Christ, the omnipotent Lord, prolonged our lives and sustained us with His grace, we might reach Jerusalem and attain complete atonement for our faults and the remittal of our sins."

Rabban Sauma tried to withhold Mark, frightening him with the labors of the journey, the perils of the itinerary, the dangers entailed by the routes, their needs and their foreign status. But Rabban Mark burned with a desire to leave; he felt in his heart that some treasure was in store for him in the West. Therefore, he continued to urge Rabban Sauma with his words, inciting him to leave. Eventually, they mutually agreed that neither would leave the other, no matter what trouble he might endure as a result; then they rose and, after distributing their meagre belongings and implements to the poor, they entered the city [of Beijing] in search of travel companions and provisions. . . .

Rabban Sauma's Journey to the Land of the Romans on Behalf of King Arghun and of the Catholicos Mar Yahballaha

Mar Yahballaha [the new name of Rabban Sauma's friend Mark], the *catholicos* ["patriarch" of the Nestorian Church of the East] was held in great esteem by the king [Arghun, Mongol Il-Khan of Persia, r. 1284–1291], and day after day he was honored more and more by the kings and queens. . . .

The king was considering the conquest and submission of the lands of Palestine and Syria, but "If the kings of the West - he thought - who are Christian, do not help me, I will not be able to attain my wish." He therefore asked the *catholicos* to recommend a wise man, suitable and capable of serving as ambassador, who could be sent to those sovereigns. Then the *catholicos* saw that no one was as eloquent in speech as Rabban Sauma, who possessed the skills to undertake the task, and summoned him.

Rabban Sauma said: "I am truly eager to carry out this mission!" King Arghun immediately had documents written out for him, addressed to the rulers of the Greeks and the Franks [Western Europeans], that is, the Romans, as also *yarlighs* [royal decrees or credentials] and letters, with gifts for each king. He gave Rabban Sauma two thousand pounds of gold, thirty good mounts and a *paiza* [a Mongol token of safe passage]. Rabban Sauma went to the patriarchal see to obtain another letter from Mar Yahballaha, the *catholicos*, and to take leave. The *catholicos* granted him leave, but when the time of separation came, was deeply sad: "What will happen? You administered the residence and know very well that with your departure my affairs will go to rack and ruin!" After more farewell talk, they parted, weeping. . . .

Rabban Sauma in Constantinople

Rabban Sauma left, accompanied by eminent priests and deacons from the residence. He arrived in the land of the Romans [Byzantine Empire], on the shores of the *dmk'* sea [Black Sea] and saw a church there. He boarded a ship, and his companions with him. On the ship were more than three hundred people, and each day he exhorted them with a discourse on faith. Most of the people on that ship were Romans who, fascinated by his eloquence, honored him a great deal.

After a few days he reached the great city of Constantinople. Before entering the city, he sent two youths to the king's gate, to announce the arrival of the ambassador of King Arghun. The king ordered people to come out and greet him and lead him into the city with great pomp and due honor. When Rabban Sauma entered the city, a house was assigned to him, that is, a residence, for the duration of his stay. Once he had rested, Rabban Sauma went before King Basileus [the Greek word *basileus* just means "king." This was Emperor Andronikos II Palaiologos, r. 1282–1328]. After greeting him, the king asked: "How are you, after the labors of the sea and the toils of the journey?" He replied: "The sight of the Christian king dispels all labors, and the toils of the journey disappear! I very much hoped to see your kingdom, may God preserve it!"

After enjoying food and drink, he requested permission from the king to see the churches, the shrines of the homes of the fathers and the saints' relics found there. The king entrusted him to the notables of his kingdom, who showed him all there was. First, he went to the great church of *asofia* [Hagia Sophia], which has 360 gates, all finished with marble. It is impossible to describe the dome that soars above the altar if you have not seen it, or to relate how tall and large it is. In that church is an image of the Holy Mother, painted by Luke the Evangelist. Rabban Sauma saw the hand of John the Baptist, relics of Lazarus and Mary Magdalene, and even the stone that was placed on the tomb of the Lord, when Joseph the counsellor [Joseph of Arimathea] took him down from the cross. Mary cried on that stone; to this day the spot where her tears fell is wet, and no matter how much it is dried up, it becomes wet again. He also saw the stone jar, in which the Lord changed water into wine in Cana of Galilee; the reliquary of a holy woman [probably Saint Theodosia], who is carried in procession every year: an ailing man, if placed under it, will be healed; the reliquary of Saint John Chrysostom. He saw the stone on which Simon Peter was seated when the rooster crowed; the tomb of the undefeated Emperor Constantine, which is red-colored; the tomb of Justinian, of greenish stone; the sepulchers of the 318 fathers: they are all in a great church and their bodies do not suffer corruption, because they held fast to their faith. They visited many other shrines of the holy fathers, saw several talismans and a statue fashioned from bronze and stone.

Finally, Rabban Sauma arrived before King Basileus and said: "May the King live forever! I thank Our Lord for being considered worthy of seeing these holy temples. Now, with the king's permission, I shall leave to accomplish the order of King Arghun, who wants me to travel to the lands of the Franks." Then the king bade him farewell benevolently and presented him with gifts of gold and silver.

Rabban Sauma in Italy and in Illustrious Rome

He left there and boarded a ship to cross the sea; on the coast he saw a monastery of the Romans, in whose treasury were two silver reliquaries, one of which contained the head of Saint John Chrysostom, the other that of the pope who baptized King Constantine [Pope Sylvester, d. 355 CE].

Then he got on a ship and, once out in the open sea, he saw a mountain [probably Mt. Etna on Sicily] from which smoke rises all day long, while at night fire appears on it; no one can venture nearby because of the smell of sulfur. People say that there is a great snake there, and that is why that sea is called Sea of the Dragon. It is indeed a scary sea, where many ships full of men have been lost.

Two months later he came ashore, after enduring much trouble, labors, and anxiety. He disembarked in a city called Naples; the king was called Irad Sharlado [French: *le roi Charles deux*; King Charles II, r. 1285–1309]. Rabban Sauma arrived before the king [more likely, the king's son Charles Martel] to let him know the aim of his visit and was received with grace and honor.

There was at the time a struggle between that king and another, called Irad Arkon [French: *le roi d'Aragon*; James, Aragonese king of Sicily], whose troops had arrived aboard numerous ships [June 23, 1287]. The army of this king was in battle formation, and they clashed. Irad Arkon defeated king Irad Sharlado, killed 12,000 of his men, and sank their ships out at sea. During the confrontation, Rabban Sauma and his companions were on the roof of their dwelling and wondered at the customs of the Franks, who did not harm anybody, apart from the combatants.

From there they proceeded overland on horseback. They rode through several cities and were marveled to see that no region was devoid of buildings. Along the way they heard that the pope had passed away [Honorius IV, d. April 3, 1287].

After a few days they arrived in the illustrious Rome; Rabban Sauma entered the church of Peter and Paul, which is the pope's residence. After the pope's demise, the holy see was entrusted to twelve men, called *kaldinnārā* [cardinals]. They were holding council to elect a pope, when Rabban Sauma had this message delivered to them: "We are ambassadors of King Arghun and of the *catholicos* of the East." Then the cardinals ordered them to enter. The Frank who had travelled with Rabban Sauma [a Genoese named Tommaso Anfossi] had taught them that, when entering the residence of the pope, there was an altar before which they were to prostrate and from there proceed to greet the cardinals. They acted accordingly, which pleased the cardinals. When Rabban Sauma arrived amongst them, no one rose before him; it was not the custom of those twelve men to do so, due to the honor of the Holy See.

But they invited Rabban Sauma to sit amongst them, and one of them asked him: "How are you after such a tiring journey?"

He replied: "Your prayers are solace and nurture for me!"

He said: "What brings you here?"

He replied: "The Mongols and the *catholicos* of the East jointly sent me to the pope, to discuss the business of Jerusalem [i.e., retaking the Holy Land], and they sent letters through me."

The cardinals told him: "Rest for now; we will discuss this together later on." They assigned a residence to him, to which he was escorted.

After three days the cardinals sent for him. When he arrived before them, they began to question him: "What part of the world are you from, and why have you come?"

And he answered to the point.

They asked him: "Where does the *catholicos* reside? And which apostle evangelized your country?"

He replied: "Saint Thomas, Saint Addai [Thaddeus of Edessa], and Saint Mari [of Edessa] evangelized our land, and to this day we follow the rules they gave us."

They asked him: "Where is the seat of the *catholicos*?"

"In Baghdad," he replied.

They asked: "What is your office there?"

He replied: "I am a deacon in the patriarch's residence, master of novices and visitor-general."

They said: "It is surprising that you, a Christian, a deacon from the patriarchal see of the East, have come as an envoy of the Mongol king!"

"My dear fathers, be aware that many of our fathers went to the lands of the Mongols, the Turks and the Chinese to instruct them, so that today there are many Christians among the Mongols. There are even sons of kings and queens who have been baptized and profess Christ's religion, and there are churches in their encampment. Christians are greatly honored, and many of the Mongols are believers. Therefore, the king, who is bound by affection to the *catholicos* and wishes to conquer Palestine and the lands of Syria, requests your assistance regarding the capture of Jerusalem. This is why I was chosen as an envoy; since I am Christian, my word will be believed by you."

They asked him: "What is your faith, and what doctrine do you follow —that which the pope now accepts or another?"

He replied: "No one came to us Orientals from the pope; we were evangelized by the holy apostles I have named, and to this day we follow what they conveyed to us."

They said to him: "What is your faith? Expound your creed."

* * * * * * * * * * * * * * * * **BRIEF EXCERPT** * * * * * * * * * * * * * * * *

Rabban Sauma's Creed, as Requested by the Cardinals

Rabban Sauma replied: "I believe in a hidden God, eternal, without beginning or end, Father, Son and Holy Spirit, three identical, unseparated *hypostases* [i.e., fundamental spiritual substances]. There is no first or last among them, neither young nor old; they are one in nature but three as *hypostases*: the generating Father, the generated Son and the proceeding Spirit; I believe that one of the *hypostases* of the regal Trinity, the Son, at the end of time garbed Himself in a perfect man, Jesus Christ, from the Holy Virgin Mary, was united to Him personally and in Him saved the world; in His divinity he is eternally generated by the Father and in His humanity, temporally, by Mary. This unity is undivided and unbroken for eternity; the unity is without mixture, confusion, or composition. This Son of unity is a perfect god and a perfect man, two natures and two *hypostases*, one person."

They questioned him: "Does the Holy Spirit proceed from the Father, the Son, or are they separate?"

He replied: "Do the Father, the Son and the Spirit share the same nature, or are they separate?"

They replied: "They share the same nature but are separate insofar as their individual characters are concerned."

Rabban Sauma asked: "What do their individualities consist of?"

They replied: "The Father is a generator, the Son is generated, the Spirit proceeds."

He said: "Who is the cause of whom?"

They replied: "The Father is the cause of the Son, and the Son is the cause of the Spirit."

He replied: "If they are equal in nature, in operation, in power and authority, and the three *hypostases* are one, how can one be the cause of the other? Necessarily, the Spirit should also be the cause of something else, and that should rather concern the learned outside the Church. We do not think this argument of yours can be proved: in fact, the soul is a cause for reason, while reason is not a cause for life. The solar globe is a cause for rays and heat, but the heat is not a cause for the rays. Therefore, we rightly believe that the Father is the cause for the Son and the Spirit, and both were caused by him. Adam similarly generated Seth and caused Eve to proceed: they are different in their generation and procession, but not distinct in their humanity."

The cardinals said: "We witness that the Spirit proceeds from the Father and the Son, but not in the way we have said to test your sanctity with argumentation."

He then said: "It is not true that two, three, or four causes exist for everything; I think this is not in conformity with our religion." Even though they refuted his argumentations with various arguments, the cardinals honored his reasoning.

He then said: "I have not come from faraway lands to dispute, nor to expose matters of faith but to receive the blessing of the pope and the temples of the saints, and to relate the word of the king and the *catholicos*. Now if you please, let us not pursue this discussion, and care to command someone show me the churches and shrines of this place; you will do your servant and disciple a great favor."

They then summoned the governor of the city and a few monks, who were ordered to show him the churches and shrines of that place. . . .

Rabban Sauma Visits Rome

First, they went to the churches of Peter and Paul: under the throne there is a chapel, where the body of Saint Peter was laid to rest, and above the throne is an altar. The altar sitting in the middle of that vast temple has four gates, each with double shutters of wrought iron. On that altar the pope celebrates Mass and no one except him ascends that platform. They then saw the Chair of Saint Peter, upon which the pope is seated during his investiture ceremony, and they also saw the shroud of pure cotton onto which Our Lord impressed his image to send it to King Abgar of Urha [King Abgar V of Edessa, d. ca. 50 CE]. The majesty of that church and its magnificence cannot be described; it rests on a hundred and eight columns, and inside it is another altar, on which their king of kings receives his investiture and is proclaimed *ampror* [French: *empereur*, i.e., the Holy Roman Emperor], that is, king of kings, by the pope. It is said that after praying the pope takes the crown with his feet and invests him, that is, he places it on his head, they say, to ensure the primacy of priesthood over royalty.

* *

Having visited all the churches and monasteries in illustrious Rome, they went to the church of Saint Paul the Apostle outside the city, whose tomb is also beneath the altar. Inside it is also the chain with which Paul was bound when they took him there. A gold reliquary was placed inside the altar, containing the head of Saint Stephen the martyr and the hand of Saint Ananias, who baptized Paul. The stick of Saint Paul the Apostle is also enshrined there. Then they went to the place where he was crowned in martyrdom: they say that when they severed his head, it bounced three times and each time it cried, "Christ!, Christ!"; from the three spots where it fell, waters gushed out, which can heal and comfort the ailing. There is a great chapel at the site, in which the bones of martyrs and illustrious fathers are preserved; Rabban Sauma and his companions received a blessing there. They then went to the Church of Saint Mary and that of Saint John the Baptist, where they saw the robe of Our Lord, the seamless one. In this church there is a board on which Our Lord consecrated His offering and distributed it among His disciples. Upon it, every year, the pope celebrates the mysteries of Easter. In the church there are also four bronze columns, each six cubits thick; they say they were brought from Jerusalem by the kings. There they saw the font of dark polished stone where Constantine, the undefeated king, was baptized. In the temple there are a hundred and forty columns of white marble; indeed, the church is massive and broad. They saw the place of the disputation between Simon Cephas and Simon Magus, where the latter fell and broke his bones. From there they went to the church of Saint Mary, where the beryl reliquary was taken out for them, containing Saint Mary's robe and the wooden board on which Our Lord slept as a child. In a silver reliquary they also saw the head of Matthew the Apostle, then again, the foot of Philip the Apostle and the arm of Jakob son of Zebedee, in the Church of the Apostles, which is also there. . . .

Rabban Sauma and his companions eventually returned before the cardinals, thanking them for having considered him worthy of visiting those shrines and receiving blessings there. Rabban Sauma requested permission to go to the kings beyond Rome. The cardinals granted him leave but said: "We cannot fulfil your request until a pope is elected."

Rabban Sauma Travels to France

From there they travelled to the land of Ṭuszkān [Tuscany] where they were honored; then they arrived in Genoa. In that place there is no king: the people choose a leader who rules them according to their liking. When they learned that an envoy had arrived from King Arghun, their leader came out with all the people and welcomed them into the city. There is a great church, dedicated to the holy Saint Laurence; the blessed body of Saint John the Baptist is preserved there, in a pure silver reliquary. They also saw a six-sided emerald basin, and were told by the locals that it was the one from which Our Lord had eaten Easter supper with his disciples; it had been brought there at the time of the capture of Jerusalem.

From there they went to the land of Onbar [Lombardy]; they saw the people there, that they did not fast on the first Saturday of Lent, and when asked: "Why do you behave so, unlike all the other Christians?", they said: "This is our custom; early in the time of our evangelization, the faith of our fathers was weak, and they could not fast: therefore their masters ordered them to fast for forty days."

Rabban Sauma in France

They later came to the place called Paris, before King Fransis [King of France, Phillip the Fair, r. 1285–1314], who sent many people out to greet them. The crowd accompanied them into the city in procession, with great honors. His lands extend for a distance of a month or more. King Fransis appointed a residence for them, and three days later he sent one of his nobles to summon Rabban Sauma. When the latter arrived, the king rose before him and honored him, asking him: "Why did you come? Who sent you?"

Rabban Sauma replied: "King Arghun and the *catholicos* of the East sent me, concerning the Jerusalem question," and disclosed all he knew, and gave him the letters he had with him and the gifts, that is, the presents he had brought.

King Fransis replied: "If the Mongols, who are not Christian, fight the Muslims to take Jerusalem, we should fight all the more. We shall come forward with an army, Our Lord willing!"

Rabban Sauma told him: "Now that we have seen the glory of your reign and admired the excellence of your power with eyes of flesh, we ask you to command that the citizens show us the churches, shrines, relics of saints, and all that is found here and not elsewhere, so that upon our return we may describe what we saw here and let it be known in our lands."

Then the king commanded his nobles, "Go and show them all the wonders found in our country; later I will personally show them what I have with me." Accordingly, the nobles left with them.

They remained for over a month in that great city, Paris, and they visited all there was to see. There are 30,000 scholars, who study both religious and secular disciplines; the interpretation and commentaries to all the Holy Scriptures and the wisdom, that is, philosophy and rhetoric, along with medicine, geometry, arithmetic and the science of planets and stars. They are constantly writing and receive a stipend from the king. They then saw a great church [Saint-Denis], in which there are coffins where the dead kings are buried; on them are their gold and silver effigies. Five hundred monks provide service to the royal burial site, whose maintenance is paid for by the king; they are assiduous in fast and prayer by the royal tombs. On the tombs are the crowns and arms of those kings, along with their clothes. In short, they saw all that was commendable and worthy.

Then the king sent for them, and they went to meet him in the church [of Sainte Chapelle]. They saw him standing by the altar and greeted him. He asked Rabban Sauma, "So, have you seen what there is to see? Is there anything else?" Sauma then thanked him. Meanwhile, he ascended with the king to a gold tabernacle. The king opened it and extracted a beryl reliquary containing the crown of thorns that the Jews put on the head of Our Lord when they crucified him.

So transparent was that beryl that the crown could be seen in the casket without opening it. There was also a fragment of wood from the Cross. "When our ancestors took Constantinople—the king told them —and looted Jerusalem, they took back these charms." We blessed the king, then begged him to grant us leave.

He then told us: "I will send with you one of the most important nobles [Gobert of Helleville] who are at my side to convey my reply to King Arghun." He then gave them presents and precious robes.

Rabban Sauma Before the King of England

They left there, that is, Paris, and went to the King Ingaltar ["England"; Edward I, r. 1272–1307], in Ksonyā [Gascony]. When, after twenty days, they reached their city [Bordeaux], the citizens came out to greet them and asked who they were.

"We are envoys," they replied, "and we come from beyond the Eastern seas. We were sent by the king and the patriarch, and the Mongol kings."

People ran in haste to inform that king, who received them with great joy; then they were escorted to him. Immediately the members of Rabban Sauma's embassy presented him with King Arghun's mandate and the gifts he had sent, along with letters from the *catholicos*. The king rejoiced greatly, even more so when the conversation came to the Jerusalem question. He said, "We kings of these lands wear the cross as a sign on our person, and we have no other thought than this. My resolve is strengthened when I hear King Arghun is of the same mind!"

He ordered Rabban Sauma to celebrate the eucharistic oblation, and he celebrated the glorious mysteries before the king and his court; the king received Holy Communion. That day the king gave a great banquet.

Then Rabban Sauma said: "We would be pleased, oh King, if you commanded every church and shrine of this place be shown to us, so we can relate that when we return to the Eastern peoples."

He replied: "You will be able to tell King Arghun and all the Eastern peoples that you saw something most commendable; namely, in the lands of the Franks [Europe] there are no two religions, only one, the one professed by Jesus Christ, and everybody is Christian." And he gave us many presents and money for the expenses.[2]

Notes

1. From William Woodville Rockhill, *The Journey of William of Rubruck to The Eastern Parts of the World, 1253-55* (Hakluyt, 1900), 52–61, 63–65, 66–83, 155–156, 200–202, with updates to modernize spelling, and a few corrections from Peter Jackson, *The Mission of Friar William of Rubruck* (Hakluyt, 1990). Headings added by the author.
2. Translation from Pier Giorgio Borbone, *History of Mar Yahballaha and Rabban Sauma* (Verlag Tradition, 2021), 69, 97–119, with minor updates to Americanize spelling and added parenthetical glosses by the author. The headings are rubrics in the original manuscript.

CHAPTER 8

DESCRIPTIONS OF THE WORLD (1225–1295 CE)

Though European scholars during the high Middle Ages were beginning to digest some Arabic knowledge, they had never been enlightened by the detailed Arabic accounts of China and India recounted in chapter 6. At the dawn of the 13th century, Europeans still imagined an Asia populated by the monstrous races of Ctesias (chapter 1) and were captivated by fantastic tales of the East from the popular *Alexander Romance*. They also vainly held out hope for a supposed Christian monarch of the Orient named Prester John. All this would change with the Mongol invasions, and the information brought back by papal envoys and missionaries like William of Rubruck (chapter 7).

On the Chinese side of Eurasia, the Tang engagement with the Indian Ocean trading world continued into the Song dynasty (960–1279 CE). After the move of the capital to the south following the cataclysm of 1127 CE, and the loss of access to the overland routes through Central Asia, ocean trade increased significantly. And while Chinese ships rarely traveled farther than the Malaysian peninsula during this period, ocean-going merchants from the Muslim world continued to make the profitable voyage to Chinese ports like Canton and Quanzhou, bringing knowledge and tales from Africa, the Middle East, and the Mediterranean. Both of the parallel authors in this chapter, Zhao Rugua and Marco Polo, sought to present a comprehensive picture of the known world beyond their homelands, though neither author personally traveled to all the of places they described. For the first time, these envisioned worlds encompassed most of the Afro-Eurasian continent, a vast landmass inter-connected through multipolar trade networks and diplomatic relationships.

INTRODUCTION TO ZHAO RUGUA

Zhao Rugua 趙汝适 (April 12, 1170–1231) was an eighth-generation descendant of Emperor Taizong (r. 976–997 CE) of the Song, born in what is now Taizhou, Zhejiang. He followed the favored track for a Confucian scholar, being awarded the *jinshi* degree in 1196. He subsequently served in a series of positions with increasing responsibilities, until he was appointed supervisor of maritime trade (*tiju shibo* 提舉市舶) of Fujian circuit in 1224, adding a simultaneous appointment in the vital port city of Quanzhou 泉州 the following year. The Tang dynasty had already established this post of supervisor of maritime trade earlier, mostly to be appointed in Canton, but the Song dynasty expanded the postings to multiple ports along the southeast coast. The supervisor of maritime trade was responsible for receiving tribute envoys from abroad, inspecting foreign shipping, managing trade, collecting duties, interdicting smuggling, procuring items for government use, managing the foreign settlements in the port cities, as well as supervising Chinese ocean-going traders.

Quanzhou was home to a bustling Muslim merchant diaspora and had superseded Canton as the center of maritime trade after the Tang. There was even a Muslim cemetery in the town. In his preface, Zhao Rugua mentions that he (or more likely his subordinates) interviewed the foreign merchants about their countries and their products. He had this information translated into Chinese, which he compiled into a book called *Account of Foreign Countries* (Zhufanzhi 諸蕃志), with a preface dated in the fall of 1225. For his compilation, Zhao copied heavily from earlier texts, most notably the *Responses to Queries on the Lands Beyond the Passes* (Lingwai daida 嶺外代答; 1178) of Zhou Qufei 周去非, but Zhao mentioned many more countries than Zhou had and added greater detail. Neither Zhou Qufei nor Zhao Rugua ever traveled outside of China, and all their information was second-hand, though some of it was fairly accurate. Although Zhao Rugua claims to have excised crude and salacious stories, a few flights of fancy remain in his compilation, as one can see from the translated excerpts below.

The *Account of Foreign Countries* is divided into two parts. The first gives the familiar catalogue-style entries for 58 foreign countries, while the second part details dozens of exotic products like aromatics, spices, and ivory that can be obtained from those places. Zhao's catalogue of countries constitutes a complete "description of the world," as known to Chinese officials and their Arab merchant informants by the year 1225 (map 6). It spans from southern Spain in the west to Japan in the East, Central Asia in the north, and to the southeast African coast in the south.

In Zhao's representation, the center of gravity in this world was the Muslim lands of the Islamic empire (Dashi), for which he mentions 24 individual "countries," ranging from southern Spain to Zanzibar to Bukhara. Even though the Abbasid Caliphate in Baghdad no longer directly controlled all these lands, several of the various splinter caliphates still gave nominal fealty to the Abbasid ruler. A few of these so-called Arabic countries were probably just small merchant diasporic communities. Starting during the 12th century, many of these diasporic Muslim merchants situated east of India were effectively cut off from direct contact with their Near Eastern homelands by the rise of Southeast Asian powers.

Map 6. The World according to Zhao Rugua and Marco Polo.

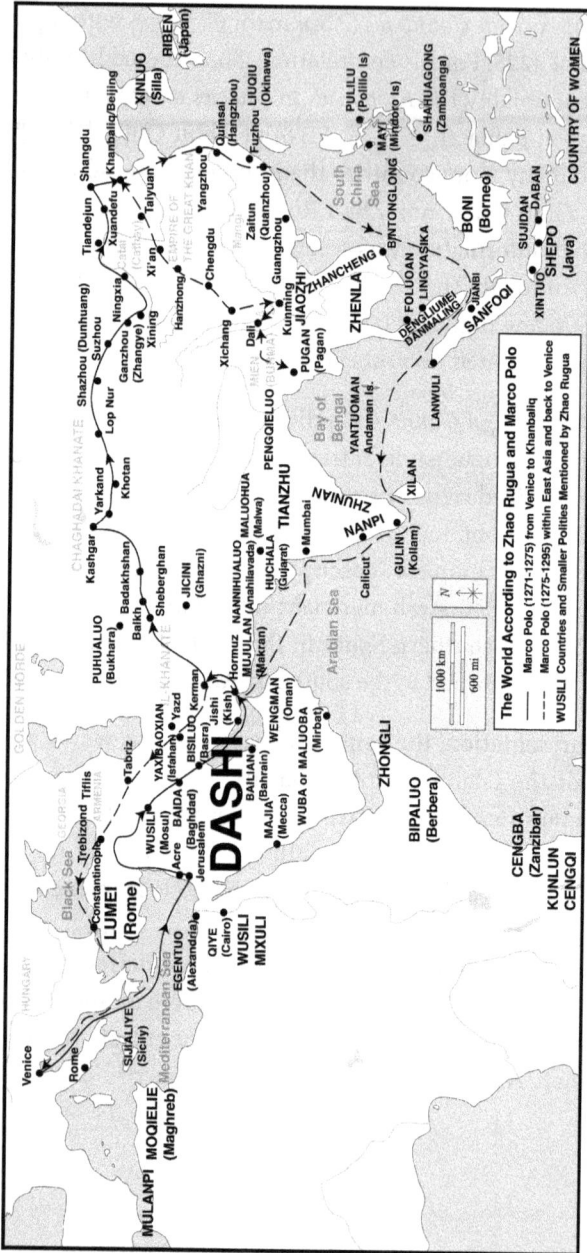

Source: Map created by Zhang Yexu and the author.

The second gravitational center of this trading world was in Southeast Asia, focused on the former domains of the trading empire of Srivijaya, which had been based on Sumatra. Because Zhao relied on earlier textual sources, his work makes it appear that the Abbasid Caliphate and the Srivijaya thalassocratic empire were still intact and supremely powerful, but their heydays had long passed. Chinese sailors and ships most certainly had traveled throughout Southeast Asia to places like Java and Sumatra for centuries, but there is almost no evidence that Chinese ships regularly traveled beyond the Straits of Malacca or the Bay of Bengal before the Yuan dynasty. But there was hardly any need to travel that far, since all the products of Arabia and Africa were traded to their doorstep in Sumatra or brought all the way home to Quanzhou, where Zhao Rugua supervised trade in 1225.

In the following translated excerpts, Zhao records wondrous animals from Somalia such as the ostrich, giraffe, and zebra, an exaggerated description of the crops and merchant galleys of Muslim Spain and North Africa, an entertaining story about Egypt that relates a different version of the story of Joseph from those found in the Bible or Qur'an, and the first mention of Sicily and its volcano Mt. Etna. Following a theme we have highlighted throughout, Zhao mentions two different "Countries of Women": one in the West near the Black Sea (which we have seen before), and a new one east of Java, retold from Arabic sailors' tales.

SOURCE TEXT, ALTERNATE ENGLISH TRANSLATIONS, AND SELECTED STUDIES

Almonte, Victoria, and Paolo De Troia. *The Historical Value of the Work Lingwai Daida by Zhou Qufei.* Aracne, 2020.

Chaffee, John W. *The Muslim Merchants of Premodern China: The History of a Maritime Asian Trade Diaspora, 750–1400.* Cambridge University Press, 2018.

Feng Chengjun 馮承鈞, ed. *Zhufan zhi jiaozhu* 諸蕃志校注. Taiwan shangwu yinshuguan, 1970. Originally published in Shanghai, 1940.

Fracasso, Riccardo M. "Ssu-chia-li-yeh 斯加里野, The First Chinese Description of Sicily." *T'oung Pao*, Second Series, 68, nos. 4/5 (1982): 248–253.

Hirth, Friedrich, and William Woodville Rockhill, trans. *Chau Ju-Kua: His Work on the Chinese and Arab Trade in the Twelfth and Thirteenth Centuries, Entitled* Chu-Fan-chï. New York: Paragon Book Reprint Corp., 1966. Originally published in 1911, St. Petersburg.

Kolnin, Ilia S. "Some Thoughts on Zhao Rugua's Biography and Zhufan zhi: Translation and Comparison of Relevant Fragments from Various Sources." *Crossroads* 17/18 (2018): 45–68.

Kuo Huei-Ying. "Charting China in the Thirteenth-Century World: The First Translation of *Zhu fan zhi* and Its Recipients in China in the 1930s." In *Knowledge in Translation: Global Patterns of Scientific Exchange, 1000–1800 CE*, edited by Patrick Manning and Abigail Owen. University of Pittsburgh Press, 2018.

Park, Hyunhee. "The Imagined among the Real: The Country of Women in Traditional and Early Modern Chinese Geographical Accounts and Maps." In *Imagining Early Modern Histories*, edited by Elizabeth Ketner and Allison Kavey. Routledge, 2016.

Schottenhammer, Angela, ed. *The Emporium of the World: Maritime Quanzhou, 1000–1400*. Brill, 2000.

Yang Bowen 楊博文, ed. *Zhufan zhi jiaoshi* 諸蕃志校釋. Zhonghua shuju, 1996.

Yang Shao-sun. "A Chinese Gazetteer of Foreign Lands; A New Translation of Part 1 of the *Zhufan zhi* 諸蕃志 (1225)." https:// storymaps.arcgis.com/stories/39bce63e4e0642d3abce6c24db470760.

EXCERPTS FROM *ACCOUNT OF FOREIGN COUNTRIES*, BY ZHAO RUGUA

Zhao's Preface

The "Tribute of Yu" [a chapter in the *Book of Documents*] records that the barbarians from the foreign islands [came to Yu the Great] wearing garments of grass and carrying silks woven in shell patterns in their baskets of tribute. Thus, we can see that trade relations between the barbarians and the Middle Kingdom were already established during ancient times. From the Han dynasty [206 BCE–220 CE] onward, the tribute in precious items has never been cut off. When it came to the Tang dynasty [618–907 CE], the supervisors of maritime trade sent trade envoys who solicited [the foreigners to come to China], so the avenue of commerce from this point forward increasingly broadened. During our current [Song] dynasty, a succession of sage sovereigns has reigned, who all consider humaneness and frugality to be genuine treasures. The civilizing influence of our sovereigns' moral power has reached these distant lands, but communication had to go through multiple stages of intermediate interpreters, before the foreigners could present their treasures. Accordingly, they set up official posts in Quanzhou and Canton to supervise the maritime trade. This was only done out of a desire to conserve the labor of the people and to give material assistance to our dynasty. There is no way that this could be referred to as "valuing exotic products" or "indulging in extravagance."

I, [Zhao Rugua], was commissioned to fulfill the post [of supervisor of maritime trade] and arrived here [in Quanzhou, 1224 CE]. In my spare time, I used to peruse the maps of various foreign countries. On these were marked such maritime hazards as the Stone Beds and the Long Sands, or barriers like the Jiao Bay [Gulf of Tonkin] or the Bamboo Island [Aur Island, off Eastern Malaysia], but when I inquired into these features in written treatises, they contained nothing about them. So I asked around among the various foreign merchants [in Quanzhou] and had them each set down the name of their country and speak about its customs and topographical features. I also had them speak about the connecting points and the distances between them, as well as the livestock and products of their mountains and marshes. I had this all translated into Chinese, excising the obscene stories and sensational rumors, but retaining those accounts that accorded with reality. I entitled it *Account of Foreign Countries.*

There are vast myriads of countries which lie overseas in the great encircling ocean. In general, all their precious and exotic products which are found for sale in China, like southern gold, ivory, rhinoceros horn, pearls, aromatics, and tortoiseshell, can be found in the pages of this book.

Alas! The "Mountains and Seas" have their *Guideways* [see chapter 1], and the "Diverse Matters" have their *Record* [i.e., the *Bowuzhi* 博物志 of Zhang Hua, ca. 290 CE], but a gentleman is ashamed if he is unaware of just one single thing, so it was good that I compiled this *Account of Foreign Countries.*

Preface of Zhao Rugua, Grand Master for Closing Court and Supervisor of Maritime Trade of the Fujian Circuit, October 4–November 1, 1225 CE. . . .

Country of Bipaluo

The country of Bipaluo [Berbera coast of Somalia] has only four towns; the rest of the population resides in villages. Each of these is ruled by strongmen who try to surpass one another. They all pray to Heaven [Allāh] and not to the Buddha. They raise many camels and sheep, and a regular meal for the people usually consists of camel meat and milk, along with baked flatbread. This land produces ambergris, large elephant tusks, and sizeable rhinoceros horns. Some of the elephant tusks weigh over one hundred Chinese pounds [63 kg], and the rhinoceros horns can weigh more than 10 pounds [6.3 kg.] The land is also rich in costus root, liquid storax gum, myrrh, and particularly thick hawksbill turtle shells. Merchants from other countries all come here to trade for these products.

The country also produces an animal called the "camel crane" [Somali ostrich]. To the top of its outstretched neck, its body measures six or seven Chinese feet [1.87–2.18 m] in height. It has wings and can fly, but not very high.

There is another wild animal called a *cula* [Arabic: *zarāfah*, giraffe], which resembles a camel and is as large as an ox. It is yellowish-brown in color, and its forelegs are five feet long [1.56 m], while its rear legs are only three feet long [93.6 cm]. Its head sits very lofty and is tilted upward, while its hide is nearly a Chinese inch thick [3.1 cm].

There is also a kind of mule with alternating reddish-brown, white, and black stripes across its body, like the warp threads on a loom [zebra]. They are a wild animal of the hills and plains. These are often thought to be an alternate species of the camel. The inhabitants of this country are fond of hunting and occasionally take down these beasts with poisoned arrows. . . .

The Country of Mulanpi
To the west of the country of the Arabs lies an enormous sea [Mediterranean Sea], and in the western portion of this sea are found innumerable small countries. But the only place where the giant merchant galleys of the Arabs can dock is this country of Mulanpi [Arabic: al-Murābiṭūn; the Berber Almoravid dynasty (1050–1147) in North Africa and Southern Spain]. After setting out from the Arab controlled country of Tuopandi [Arabic: Dumyāṭ, port of Damietta in Egypt], one arrives in this country after sailing due west for over a hundred days. Just one of these merchant galleys can hold several thousand men, and aboard ship they have set up wine taverns, marketplace stalls with prepared food, weaving workshops, and other such amenities. When one speaks about giant ships, none can compare with these galleys bound for Mulanpi.

The products of this country are quite extraordinary. The kernels of wheat grow to three Chinese inches in length [9.36 cm]. Their melons have a circumference of six Chinese feet [1.87 m], and a single melon can feed 20 or 30 men. Their pomegranates can weigh five Chinese pounds [3.17 kg], the peaches two pounds [1.27 kg], and their citrons weigh more than twenty pounds [12.7 kg]. Each head of lettuce can weigh more than 10 pounds [6.3 kg], with leaves of three or four feet in length [94–125 cm]. They excavate underground silos for storing grain and wheat, which can keep for several decades without spoiling.

They produce a breed of foreign sheep that is several Chinese feet tall and has a tail as large as a fan [the North African fat-tailed sheep]. Each spring, herders slit open their bellies and extract several dozen pounds of fat, then sew the sheep back up again, so they continue living. It they do not extract this fat, the sheep would become so bloated and obese that they would perish.

If one journeys overland for 200 days [to northern Europe], he will arrive in a land where the hours of daylight shrink to only three Chinese double hours.

During the autumn months [in the Sahara Desert], windstorms sometimes arise suddenly. Men and beasts must immediately find water to drink to remain alive. If they delay even for a little while, they will die of thirst.[1]

* * * * * * * * * * * * * * * **BRIEF EXCERPT** * * * * * * * * * * * * * * * *

Country of Wusili

The country of Wusili [Arabic: Miṣr, Egypt] is under the control of the country of Baida [Baghdad, the Abbasid Caliphate]. The king of the country has a light-skinned complexion. He wraps a turban around his head, wears a foreign-style upper garment, and dons black riding boots. When the ruler rides out from his palace, he is preceded by three or four hundred attendant riders, whose saddles and bridles are completely decorated with gold and precious gems. He is also accompanied by 10 tigers, leashed with iron chains. Each is subdued by 100 tiger tamers, with 50 men just to tend to the chains. The entourage also includes 100 cudgel-wielding enforcers and 30 men with hawks perched on their arms. The king is surrounded by 1,000 mounted guards and 300 trusted personal slaves, each fully armored and brandishing a sword. Two men bear the royal armaments leading the procession before the king, while 100 drummers on horseback bring up the rear. The entire honor guard procession is quite wonderful.

The people here eat only bread and meat; they do not eat rice.

The climate of the country is predominantly dry. It governs 16 provinces, covering a territory with a perimeter requiring more than 60 days to traverse. Paradoxically, when there is rainfall, it washes away the people's crops and ruins them. There is a river in this country [the Nile] with extremely clear and fresh water, but nobody knows from what source it springs. During years of drought, the rivers in all the other countries will be greatly diminished, but only this river will remain constant in its flow, providing ample water for farm fields. The peasants rely on it for their irrigation and cultivation, and every year it is generally like this. There are men who have reached seventy or eighty years of age who cannot recall it ever having rained here.

An old tradition says that when the third-generation descendant of Puluohong [Abraham] named Shisu [Joseph] seized control of this country, he feared that it would suffer from a great drought on account of it never raining. Therefore, he selected lands adjacent to the Nile and established 360 new villages. All the villages were required to plant wheat, which provided the daily food needs of the people year after year. If each village provided enough food to supply the country for one day, then the 360 villages together would be able to contribute sufficient food for an entire year.

Furthermore, the country has a prefecture called Qiye [Arabic: al-Qāhirah, Cairo] which lies adjacent to the river.

Every second or third year, an "old man" invariably emerges from the river. He has short black hair and pure white whiskers and seats himself on a rock in the river, revealing only half his body. He washes his face with water and trims his nails. When the people of this country first saw him, they realized he was a supernatural being. They approach him and respectfully ask, "Will the people have good or bad luck this year?" If he does not speak, but only smiles, then there will be a bountiful harvest, and the people will not be burdened with epidemics. If he knits up his brows, then in the current year, or the following one, there will certainly be crop failures and plague. After sitting there on his rock for a long while, he will then submerge into the water and disappear.

In this river, there are also "water camels" [possibly rhinoceros] and "water horses" [hippopotamus], which occasionally climb up on the riverbank and chew on the grasses, but as soon as they catch sight of a man, they will submerge back into the river.[2]

* *

Miscellaneous Countries upon the Oceans

The Country of Women

Even farther to the southeast [of Indonesia], there is the Country of Women. The current here flows perpetually eastward. Every few years, the water overflows in a tidal surge from which sometimes emerge lotus seeds more than a Chinese foot in length and peach pits two feet long. When people acquire these, they present them as tribute to the queen of this country.

In the past, whenever a ship drifted onto these shores, hordes of women would seize all the crew and return home with them [to satisfy their sexual cravings]. Within a few days, all had died [of sheer exhaustion]. But once there was a cunning captive who fled by stealing a boat at night, and so he managed to escape with his life and passed down this tale.

When the south wind blows at full gale, the women of this country face into the direction of the breeze and expose their naked bodies and so are impregnated by the wind. They only give birth to female offspring.

In the Western Sea [Mediterranean and Black Sea], there is another Country of Women, where there are five men for every three women, but the monarch of the country is a woman, and women serve in all the civil posts, whereas the men serve in the military. Women are highly honored here, and so have many male servants, but men are not allowed to have any female servants. When a child is born, it takes the mother's family name. The climate is mostly cold, and they subsist on hunting with bows and arrows. They carry on trade with Da Qin [Byzantine Empire] and India from which they make a profit of several hundred percent. . . . [3]

Island Country of Sijialiye

The Country of Sijialiye [Sicily] lies near the border of Lumei [Rome]. It is an island in the sea that measures 1,000 Chinese miles [approx. 500 km] in breadth. The clothing, customs, and spoken language are the same as those of Lumei. There is an extremely deep mountain cave on this island [Mt. Etna], which emits flames during all four seasons a year. When viewed from a great distance, it appears to emit smoke in the morning and flames in the evening. When one observes it from close up, the power of the flames is quite intense. Sometimes, the inhabitants of this country will work together using poles to lift a huge boulder, some 500 [313 kg] or 1,000 Chinese pounds [633 kg], to the mountaintop and cast it into the crater. A short while later, there will be a tremendous explosion, blasting the stone into fragments of pumice. Once every five years, flames and rocks spew forth from the mountain, flowing all the way down to the seashore before retreating. The lava does not burn down the forests it passes through, but if it encounters any rocks, it will burn them up into ash. . . . [4]

INTRODUCTION TO MARCO POLO

According to his book's prologue, Marco Polo's (ca. 1254–January 8, 1324) father, Niccolò, and his uncle Maffeo were Venetian traders in the Black Sea area. On one trading trip, they ventured as far as Sarai, the capital of the Mongol Golden Horde, near present-day Volgagrad. A war between rival Mongol khanates prevented their return, so they continued eastward toward Bukhara, eventually joining the entourage of a Mongol envoy and reaching the court of the Great Khan Kublai in North China. Kublai entrusted the two Polos with a diplomatic and religious mission to ask the pope for a delegation of learned Christians to adorn his court and for oil from the Church of the Holy Sepulcher. They returned to Europe, and while waiting for a new pontiff to be elected went home to Venice and retrieved Niccolò's 15-year-old son, Marco.

Their long return trip to China (1271–1275 CE) can be tentatively reconstructed from itineraries and descriptions in Marco's book (see map 6). After passing through the Holy Land, they journeyed overland through Mosul and Baghdad, then probably traveled briefly by sea to Hormuz, then up through Kerman in Persia. After traversing Bactria in Afghanistan, they eventually reached Kashgar in western Xinjiang, traveled the southern Silk Road through Khotan, then passed through Lop Nur and the towns of Gansu, including Dunhuang. Eventually, they reached the Mongol seasonal capital of Shangdu, where they again had an audience with Kublai. After 17 eventful years in China, the Polos returned to Europe, following one final mission by sea to Persia, arriving in Venice in 1295. Marco was later captured in a naval battle in 1298 and spent time under house arrest in Genoa, where he collaborated with the author Rustichello of Pisa to write a book, probably called *The Description of the World* (Le devisament dou monde).

Marco Polo claimed that he had mastered several written languages of Mongol administration and had been a favorite of Kublai, who sent him on important oversight missions. Marco's book contains some accurate reporting on geography, architecture, products, and revenues, along with a good dose of hearsay, exaggeration, and blatant fabrications, often inflating his own importance and involvement. Marco's account is strangely impersonal, for he often erases himself from his travel itineraries, so we don't hear much of the hardships of his adventurous journeys.

The original manuscript collaboration by Rustichello and Marco written in Genoese confinement is now lost. What has come down to us are over 150 manuscripts and early printed versions, translated into every European language, each somehow incomplete or corrupt, with both subtle and substantial differences. Scholars have grouped them into two categories: an A-group in Franco-Italian, courtly French, Tuscan, and Venetian, descended from Rustichello's original, along with a much smaller B-group, which are more detailed in certain sections (such as regarding Quinsai), possibly expanded by Polo himself later in life.

Polo's book defies easy categorization for our anthology, due to its uniqueness. Marco certainly displays a mercantile interest, like the author of *Navigating Around the Red Sea* (chapter 3), for he was keen to notice trade nexuses, commercial duties, and valuable commodities. Though not an ordained monk like William of Rubruck (chapter 7), Marco still did show some sensitivity towards religious differences, describing Indian ascetic practices and the life of Siddhartha Gautama. Moreover, while Marco himself never intended to be a diplomat, Kublai sent him on several sensitive missions to other polities, including Myanmar and Vietnam. The nature of his book is best summed up by the earliest of its many titles, *The Description of the World,* for it is a comprehensive empirical, anthropocentric geography of Central, South, and East Asia. It is still loosely structured under the old literary tradition of Marvels of the East, but most of Marco's marvels of urbanism and empire are now quite real, though they are almost more incredible.

Doubts that Marco Polo actually went to China were first expressed during the 18th century, due to omissions in his book, where he failed to notice things that seemed remarkable to contemporary China observers, such as the Great Wall, foot binding, Chinese characters, tea, and chopsticks. However, those things would not have been obvious or evident to Marco at the Mongol court or in the Persian-speaking circles in which he moved. He did write accurately about Chinese paper money, coal, Buddhism, and public charity. His book was also the earliest to mention Japan (Cipangu, i.e., Ribenguo 日本國) and how it rebuffed an attempted Mongol invasion.

Certain contradictions and errors in Marco's book also raised suspicions. One manuscript of his book claims that Marco had been the "governor" of the major city of Yangzhou for three years, but Chinese sources fail to support this. However, we know from the famous tombstones of Katerina and Antonio Vilioni that other Italians were indeed living in Yangzhou during the Mongol period. Traditionally, the Chinese government assigned headmen to supervise such foreign groups, selected from among their compatriots. Perhaps Marco served as such a headman of the foreign merchant community in Yangzhou, not as governor of the entire city.

Even more problematic, Marco claims that his family and their associates helped instruct the Mongols in making catapults for the siege of the important Song stronghold of Xiangyang. Yet Chinese accounts record this siege for the years 1268–1273 CE, more than a year before the three Polos could have arrived in China and note that engineers sent from Persia designed the catapults for the Khan. The Chinese scholar and travel writer Shan Shili (chapter 13) was one of the first to point out this anachronistic falsehood in Polo's account.

Finally, the titles and personal names in Polo's book are often derived from Persian, not Chinese, pronunciation, suggesting either that Marco consorted with many Persians during his stay or that he copied parts of his account from now-lost Persian sources. At the time, Persian was a language commonly heard at the Mongol court, and Marco obviously never learned Chinese. For example, he calls the former Southern Song emperor Duzong (r. 1264–1274) "Facfur," a name which derives from the Persian form *baghpūr,* meaning "son of God," rather than using the emperor's reign name or the proper Chinese title, *tianzi* 天子 ("Son of Heaven").

But the textual and physical evidence that Marco did sojourn in China for 17 years far overwhelms such doubts. This includes unique information, not to be found in other earlier or contemporary Persian or Western accounts, which can be corroborated by Chinese sources, namely his descriptions of the Mongol capitals and the Grand Canal, the ocean voyage to escort Princess Kökechin to Persia, commercial duties and revenues at Quinsai, the Mongol campaigns against Myanmar and Vietnam, the postal service, detailed information about salt production and revenues, and the funerary custom of burning paper effigies. Physical proof also comes from the *paizi* ("tablet of command") given by the Great Khan, a Mongol silver belt, a lady's headdress, and a Mongol slave, which were enumerated in Marco's will of 1324 and in a separate inventory of his possessions.

Polo's glorifying portrait of Kublai Khan ("the most powerful man in people and in lands and in treasure that ever was in the world") was a not-so-subtle indictment by comparison of the self-important kings and dukes of Western European monarchies and Italian city states. Along with his power and wealth, Kublai's fairness and astuteness in the employment of worthy men could be seen as an encouragement for European rulers to act similarly to employ a man of worth and knowledge like Marco himself. His portrayal of Kublai as the ideal monarch required Polo to cover up his obesity, alcoholism, and gout.

Only two cities in China proper receive detailed descriptions in Polo's book, the Khan's new capital of Khanbaliq (Beijing) and the most populous city in the world, Quinsai (present-day Hangzhou). The dozens of other cities in north, east, and southwest China that he describes on itineraries are passed over quickly in formulaic language ("They are idolators, subject to the Great Khan, and use paper money"), suggesting to some scholars that they were not visited by Polo at all and were being read off some administrative document in his possession. However, his extended account of Quinsai (excerpted below) is one of the centerpieces of the book, describing a bustling, glittering city with canals like Venice, home to over 1.6 million residences, a number close to the population estimated from Chinese sources. Though the core of that fascinatingly accurate account is present in all the older A-tradition manuscripts, it is much longer and detailed in the B-tradition manuscript in Italian published posthumously by Giovanni Battista Ramusio in 1559. The core account written by Rustichello and Polo in 1298 had either been supplemented by other experienced visitors, or by Polo himself, sometime after his initial captivity in Genoa.

These two aspects of China, an all-powerful wise monarch and a bustling urban civilization, constituted the great "Revelation of Cathay" to Western Europe. Though not everyone believed Marco's tales, it was convincing enough to alter the worldview of famous cartographers (see appendix) and to inspire men like Christopher Columbus and Vasco da Gama to seek the riches and spices of the East through various routes.

SOURCE TEXT, ALTERNATE ENGLISH TRANSLATIONS, AND SELECTED STUDIES

Haw, Stephan. *Marco Polo's China: A Venetian in the Realm of Khubilai Khan.* Routledge, 2006.

Larner, John. *Marco Polo and the Discovery of the World.* Yale University Press, 1999.

Olschki, Leonardo. *Marco Polo's Asia.* University of California Press, 1960.

Moule, A .C. *Quinsay with Other Notes on Marco Polo.* Cambridge University Press, 1957.

Moule, A. C., and Paul Pelliot, trans. *Marco Polo: The Description of the World.* 2 vols. George Routledge & Sons, 1938.

Pelliot, Paul. *Notes on Marco Polo.* 3 vols. Imprimerie nationale, 1959–1973.

Vogel, Hans Ulrich. *Marco Polo Was in China.* Brill, 2013.

Wood, Frances. *Did Marco Polo Go to China?* Secker & Warburg, 1995.

Yule, Henry, and Henri Cordier. *The Travels of Marco Polo*. 2 vols. Dover, 1993.

EXCERPTS FROM *THE DESCRIPTION OF THE WORLD*, BY MARCO POLO (1298)

Kublai Khan and his Women

Now I wish to begin to tell you, in this part of our book all the very great doings, and all the very great marvels of the very great Lord of the Tartars, namely the great Kaan [Khan] who now reigns, who is called Cublai Kaan [Kublai Khan, r. 1260–1294 CE], which "Kaan" means to say in our language, the great lord of lords, emperor, and this lord, who now reigns, indeed he really has this name of lord of lords by right, because everyone knows truly that this great Kaan is the most powerful man in people and in lands and in treasure that ever was in the world or that now is, from the time of Adam our first father till this moment; and under him all the people are set with such obedience, as has never been done under any other former king. . . .

The great lord of lords, that is of all those of his dominion, who is called Cublai Kaan is like this. He is of good and fair size, neither too small nor too large, but is of middle size. He is covered with flesh in a beautiful manner, not too fat, nor lean; he is more than well formed in all parts. He has his face white and partly shining red like the color of a beautiful rose, which makes him appear very pleasing; and he has the eyes black and beautiful; and the nose very beautiful, well made and well set on the face.

And he has four women whom he holds always as his true wives, and the eldest son that he has of the first of these four women ought to be the lord of the whole empire by right when the great Kaan the father should die. They are called empresses, and each of those four women is called also by her other proper name. And each of these four ladies holds a very fine royal court by herself in her own palace, for there is none of them who has not three hundred chosen girls very fair and amiable. They have very many valets, eunuchs, and many other men and women, so that each of these ladies has in her court quite 10,000 persons. And whenever he wishes to lie with any one of these four women, he makes her come to his room; and sometimes he goes to the room of his wife. And he has also very many other concubines, and I will tell you in what way.

It is true that there is a province in which dwells a race of Tartars who are called Ungrat [Onggirat], and the city likewise, who are very handsome and fair-skinned people; and these women are very beautiful and adorned with excellent manners. And every second year, 100 maidens, the most beautiful that are to be found in all that race, are chosen and are brought to the great Kaan as he may wish. The great Kaan sends his messengers to the said province that they may find him the most beautiful girls, according to the standard of beauty which he gives to them, 400, 500, more and less, as they think right. And these girls are judged in this way. When the messengers come, they make all the girls of the province come to them. And there are judges deputed for this purpose, who seeing and considering all the parts of each separately, that is the hair, the face, and the eyebrows, the mouth, the lips, and the other limbs, that they may be harmonious and proportioned to the body, value some at 16 *carats*, others at 17, 18, 20, and more and less according to as they are more or less beautiful. . . .

And when they have come into his presence, he has them valued again by other judges. And of them all he has 30 or 40 who are valued at the most *carats* chosen for his own room. And he has them kept by the elder ladies of the palace, one to each of the wives of the barons, who use diligent care in watching them, and make them lie with them in one bed to know if she has good breath and sweet, and is clean, and sleeps quietly without snoring, and has no unpleasant scent anywhere, and to know if she is a virgin, and quite sound in all things. And when they have been carefully examined, those that are good and fair and sound in all their limbs are sent to wait on the lord in such a way as I shall tell you.

It is true that every three days and three nights, six of these girls are sent to wait on the lord when he goes to rest, and when he gets up, both in the room and in the bed, and for all that he needs; and the great Kaan does with them what he pleases. And at the end of these three days and of three nights come the second six girls in exchange for these, and those depart. And so it goes all the year that every three days and three nights, they are changed from six to six girls until the number of those hundred is completed and then they begin again another term. . . .

The Magnificent City of Quinsai

And when one is gone riding these three days journeys [from Chang'an zhen 長安鎮, Zhejiang] then one finds the very most noble and magnificent city which for its excellence, importance, and beauty is called Quinsai [probably derived from the Persian pronunciation of the term *xingzai* 行在 "temporary residence of an emperor," a name for Lin'an 臨安, the capital of Southern Song China, present-day Hangzhou], which means to say in French, the "City of Heaven," just as I told you before, for it is the greatest city which may be found in the world, where so many pleasures may be found that one fancies himself to be in Paradise.

And since we are come there, so will we tell you all its great nobility, because it does well to relate, that it is without fail the most noble city and the best that is in the world, and it is the principal in the province of Mangi [Nanjia 南家 "Southerners," i.e., China south of the Yangzi]. . . .

Master Marco Polo was in the city many times and determined with great diligence to notice and understand all the conditions of the place, describing them in his notes. . . . First of all that the city of Quinsai is so large that in circuit it is in the common belief 100 miles around or thereabout, because the streets and canals in it are very wide and large. Then there are squares where they hold market, which on account of the vast multitudes which meet in them are necessarily very large and spacious. And it is placed in this way, that it has on one side a lake of fresh water [West Lake, Xihu 西湖], which is very clear, and on the other there is an enormous river [Qiantang 錢塘 River] which, entering by many great and small canals which run in every part of the city, both takes away all the impurities and then enters the said lake, and from that runs to the ocean. And this makes the air very wholesome, and one can go all about the city by land and by these streams. And the streets and canals are very wide and great so that boats are able to travel there conveniently and carts to carry the things necessary for the inhabitants (figure 2).

And there is a story that it has 12,000 bridges, between great and small, for the greater part of stone, for some are built of wood. And for each of these bridges, or for the greater part, a great and large ship could easily pass under the arch of it, and for the others smaller ships could pass. But those that are made over the principal canals and the chief streets are arched so high and with such skill that a boat can pass under them without a mast, and yet there pass over them carriages and horses, so well are the streets inclined to fit the height. And let no one be surprised if there are so many bridges, because I tell you that this town is all situated in water of lagoons as Venice is, and is also all surrounded by water, and so it is needful that there may be so many bridges for this, that people may be able to go through all the town both inside and out by land; and if they were not in such numbers you could not go from one place to the other by land, but only by boats. . . .

Figure 2. "The City of Quinsai [Hangzhou]."

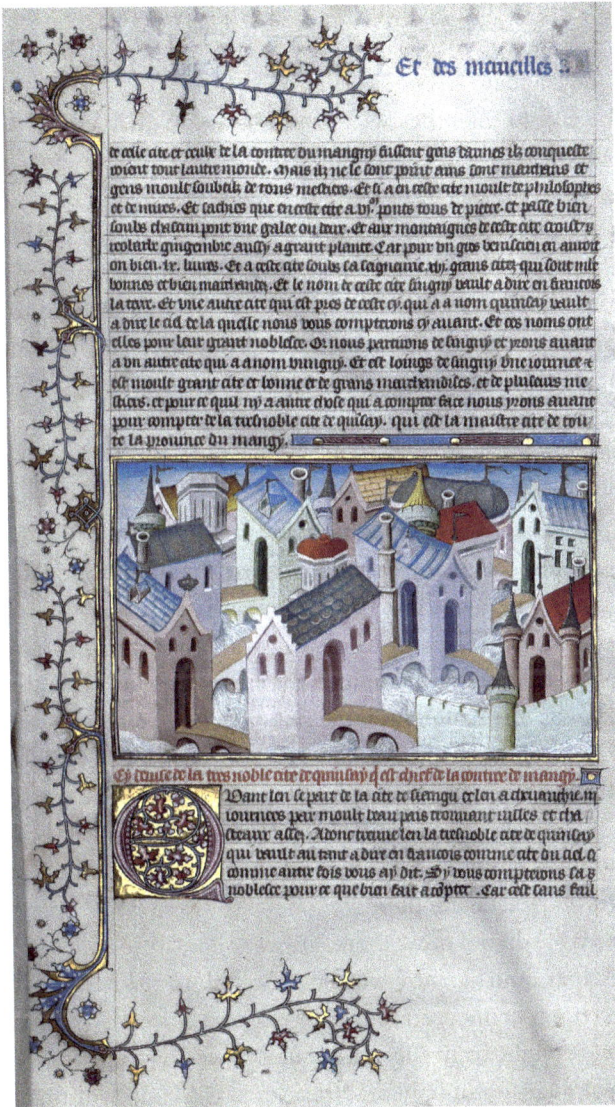

Source: Marco Polo, *Livre des merveilles*. Bibliothèque nationale de France, gallica.bnf.fr, Département des Manuscrits. Français 2810, 67 recto.

Public Squares and Markets in Quinsai

There are 10 principal open spaces, beside infinite others for the districts, which are square, that is half a mile for a side. And along the front of those there is a main street 40 paces wide, which runs straight from one end of the city to the other with many bridges which cross it level and conveniently; and every four miles is found one of these squares such as have two miles (as has been said) of circuit. There is in the same way a very broad canal which runs parallel to the said street at the back of the said squares, and on the near bank of this there are built great houses of stone where all the merchants who come from India and from other parts deposit their goods and merchandise that they may be near at hand to the squares. And on each of the said squares three days a week there is a concourse of from 40,000 to 50,000 persons who come to market and being everything you can desire for food, because there is always a great supply of victuals; of game, that is to say of roebuck, red deer, fallow deer, hares, rabbits, and of birds, partridges, pheasants, francolins, quails, fowls, capons, and so many ducks and geese that more could not be told, for they rear so many of them in that lake that for one Venetian silver groat may be had a pair of geese and two pair of ducks. There are too the shambles where they slaughter the large animals like calves, oxen, kids, and lambs, the which flesh the rich men and great lords eat. But the rest who are of low position do not abstain from all the other kinds of unclean flesh, without any respect. There are always on the said squares all sorts of vegetables and fruits, and above all the rest immense pears, which weigh 10 pounds apiece, which are white inside like a paste, and very fragrant; peaches in their season, yellow and white, very delicate. Grapes for wine do not grow there, but very good dried ones are brought from elsewhere, and likewise wine, of which the inhabitants do not make too much count, being used to that of rice and spices. Then there comes every day, brought from the ocean sea up the river for the space of 25 miles, great quantity of fish; and there is also a supply of that from the lake (for there are always fishermen who do nothing else), which is of different sorts according to the seasons of the year and because of the impurities which come from the city, it is fat and savory. Whoever saw the quantity of the said fish would never think that it could be sold, and yet in a few hours, it has all been taken away, so great is the multitude of the inhabitants who are used to live delicately; for they eat both fish and flesh at the same meal. All the said 10 squares are surrounded by high houses, and underneath are shops where they work at all sorts of trades, and sell all sorts of merchandise, and spicery, jewels, pearls; and in some shops nothing else is sold but wine made of rice with spices, for they continually go making it fresh and fresh, and it is cheap.

Courtesans

In other streets are stationed the courtesans, who are in so great number that I dare not say it, and not only near the squares, where places are usually assigned to them, but all over the city. And they stay very sumptuously with great perfumes and with many maidservants, and the houses all decorated. These women are very clever and practiced in knowing how to flatter and coax with ready words and suited to each kind of person, so that the foreigners who have once indulged themselves with them stay as it were in an ecstasy and are so much taken with their sweetness and charms that they can never forget them. And from this it comes to pass that when they return home, they say they have been in Quinsai, that is the City of Heaven, and never see the hour that they may be able to go back there again. . . .

Along the principal street of which we have spoken, which runs from one end of the city to the other, there are on one side and on the other houses, very large palaces with their gardens, and nearby them houses of artisans who work in their shops, and at all hours are met people who are going up and down on their business, so that to see so great a crowd anyone would believe that it would not be possible that victuals are found enough to be able to feed it; and yet on every market day, all the said squares are covered and filled with people and merchants who bring them both on carts and on boats, and all is disposed of. . . .

Mansions and Pleasures of the Wealthy

They have their houses very well built and richly worked, and they take so great delight in ornaments, paintings, and buildings, that the sums they spend on them are a stupendous thing. The native inhabitants of the city of Quinsai are peaceful people through having been so brought up and habituated by their kings, who were of the same nature. They do not handle arms or keep them at home. Quarrels or any difference are never heard or noticed among them. They do their merchandise and arts with great sincerity and truth. They love one another so that a district may be reckoned as one family, on account of the friendliness which exists between the men and the women by reason of the neighborhood. So great is the familiarity that it exists between them without any jealousy or suspicion of their women, for whom they have the greatest respect; and one who should dare to speak improper words to any married woman would be thought a great villain. They are equally friendly with the foreigners who come to them for the sake of trade, and gladly receive them at home, saluting them, and give them every help and advice in the business which they do. On the other hand, they do not like to see soldiers or those of the great Kaan's guards, as it seems to them that by reason of them they have been deprived of their natural kings and lords.

Villas at West Lake

And again I tell you that towards midway from the city, that is to say inside the city, is a lake [West Lake] very beautiful and great which is quite 30 miles round, and all round this lake are built many very beautiful and great palaces and many beautiful houses so wonderfully made that they could not be better devised nor made, nor more richly, which belong to gentlemen and to the great of the city, and the they are marvelously adorned inside and outside. And again, there were many abbeys in that place round the lake and many monasteries of idols [i.e., Buddhist temples], which are in the very greatest numbers, where stay a large number of monks who serve them. And again, I tell you that in the middle of the lake are two little islands on which there are, on each one, very wonderful palaces very great and noble and rich, so well made and so ornamented that they are really like some emperor's palaces, with so many rooms and galleries that it could not be believed. And so when some notable one wishes to make a great wedding or any great banquet in a smart place, they go to one of these palaces and there with dignity can make their wedding and their feast. And they find there all the furniture that is needed for the banquet, that is of plate and of linen and of dishes and everything else which they need, according to their usages, which are all kept in the said palaces for the people of the said city for this purpose, because they were built by them. And sometimes there would be 100, and some would wish to make feasts and others, weddings, and yet all would be accommodated in different rooms and verandas with such order that one does not inconvenience the others. Besides this, boats or barges are found on the said lake, in great numbers, large and small, to go for enjoyment and to give oneself pleasure; and in these there can stay, 10, 15, and 20, and more persons, because they are 15 to 20 paces long with broad and flat bottoms, so that they sail without rocking on either side. And everyone who likes to enjoy himself with women, or with his companions, takes one of the boats like these, which are always kept adorned with beautiful seats and tables, and with all the other furniture necessary for making a feast. Above they are covered and flat, where men stand with poles which they stick into the ground, for the said lake is not more than two paces deep and guide the said barges where they are ordered. The covering on the inside part is painted with different colors and patterns, and likewise, all the barge; and there are windows round about, which they can shut and open, for those who stay seated at the meal at the sides may be able to look this way and that and delight the eyes with the variety and beauty of the palaces to which they are taken. Here come the best wines, hence are brought perfect confections; and in this way, those men go about this lake rejoicing together, for their mind and care is set on nothing else but bodily pleasure and enjoyment in feasting together. . . . And barges like these are found on the said lake at all times with people who go for enjoyment; for the inhabitants of this city never think of anything else but after that they have done their work or business

to spend part of the day with their ladies, or with courtesans, in
giving themselves to pleasure. . . .

* *

Paper Money and Silk

Moreover, I tell you that the people of this city are idolaters and since it
was conquered are subject to the rule of the great Kaan and have Tartar
money of notes by the commandment of the great Kaan. One makes the
money thus. One takes the innermost bark of a mulberry tree and lays it
together and makes of it, the same as one does with us, paper of which
one makes sheets, as one does our paper. The sheets one tears after the
shape of a penny on which one prints the stamp and mark of the great
Kaan. The money is taken for everything which one will buy and sell.

And the men as well as the women are fair and handsome and always
dress for the most part in silk, because of the great abundance which
they have of that material that is produced in the whole territory of
Quinsai, besides the great quantity which is continually brought in from
other provinces by merchants. . . .

Mongol Garrison at Quinsai

And know that the great Kaan has this town very well guarded and with
a very great people because it is the head and see of the kingdom and
of all the province of Mangi, and because there is also much property
and great treasure in this city and on the other hand, the great Kaan
has great revenue and great duties from it, so great that whoever heard
it said could hardly believe it, and so greater and more anxious care is
spent on it. And again, the great lord has it guarded so well and with so
any people for fear that they may rebel. . . .

Harbor for Ocean-Going Trade

And again I make you to know that 25 miles distant from this city of
Quinsai between the Greek wind and the sunrising is the Ocean sea of

India, and there on the sea is a city which is called Gampu [Ganpu 澉 浦, on north shore of Hangzhou Bay], and there is a very good harbor, and all the very large ships with very great merchandise of many kinds and of great value come and go there from India and from other foreign parts in very great number, which add to the value of the city. And from this city of Quinsai to the harbor is a great river [Qiantang River], which makes this harbor as it enters the sea, by which the ships are able to come up to the city; and all the day the ships of Quinsai go up and down with merchandise, at their pleasure, and there they load on to other ships that go through different parts of India and of Cathay. . . .

Guarding Against Rebellion

Moreover, I shall tell you one thing more at which you will be much astonished. For I tell you there is no doubt that in the vast province of Mangi are altogether quite 1,200 cities, besides the castles and towns of which there is great quantity, all fair and rich, all dwelt in by a great multitude of rich and industrious people, and in each is a guard for the great Kaan as large as I shall tell you, according to the extent and need of the case, to guard them against rebellion. . . .

And if it happens that some cities rebel (for the men, overtaken by some madness or intoxication, often murder their rulers), the moment the event is heard the neighboring cities send so many men of these armies who destroy those cities which have made the mistake; for it would be a long affair to wish to make an army come from another province of Cathay, which would require two months' time. But do not understand that these men are all Tartars, but they are from Cathay, good men at arms. For the Tartars are horsemen and do not stay except near the cities that are not in marshy places, but in those situated in firm and dry places where they can take exercise on horseback. And these men who guard these cities are not all on horseback, but there is a great part on foot according as the protection of each place demands. Into these cities of marshy places, he sends there Cathayans and some of those of Mangi who are men that bear arms; for all are men of the armies of the great

Kaan. For of all his subjects he has every year those who seem to be fitted for arms chosen out and enrolled in his army, and they are all called trained men. And the men who are taken from the province of Mangi are not set to the guard of their own cities but are sent to others which may be 20 days march of road distant, where they stay from four to five years and then return home, and some of the others are sent in their place. And the Cathayans and those of the province of Mangi observe this rule. . . .

Funerary Customs

They burn the corpses of the dead. And again, I tell you that they have likewise for a habit that when any great rich master dies, when the dead bodies are carried to burn, all the relations and the friends make very great mourning, and the relations, women and men, dress themselves cheaply in hemp for mourning and go behind with the body, which is carried to the place where they wish to burn him, and take with them their instruments, many and different, and go playing, and singing idol prayers aloud. And when they come to the said place where the body must be burned, they stopped themselves and cause horses to be made, and slaves or servants, male and women, and camels, saddles, trappings, and cloth of gold and of silk and money of gold and silver in very great abundance. And all these things they make to be painted on sheets of cotton paper. And when they have done all this, they make the great fire and burn the body with all these things and say that the dead man will have all those things in the other world, alive of flesh and of bones and the money of gold, cloths of gold and of silk. And the burning finished, they sound all the instruments together with great cheerfulness continually singing, for they say that all the honor that they do him when he is being burned, just such another will be done him in the other world by their gods and by the idols, and that the instruments that they have sounded and the songs of the idols will come to meet them in the other world and that the idol himself will come to do him honor, and that he is born again in the other world and begins a life anew. And because of this sort of faith, they fear not nor care about death, provided that honor be shown them in it as is said before, firmly trusting to be honored in like manner in the other world. . . .

Population and Its Registration in Quinsai

And again you may know quite truly that having found himself in this city of Quinsai, when account of the revenues and number of the inhabitants was given to the factors of the great Kaan, Master Marco saw that in this city of Quinsai are registered 160 *toman* of fires in Italian idiom, or families, counting for a fire the family which lives in one house, that is to say 160 *toman* of houses. And I tell you that a *toman* is "ten-thousand" [from Mongolian, *tümän,* "ten-thousand"] and then you must know that there are altogether 1,600 thousands [1.6 million] houses, among which are a great quantity of rich and very beautiful and large palaces. In this city there is, in so great a number of people, no more than one very beautiful church of Nestorian Christians only.

And since I have described to you about the city, so will I tell you another thing that does well to relate. Now you may know that all the burghers of the city, fathers of families, and also of all the others in the whole province of Mangi, have a custom and usage like this. For each has his name and the names of his wife and of his sons and of the wives of his sons and of his slaves and maids, and of all those of his house written on the door of his house. And it is also written there how many horses he keeps. And if it happens that any of them dies, they have his name taken away; and if any is born there or is received, his name is also added there, likewise with the others. And in this way, the lords and rulers of each city know all the people whom they have in their towns. And it is done also through all the province of Mangi and of that of Cathay.

And again, I will tell you another fine custom that there is. For you may know that all those who keep inns and who lodge wayfarers in these two provinces, they write on the door of the house all those who come to lodge in their inns by their names and what day of what month and the week he lodges there, with the day and the hour that they leave, so that all through the year, the great Kaan can know who goes and who comes throughout his land. And it is indeed a thing that befits wise men. . . .

Quantity of Pepper Consumed

And to speak of comparison of the pepper which is consumed in the city, so that from this, you may be able to estimate the quantity of the victuals, flesh, wines, spices, which are provided for the general expenditure which they make; Master Marco watched the reckoning made by one of those who attend at the customs of the great Kaan, that in the city of Quinsai, it has been found by inquiry that on any day you please 43 loads of pepper are expended for its use, and every load is of the weight of 223 pounds; and so from this, you can calculate how many are the other spices which are expended there, and also how many necessaries are required for the whole expenses. . . .

Tax Revenue

I wish to speak and tell you something of a very great amount of revenue which the great Kaan has each year from the city of Quinsai, aforesaid, and from the other lands which are under her domain, which is one of the nine parts, or the ninth kingdom, of the province of Mangi. Now I will tell you first of the salt, because it is of more value for revenue. Now you may know, therefore, quite truly that the salt of this town pays each year as a rule 80 *toman* of gold altogether and that each *toman* is 70,000 *saggi* of gold, so that the 80 *toman* amount to 5,000 thousands and 600 thousand *saggi* of gold [5.6 million]; while each *saggio* [approx. 4.9 g in Venice; for Polo, equivalent to one Chinese *qian* 錢, 4 g] is worth more than one florin [the Florentine gold coin of 3.5 g] of gold or than one ducat [the Venetian gold coin of 3.5 g] of gold. And this is quite a marvelous thing and a very great sum of money. And the revenue of salt is so great in this city for this reason, because it is in the borderland next to the sea, where there are many lagoons or marshes, where the water of the sea is condensed in the summer, and a great quantity of salt is made, and they take thence so great a quantity of salt that from this city, quite five other kingdoms of Mangi are supplied with salt for their uses.

And since I have told you of the salt, now I will tell you of the other things and merchandise also, for the duty which they pay. I tell you that sugar, which pays three and a third percent, grows and is made in this city and its dependencies, which is one of the nine parts of the province of Mangi, and it is also made in the other eight parts of the said province everywhere, and so it is made in this province in very great quantity and more than double that which is made in all the rest of the world; many people say it in truth; and this is again a very great source of revenue. And then the spicery, which is there without measure. . . . For you may know that all spiceries pay three and a third percent; and all goods they pay also three and a third percent to the king. And from the wine, which they make of rice and of spices, they have very great revenue also, and from charcoal. And from all the twelve crafts of which I have told you above that they have, each craft, 12,000 stations, and from these crafts, they have very great revenues, for they pay duty on everything. But what? All the merchants who carry merchandise to the city by land and carry them away from it to other parts, and those also carry them away by sea, pay in the same way a thirtieth of the goods themselves, namely, one of 30, which takes three and a third percent; but those who carry merchandise to it by sea and from far countries and regions as from the Indies, give 10 percent. Moreover, of all the things which grow in the country, produce coming both from animals and from the land, and silk, a 10th part is applied to the lord's government. And the duty on the silk of which they have so great abundance that it is a wonderful thing is very exceedingly great. . . . You may know that of the silk, 10 percent is given, and this amounts to untold money. And there are many other things which also pay 10 percent. So that all these revenues amount to a sum of money so great and so immeasurable that it is an incredible thing to hear; and it is every year; and again, it is only the ninth part of the province of Mangi. And that you may know the sum of it, I will tell you that I, Marco Polo, who several times was sent by the great Kaan to see, and heard the count of the annual revenue, which the lord had from all these things, without the salt (of which we have told you before) made, say that it is usually worth 210 *toman* of gold for each year, and each *toman*, as has been said above is worth 70,000 *saggi* of gold, which are worth 14,000 thousand and 700 thousand [14.7 million] *saggi* of gold [approx. 58,800 kg]. And this is quite one of the most great and incalculable amounts of revenue of money that was ever heard tell. And this is one of the nine parts of the province of Mangi. So you can well see that when the lord has so great revenue from the ninth part of the country, that the revenue from the other eight parts can be worth much; but indeed this is the largest and the most profitable part. And for the great profit which the great lord has from this country, he loves it much and does much to guard it carefully and to keep those who dwell there in great peace. But yet the great Kaan has all these revenues spent on the armies that are guarding the cities and countries, and to remove the poverty of the cities.[5]

APPENDIX TO CHAPTER 8

The *Catalan Atlas* (1375 CE)

The *Catalan Atlas*, one of the most important world maps from the Middle Ages, is the oldest surviving atlas to show the influence of Marco Polo's account of East Asia. Attributed to the Jewish Mallorcan creator of maps, clocks, and compasses, Cresques, son of Abraham (a.k.a. Elisha ben Abraham Cresques; 1325–1387), along with his son Jehuda, internal evidence suggests the map was completed in 1375. It was originally painted on vellum and mounted to six wooden panels (65 cm high by 50 cm wide). Some of the panels have a north orientation, while others have a south orientation. The panels may have originally been arranged from east to west. The first two panels (not depicted here) portray a schematic diagram of a geocentric cosmos.

It combines elements of the schematic *mappamundi* tradition, seen on examples like the *Hereford Atlas* (ca. 1300), such as placing Jerusalem at the center and putting apocalyptic imagery at the margins, along with the rectangular layout and information drawn from the *portolano* navigation charts used by sailors of the Mediterranean, an area that is represented on the atlas with great fidelity. Beyond this core area, however, the geography becomes more schematic and distorted, influenced by Classical and medieval conceptions of the world.

The *Catalan Atlas* is inscribed with more than 2,300 placenames, along with longer textual captions and illustrations of historical figures and monstrous races. Several of the monstrous races from the writings of Ctesias (see chapter 1) and Pliny, such as pygmies and naked fish-eating savages, have been relegated to East Asia or North Asia.

A caravan on horseback and camel depicted at the top of the sixth panel (see map 7, inset) is often assumed to represent Marco Polo and his father and uncle heading toward China. However, the Polos are not named in the caption, which only mentions a journey to Catayo (Cathay) from the empire of Sarra (Sarai), a similar route taken by the Polos.

The final panel of the map represents Mongolia and China, as well as Ceylon and Southeast Asia. Captioned depictions of Kublai Khan and his capital of Khanbaliq (Beijing), taken nearly verbatim from Polo's book, are placed below images of the Antichrist, Alexander the Great confronting a devil, and the peoples of Gog and Magog. More than a dozen other Chinese towns along the coast are named, reproducing Polo's corrupted spellings. However, the map's information is outdated, for Kublai's dynasty had been overthrown in 1368.

SCHOLARLY EDITION, ONLINE EDITION, AND RELEVANT STUDIES

Cresques, Abraham. *Atlas Catalan.* Bibliothèque nationale de France. Département des Manuscrits. Espagnol 30. https://gallica.bnf.fr/ark:/12148/btv1b52509636n.

Grosjean, Georges. *Mappamundi: The Catalan Atlas of the Year 1375.* Urs Graf, 1978.

Kogman-Appel, Katrin, and Charles Burnett. "The Geographical Concept of the Catalan Mappamundi." In *Knowledge in Translation: Global Patterns of Scientific Exchange, 1000–1800 CE*, edited by Patrick Manning and Abigail Owen. University of Pittsburgh Press, 2018.

Map 7. The *Catalan Atlas*.

The caravan set out from the empire of Sarra [i.e., the Golden Horde] to go to Catayo [Cathay, i.e., China].

The most powerful prince of all the Tartars is named Holubeim [Kublai], which means Great Khan. The emperor is far wealthier than any other monarch in the whole world. This emperor is guarded by 12,000 horsemen and these again have four captains. These 12,000 horsemen perform the duties of serving at court in relays, each of the four captains with his troop serving for three months in the year.

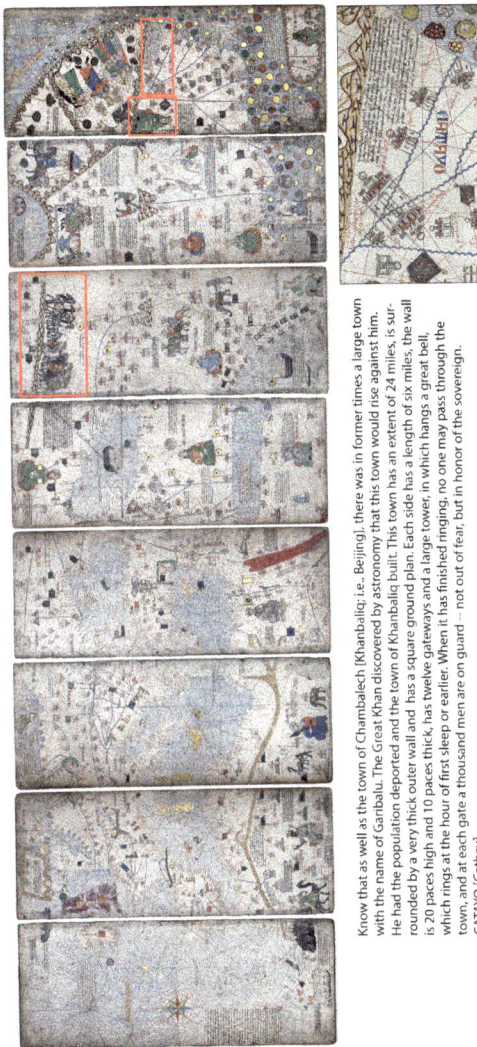

Know that as well as the town of Chambalech [Khanbaliq; i.e., Beijing], there was in former times a large town with the name of Garibalu. The Great Khan discovered by astronomy that this town would rise against him. He had the population deported and the town of Khanbaliq built. This town has an extent of 24 miles, is surrounded by a very thick outer wall and has a square ground plan. Each side has a length of six miles, the wall is 20 paces high and 10 paces thick, has twelve gateways and a large tower, in which hangs a great bell, which rings at the hour of first sleep or earlier. When it has finished ringing, no one may pass through the town, and at each gate a thousand men are on guard — not out of fear, but in honor of the sovereign. CATAYO [Cathay]

Source: Bibliothèque nationale de France, gallica.bnf.fr, Département des Manuscrits; Espagnol 30. Caption translations by Georges Grosjean, with modification by the author.

NOTES

1. Yang Bowen 楊博文, ed., *Zhufan zhi jiaoshi* 諸蕃志校釋 (Zhonghua shuju, 1996), 117–120. Translation by the author.
2. Yang Bowen, *Zhufan zhi jiaoshi*, 120–123. Translation by the author.
3. Yang Bowen, *Zhufan zhi jiaoshi*, 130–131. Translation by the author.
4. Yang Bowen, *Zhufan zhi jiaoshi*, 133–134. Translation by the author.
5. Translation from A. C. Moule and Paul Pelliot, *Marco Polo: The Description of the World* (George Routledge & Sons, 1938), 192, 204–206, 326–342, with minor updates to Americanize the spelling and some emended calculations from Hans Ulrich Vogel, *Marco Polo Was in China*. Subheadings added by the author.

CHAPTER 9

EPIC VOYAGES DURING THE GLOBAL AGE OF DISCOVERY (1405–1560)

The Pax Mongolica that enabled missionaries and envoys like William of Rubruck and Rabban Sauma (chapter 7) or merchants like the Polos (chapter 8) to cross Eurasia during the second half of the 13th century with limited hindrance quickly broke down. During the 14th century, the four major khanates of the Mongol world continued to squabble and disintegrate, with conflicts routinely blocking trade routes. By the middle of the century, the Black Death, which had been brought into Central Asia by the Mongols from their incursions into the Himalayan foothills, spread over trade routes to both Europe and China, killing a third to a half of the population of both regions and destroying economic growth and trade, just as the Plague of Justinian had accomplished seven centuries previously. Kublai's Yuan Dynasty (1271–1368) in China devolved into factional infighting and collapsed a few generations after his death. The Mamluk Sultans (1250–1517 CE) took over Egypt and the Levant, monopolizing the spice trade and blocking access to the Indian Ocean trading routes for Western powers, and the Turks conquered Constantinople in 1453, ending access to trade colonies for Italian merchants.

East Asia recovered first from the great upheavals of plague and civil war, with the establishment of the native Ming Dynasty in 1368. The first Ming emperor was xenophobic and opposed long-distance trade, but his energetic son, the Yongle 永樂 Emperor (r. 1402–1424), sponsored the first great voyages of the Global Age of Discovery. Decades later, the Portuguese, sailing from their Atlantic-facing ports, began to explore the west coast of Africa, looking for gold and a way to bypass Egypt's monopoly on the pepper trade.

INTRODUCTION TO MA HUAN

More than 80 years before Vasco da Gama (1469–1524) rounded the Cape of Good Hope and arrived on the west coast of India (May 1498), marking one of the foremost milestones in the Global Age of Discovery, the Chinese eunuch admiral Zheng He 鄭和 (born Ma He 馬和; 1371–ca. 1433), personally led or indirectly supervised seven voyages across the Indian Ocean to conduct diplomacy and trade (1405–1433), visiting Southeast Asia, India, the Persian Gulf, Arabia, and East Africa (map 8).

Map 8. Zheng He's seventh voyage.

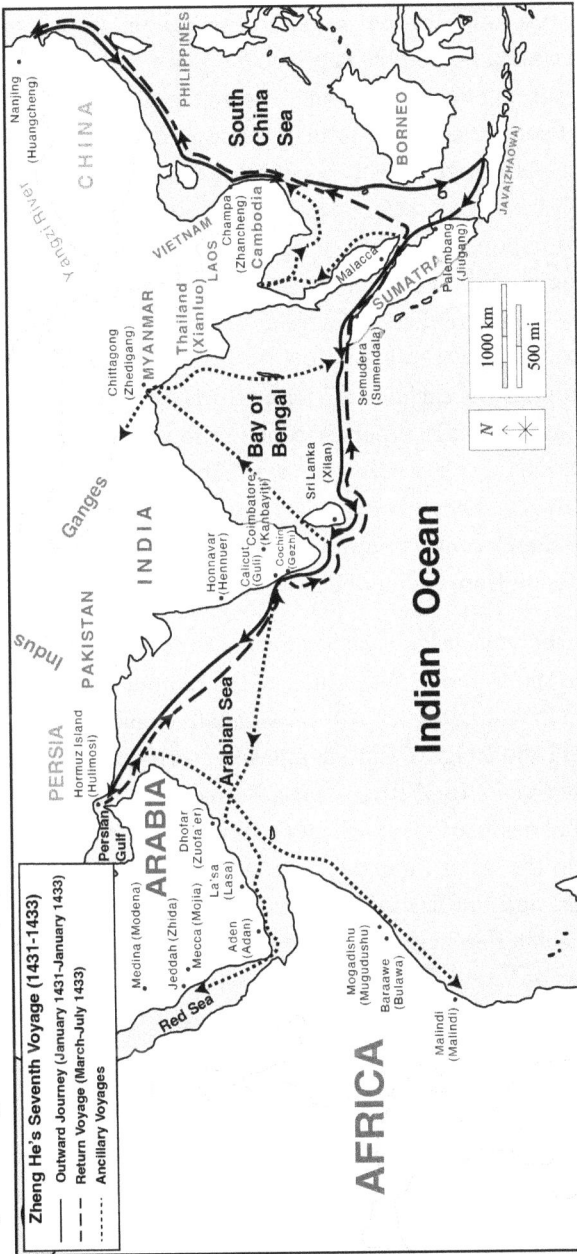

Zheng He's Seventh Voyage (1431-1433)
— Outward Journey (January 1431–January 1433)
– – Return Voyage (March–July 1433)
······ Ancillary Voyages

Source: Map created by Zhang Yexu and the author.

The voyages were initially sponsored by the Yongle Emperor, who had rebelled against and killed his nephew, the Jianwen 建文 Emperor (r. 1398–1402), usurping the Ming throne. The voyages were designed to display his might and to encourage foreign acknowledgment that he was the legitimate Son of Heaven. They were not specifically devised as trading missions, though trade was certainly conducted, nor were they meant to be voyages of conquest and colonization, although overwhelming military force was applied on multiple occasions to enforce the will of the Chinese emperor. Nor were these voyages of "discovery" in the Western sense, for Chinese rulers were aware of nearly all these foreign countries through earlier diplomatic exchanges or from the accounts of merchants (see chapter 8), and Chinese sailors had plied the routes through the eastern Indian Ocean for over a millennium (either aboard foreign vessels or on Chinese ships) and knew how to get to India and the Persian Gulf. Recall that more than a century earlier (1295), Marco Polo had voyaged on an official mission from China to Persia.

There were seven voyages in total, six during the reign of the Yongle Emperor and one sponsored by his grandson, the Xuande 宣德 Emperor (r. 1425–1435). The roundtrip expeditions usually took two years to accomplish, taking advantage of shifting monsoon winds, and the normal itinerary included visits to Vietnam, Java, Sumatra, Malacca, and Sri Lanka, with a final destination of Calicut (Kozhikode) on the southwest coast of India. On the latter three voyages, the fleet would separate at Calicut or earlier, and flotillas would travel separately to Bengal, the Persian Gulf, Arabian Peninsula, or East Africa (see map 8). Compared to the puny fleets of Vasco da Gama or Columbus, these were sizeable armadas. A typical Zheng He voyage brought nearly 300 ships, some as long as 137 meters with nine masts, and carrying as many as 28,000 officers and seamen in total.

Premodern Chinese diplomacy was often organized under a tributary system. On his voyages, Zheng He was deputized to present seals and garments of office to foreign rulers, who in turn were expected to come to the Chinese court and kowtow before the emperor, presenting local tribute from their country. In exchange, they would be granted generous gifts from the emperor and returned on the next voyage. As seen in Ma Huan's account of Calicut (excerpted below), Zheng He's fleet also engaged in significant trade in luxury goods, exchanging Chinese textiles and other manufactured goods for jewels and aromatics at negotiated prices. This Chinese tributary diplomacy was very insistent and heavy-handed, and one could not say that the voyages of Zheng He were entirely peaceful in their intent or actions. Those rulers who did not submit to the Chinese diplomatic overture and offer tribute were cowed by force. Zheng He captured and executed a Chinese pirate king on Sumatra in 1407 and installed a Chinese governor, deposed an unfriendly ruler on Sri Lanka in 1411 and a usurper in northern Sumatra in 1415, and probably used siege engines in southern Arabia in 1418.

Though Zheng He left behind several inscriptions detailing his diplomatic and military exploits, he does not seem to have kept a log of his journeys. Fortunately, several of his sailors recounted the places they visited in published works.

Ma Huan 馬歡 was a Chinese Muslim from Kuaiji, near present-day Shaoxing, Zhejiang. Because he could read the Arabic script and possibly speak both Persian and Arabic, he was selected to be a translator and interpreter on the fourth, sixth, and seventh voyages, which for the first time had the Persian Gulf and Arabia as intended destinations. His *Wondrous Sightseeing Along the Great Ocean's Shores* (Yingyai shenglan 瀛涯勝覽) was written in stages between 1406 and 1434 and finally printed in 1451, after having circulated informally for decades. One preface mentions that he worked with a collaborator named Guo Chongli 郭崇禮. Zheng He's private secretary Gong Zhen 鞏珍 (1434) and another sailor, Fei Xin 費信 (1436), also published accounts of the voyages, but much of their material seems to have been lifted from Ma Huan's account. In 1597, Luo Maodeng 羅懋登 published a fictional novel based on the voyages,

The Grand Director of the Three Treasures Goes down to the Western Ocean (Sanbao Taijian xia xiyang 三寶太監下西洋).

Ma Huan's book is the oldest surviving first-hand account from a Chinese observer of late medieval India and the early 15th century Arab world, before the arrival of the Portuguese. He includes 20 countries in his book, almost all of which he visited in person. The translated excerpts include his detailed accounts of Calicut and Mecca.

In his notes on the countries that he visited, Ma Huan mostly adheres to the genre of reports on foreign countries prevalent in China since Zhang Qian's (see chapter 2). He describes each country's topography, climate, people, animals, and local products, along with some strange or shocking customs, like ritual *suttee* in Java, matrilineal royal descent in Calicut, or the ox-dung smearing rituals of India. But in some instances, such as in his report on Calicut, Ma Huan deviates from the genre to provide some fascinating insights into 15th century Indian Ocean trade, including the logistics of trade negotiations, currency purity and exchange, weight conversion, and sailing directions.

When Faxian and Xuanzang resided in India during the fifth and seventh centuries, they viewed Central and South Asian landscapes through Buddhist lenses (chapter 5). They imagined a terrain marked by sites associated with the historical Buddha or with the Buddhist kings Ashoka and Kanishka. They were attentive to minor differences in Buddhist practice and highly critical of Brahmins, Jains, ascetic yogis, and other non-believers. The southwest India that Ma Huan encountered during the early 15th century was a very different world. Buddhism had been almost completely eclipsed, and Hindu and Islamic beliefs predominated. While Ma Huan recognized and appreciated that several high officials in Calicut were fellow Muslims, he failed to recognize that the king of Calicut and the majority of the population worshipped Hindu deities like Vishnu, calling them all Buddhists instead. It is probable that as a Muslim, he viewed all idolaters the same, so a statue of Vishnu was just another Buddhist idol to him. His explanation for why "Buddhists" (Hindus) in

India venerated the ox and the elephant is one of the most entertaining sections of his account; he retells a version of the "Moses and the Golden Calf" story, based on the one found in the Qur'an, but the golden calf in Ma's version shits gold and his Samiri/Aaron escapes Moses's wrath by riding away on an elephant!

His pro-Muslim bias is most evident during the time he supposedly spent in Mecca, which he calls a "realm of supreme happiness." Ma describes the men there as "handsome and stalwart" and asserts that poverty was nonexistent. Furthermore, the people were all peaceful and law abiding. These utopian assertions are implicit criticisms of Ming China.

We can forgive Ma Huan his cultural blind spots such as these, but his accounts of Mecca and Medina contain glaring errors, leading some scholars to suspect that he never visited those places himself. Similar mistakes in Marco Polo's account of China caused some to doubt he went there either (see chapter 8). Although Ma Huan gives an excellent description of the Kaaba shrine, its *kiswa* covering, and Ishmael's tomb, he gives the wrong direction for travel from Jedda to Mecca, claims that the Kaaba is a half day's journey outside of Mecca, and reports a completely wrong distance and direction for the trip from Mecca to Medina. He also places the famous Well of Zamzam in Medina rather than near the Kaaba. As with the criticism of Polo's account, Ma Huan's use of transliterated Persian words in his report, rather than the proper Arabic ones, has led some scholars to suspect he was just relaying the hearsay of Persian travelers.

PRIMARY SOURCE AND ALTERNATE ENGLISH TRANSLATION

Feng Chengjun 馮承鈞. *Yingyai shenglan jiaozhu* 瀛涯勝覽校注. Shangwu yinshuguan, 1935.

Mills, J. V. G. *Ying-Yai Sheng-lan: The Overall Survey of the Ocean's Shores.* Cambridge University Press, 1970.

EXCERPTS FROM *WONDROUS SIGHTSEEING ALONG THE GREAT OCEAN'S SHORES*, BY MA HUAN (1416–1434)

Ma Huan's Preface [dated November 19, 1416]

In the past, I had perused the *Brief Account of the Island Barbarians* [by Wang Dayuan, ca. 1350], which records variations in seasons and climate and differences in topography and people. I heaved a great sigh and said, "How can there be this much diversity in the world?"

Then, in the 11th year of the Yongle reign, the Grand Ancestor, the cultured emperor of the Ming [r. 1402–1424 CE], issued a decree that commanded the official envoy, the palace eunuch Zheng He [d. ca. 1433], to take overall command of the treasure ships and travel to the various foreign countries of the Western Ocean to publicly read out the emperor's edict and bestow rewards. Though unworthy, I was sent to accompany the envoy in a peripheral capacity as an interpreter and translator of foreign documents. I followed him wherever he went, through vast expanses of huge ocean waves, along a route an unknowable number of tens of millions of Chinese miles in length. We passed through the various countries, and I witnessed with my own eyes and experienced for myself their seasons, climate, topography, and peoples. So later, I understood that what was written in the *Brief Account of the Island Barbarians* was not fabrication, and that, remarkably, even greater strange and wondrous things existed in the world than were recounted in that book.

Thereupon, I collected notes on the beautiful or ugly appearance of the people of each country, the variations in their soil and customs, along with a separate listing of their local products and their systems of territorial control and arranged them in geographic order into a bound book, which I have entitled *Wondrous Sightseeing Along the Great Ocean's Shores*. It enables the attentive reader, in just a moment's perusal, to understand all the essential facts about the various foreign countries. It especially allows one to see how far the sagely civilizing influence of our emperor has spread to an extent that no previous dynasty could match.

I am ashamed of being so ignorant, a mere insignificant peasant. But I was given the privilege of accompanying the special envoy and with him experienced such wondrous sightseeing. It was truly a fortuitous adventure that occurs once in a thousand years! As for this book, when concentrating on the words I would hand down to posterity, I was incapable of producing polished prose. I could only write down a straightforward account of these matters and nothing more. I hope that the reader will not reproach me for its superficiality. This notice shall serve as a preface.

Calicut

The country of Guli [Calicut; present-day Kozhikode, India] is the great country of the Western Ocean. You set sail from the harbor of the country of Gezhi [Kochi, India] and go toward the northwest and will arrive after three days journey. The country lies along the seacoast, though east over the mountains across a distance of 500 to 700 Chinese miles [250–350 km], it communicates with the faraway country of Kanbayiti [Coimbatore, India]. To the west, it overlooks the great [Arabian] sea; to the south, it adjoins the border of the country of Gezhi; and its northern frontier connects with the region of Hennuer [Honnavar, India]. The great country of the Western Ocean is precisely this place.

During the fifth year of the Yongle reign [1407], the imperial court commanded the principal envoy, the palace eunuch Zheng He and others, to present to the king of this country [the Zamorin] an imperial edict bestowing upon him a royal investiture and granting him a silver seal of office and rewarding all of his subordinate chieftains with promotions to various ranks, along with appropriate caps and sashes of office. Zheng He arrived there in general command of the great fleet of treasure ships. He built a pavilion for sheltering stelae and erected an inscribed stone that read, "This country is more than 100,000 Chinese miles distant from the Middle Kingdom. The people and products all suitably conform to their intrinsic nature and are appropriate to the seasons and climate. The people are cheerfully contented, and their customs are largely identical.

We have inscribed this stone here to perpetually make this known for 10,000 generations."

The king of the country belongs to the indigenous Nanpi [Nambudiri] people. He is a believer in the Buddhist religion and reveres the elephant and the ox. The people of the country fall into five classes: the Huihui [Muslims], Nanpi, Zhedi [Chetty, merchant castes and monied property owners], Geling [Kling, a Malaysian term for southern Indians], and the Mugua [Mukkuvar, an ethnic group of fishermen and porters]. The king and the people of this country all refrain from eating beef, and the great chieftains are Muslims and all refrain from eating pork. Formerly, the king had pledged an oath with the Muslims, determining that "you will not eat beef, and we will not eat pork," placing a reciprocal taboo on both. It has been this way up until the present.

The king has cast an image of the Buddha out of bronze, and it is called Naina'er [probably Nārāyaṇa; a form of Vishnu]. He first constructed a Buddha Hall, casting roof tiles out of bronze to cover it. Beside the base of the Buddha statue, he had a well dug. Each daybreak, the king comes to draw water and bathe the Buddha statue. When his prayers are completed, he orders people to collect the purified dung of a yellow ox, mix it with water in a bronze basin until it is like paste, and then smear it everywhere inside the Buddha Hall, including on the floors and walls. Furthermore, he has commanded the chieftains, along with the wealthy households, to also smear other Buddha Halls with ox dung every morning. Moreover, he takes the ox dung, bakes it until it is reduced to white ash, then grinds it into a fine powder. He has a small pouch made out of fine hemp and fills it with this ash, always carrying it on his person. Each dawn, after he has finished washing his face, he takes some of the ox-dung ash, mixes it with water, and smears it on his forehead and between his inner thighs, three times in each place. This shows the sincerity with which he reveres the Buddha and the ox.

Local tradition says that once upon a time, there was a holy man named Moxie [Moses] who established a religious teaching. People realized he was a genuine saint, so they all revered and followed him. Later, for some reason, the holy man went somewhere else and ordered his younger brother, named Samoli [Samiri; the wicked idolater in the Qur'an, here conflated with Aaron], to govern the believers. His younger brother became arrogant and reckless. He cast a golden calf, saying, "This is your Holy Lord. If you strike your forehead on the ground before it, you will receive a miracle." The believers followed his orders and worshipped his golden calf, saying that it always shits out gold. Since people could obtain gold, they grew covetous in their hearts and forgot all about the way of heaven, for they all regarded this calf as the real Lord. Later, the holy man Moxie returned and saw how the masses had been misled by his younger brother Samoli into ruining the holy way, so he destroyed his calf and wanted to punish his younger brother, but he fled, riding away on an elephant. Afterward, people thought longingly of him and eagerly awaited his return. If it were the beginning of the month, they would say that he would certainly arrive by mid-month. When the middle of the month came, they would say once more that he would certainly arrive by the close of the month. Even until today they endlessly hope for his return. This is the reason why the Nanpi here worship the elephant and the ox.

* * * * * * * * * * * * * * * **BRIEF EXCERPT** * * * * * * * * * * * * * * * *

The king has two great chiefs who manage the affairs of the country for him, and both are Muslims. The majority of the people in the country profess the Muslim religion. There are mosques in about 20 or 30 locations. Once in every seven days, they travel to one to worship. When that day comes, everyone in the household will fast and bathe themselves and will not engage in any work. Between 9:00 a.m. and 1:00 p.m., men and boys will arrive at the mosque for services, and around 3:00 p.m., they will disperse and return home. Only then will they engage in buying and selling or manage household affairs.

The people are very honest and trustworthy, and quite handsome and splendid in appearance.

The two chieftains received promotions and rewards from the court of the Middle Kingdom. When the treasure ships arrive there [from China], the management of the buying and selling relies entirely on these two men. The king dispatches these chieftains along with a Chetty broker to make a full accounting record and conduct financial calculations for the government. When the brokers arrive, they meet with the high official in charge of the fleet and select a certain day to bargain over prices. When that day arrives, they first carry out the polychrome silk brocades, patterned tabby silks, and other articles that have been brought there from China and one by one negotiate the price. When this has been determined, they then write up a contract specifying the price and quantity of pieces to be bought, and each party retains a certified copy. The chieftain and the Chetty broker then publicly clasp hands with the chief eunuch, and the broker proclaims, "On such and such a month and day, we have struck our palms together in the presence of the multitude. It is all settled. Whether the prices are high or low, we shall not later regret it and try to alter them." After this, the Chetty traders and other wealthy householders bring precious gems, pearls, corals, and other items for us to see and negotiate prices. These cannot be all determined in one day. If it goes rapidly, it might take one month, or if more leisurely, it might take two or three months. Once a price in coins has been determined through consultation and negotiation, if one buys a pearl or other items from a particular owner, the price which must be paid is some amount that had been calculated by the chieftain or merchant who originally handled the negotiations. And as for the quantity of hemp, silk, or other commodities that must be given in exchange for something, merchandise can only be handed over according to the original handshake deal. No changes are allowed. Their method of calculation does not employ an abacus. They only use the 20 digits of their hands and feet to calculate figures, but there is never the slightest discrepancy. It is quite extraordinary.

The king uses gold of 60 percent purity to cast a coin and puts it into circulation. It is called the *banan* [gold *fanam*]. The diameter of the face of each coin is three-tenths and eight-hundredths of an inch [1.24 cm] in terms of our official measure. The front and back faces both display patterns. It weighs one-hundredth of a Chinese ounce [0.37 g.] on our official scales. He also uses silver to mint a coin called a *da'er* [silver *tar*]. Each one weighs about three-thousandths of an ounce [approx. 0.11 g], and these coins are employed as pocket money for minor transactions. . . .

* *

Western Ocean cloth is called *cheli* [*shālyēt*; named for a town near Calicut] cloth in this country and comes out from the neighboring territory of Kanbayiti and several other places. Each bolt is four Chinese feet and five inches [153 cm] in breadth and two staves and five feet [8.5 m] in length. It sells for eight or ten of their local gold coins. The people of this country also take raw silk thread, boil and scour it, then dye it various colors for weaving into kerchiefs with striped decoration. They are four or five feet [1.36–1.6 m] in breadth and one stave and two or three feet [4.08–4.42 m] in length. Each strip can be bought for 100 local gold coins.

As for pepper, the people who live in the hill villages have established plantations that extensively cultivate it. During the 10th month, the pepper ripens, is picked, dried in the sun, and sold. Naturally, there are major pepper harvesting agents who come to gather it. They forward it to a government warehouse, where it is taken in and stored. If there is a buyer, the government issues it from the warehouse to be put on sale. They make a calculation based on the quantity, assessing a tax in cash that must be paid to the government. One *bahār* [236 kg] of pepper is sold for 200 of their local gold coins. The Chetty merchants frequently purchase every color of precious stone and pearls, as well as manufactured coral beads and other items. When foreign ships arrive here from each land, the king of Guli dispatches a chieftain along with someone who can write to jointly inspect the sales, and then assesses a tax in coins that is paid to the government.

Wealthy households often plant coconut trees, some, 1,000 trees, and others, as many as 2,000 or 3,000 trees as a productive enterprise. Their coconuts have 10 different uses. The young tender ones produce a syrup that is very sweet and tasty and can be fermented into wine. The older coconuts have a flesh from which you can express oil, make sugar, or produce a food for eating. The husk fiber that envelopes the outside can be pounded into the ropes used in ship building, and the shell can be used to make bowls and cups. It is also good for burning into an ash that can be used [as an adhesive] for the delicate work of inlaying pounded sheets of gold and silver. The timber from the tree is good for constructing houses and the leaves are well suited for thatching roofs.

For vegetables, they have Indian mustard, fresh ginger, radishes, coriander, scallions, garlic, calabash, eggplant, Oriental pickling melon, and wax gourd, and they have all these during each of the four seasons. They also have a type of small melon, only as large as a finger, two Chinese inches [6.8 cm] in length, and perhaps it tastes a little like a cucumber. Their onions have a purple skin and resemble a large head of garlic, with small leaves. They are weighed by the Chinese pound [590 g] when sold. Jackfruit and Japanese bananas are widely sold. The gac plant can grow taller than 10 staves [34 m]. It forms a fruit that looks like a green persimmon, which holds 30 or 40 seeds inside. When it ripens, it falls on its own. Bats as large as hawks all hang upside down from this tree to rest.

They have both red and white rice, but they have neither barley nor wheat. Their wheat flour is all imported from other places for sale. Chickens and ducks are extensively raised, but they have no geese. Their goats have tall legs and are grey in color, like the foal of a donkey. Their water buffalo are not very big, and their cattle weigh three or four hundred Chinese pounds [177–236 kg]. People do not eat their flesh but consume only their milk and butterfat. If people don't have butter, they will not eat food. They care for their oxen until they die of old age and then bury them.

The price of every variety of ocean fish is extremely cheap. Deer and rabbit taken from the hills are also on sale. Many households raise peacocks. As for other game birds, they have crows, goshawks, egrets, and swallows, but besides these, they don't have any other type of large or small game birds.

The people of this country also gather to play stringed instruments and sing. They take the shell of a calabash to make the musical instrument and fashion the strings from copper wire [the South Asian *veena*]. They sing their foreign ballads to harmonize with the plucking [of the *veena*], and the melodies are tolerable to listen to.

Regarding the ritual customs for marriage and funerals among the people, the Suoli people [Chola, native South Indian] and Muslim people each follow the form of ritual for their class, which are dissimilar.

The royal throne is not handed down to the king's own son but to his sister's son, because only a child born from the womb of a woman of the maternal royal line is considered to be from the legitimate lineage. If the king does not have an older or younger sister, the throne will be handed down to his younger brother. If he has no younger brother, he will abdicate and cede the throne to a man of moral worth. The succession has always been like this.

The royal laws contain no punishment of flogging or caning. For minor crimes, they sever a hand or a foot, and for heavier crimes, they impose a fine in gold or execute the person. For heinous crimes, they make an inventory of the criminal's property, confiscate it, and exterminate his whole lineage. When a person violates the law, they detain him and bring him before an official, where he promptly admits to his crime. If there is some extenuating circumstance to the case or the accused believes he has suffered injustice and will not admit to the crime, then they will place an iron cauldron in front of the king or a major chieftain and fill it with four or five Chinese pounds [2.36–2.95 kg] of oil, bringing it to a rolling boil. They first take a tree leaf and toss it into the cauldron to see if it makes crackling and popping sounds. Then they command the accused to fry two fingers from his right hand in the oil for a short while. They wait for them to become scalded before he is allowed to lift them out. They wrap the hand in a cloth and affix it with an official seal that is recorded. The accused remains imprisoned in the judicial office. After two or three days, they open the seal and inspect the hand before an assembled crowd. If the hand is festering and ulcerated, this indicates that there was no injustice in the matter, and the prescribed punishment is imposed. If the hand is just like before, with no injury, they release him. The chieftain and others will ceremonially escort this person back to his home, to the accompaniment of drums and music. His various relatives, neighbors, and friends will present him with gifts of food in congratulation, drinking wine and making merry to celebrate this matter. It is the strangest thing!

When the principal envoy [Zheng He] returned, the king of Calicut desired to send tribute to the Ming court, so taking 50 Chinese ounces [1.85 kg] of the purest gold, he ordered the foreign craftsmen to draw out gold threads as fine as human hair and then wind them into ribbons. He then created a bejeweled treasure belt by inlaying it with every color of precious stone and large pearls. He dispatched his chieftain, Naina [Narayana], to present it as tribute to the Chinese court.

The Country of the Heavenly Cube
This country is the kingdom of Mojia [Mecca]. Setting sail from Guli [Calicut], you launch on a southwesterly bearing toward the *shen* position on the compass (240°), and after a voyage of three months, you arrive at the anchorage of this country, whose foreign name is Zhida [Jeddah, Saudi Arabia]. A chieftain is responsible for the defense of the port. From Zhida, you travel overland to the west [Ma Huan meant "east"] for one day and arrive at the city where the Sharif resides, which is named Mojia. Everyone believes in the Muslim religion. A holy man [the Prophet Muhammad] first began to expound and propagate the doctrine in this

country, and up until the present day, all the countrymen observe the religious rules and regulations in their actions. No one dares to violate them in even the slightest detail.

The people in this country are handsome and stalwart, and their limbs and features have a deep purple complexion. The men wrap their heads in turbans, dress in long robes, and wear leather shoes. The womenfolk all put a scarf over their heads, so you cannot see their faces. They all speak the Arabic language. The laws of the country prohibit alcohol, and their folkways are peaceful and commendable. There are no poverty stricken families there. Everyone observes the religious rules, and those who violate the laws are extremely rare. It truly is a realm of supreme happiness. For marriage and funerary rites, they all carry them out according to the precepts of their religion.

If you travel for more than half a day's journey, you arrive at the Mosque of Heavenly Paradise; the foreign name for this hall is Kai'abai [Kaaba]. It is surrounded on all sides by an enclosure wall outfitted with 466 openings. Each opening is supported on either side by a column of white jade [marble]. Altogether there are 467 columns in the colonnade. . . . The main hall of the Kaaba is constructed from laying courses of multicolored stones and takes the form of a flat-roofed cube. On the interior, they used five enormous beams of agarwood for the rafters and solid gold to construct a raised shelf. The interior walls of the hall were constructed of clay mixed with rosewater and ambergris, so the fragrance never diminishes. The top of the Kaaba is enshrouded by a mantle [the *kiswa*] made of black-dyed hemp and silk. They keep two black lions to guard the entrance door. Every year on the 10th day of the 12th month [Dhu al-Hijja], Muslims from every foreign country arrive to pray inside the mosque, sometimes after a journey of one or two years. They all cut off and retain a small piece of the hemp-silk mantle as a memento and proof of their experience before they depart. When the mantle has been completely cut up, the king again covers the Kaaba with a new mantle that he has had woven in advance. This remains the same, year after year, without interruption. To the left of the mosque is the tomb of the holy man Simayi [Ishmael]. The tomb was constructed entirely of green *sabuni* [whitish-green emerald] gemstone and measures one Chinese stave and two feet [approx. 3.84 m] in length, three feet in height [approx. 96 cm], and five feet [1.6 m] broad. The enclosure wall around the tomb is made of courses of purple topaz and is more than five feet [1.6 m] in height. Inside the enclosure wall of the mosque, they constructed four towers [minarets], one at each of the four corners. Each time for worship, someone would climb these towers, shouting out broadly and chanting the ceremonial call to prayer. To the left and right of the mosque are located the halls where all the founding clerics of the different schools expound their doctrines. These were also built with courses of stone and were all decorated most magnificently.

The climate of this place during the four seasons is always just as hot as in midsummer, and there is never any rain, lightning, frost, or snow. But at night, the dew is very heavy, so the plants and trees can rely on the moisture from the dew for nourishment. If you put out an empty bowl at night, and allow it to receive the condensation until dawn, there will be three-tenths of a Chinese inch [1 cm] of water in the bottom of the bowl. As for the products of the soil, rice is exceedingly rare, but they all cultivate foxtail millet, wheat, black broomcorn millet, and varieties of melons and vegetables. They even have some watermelons and muskmelons that are so large that it takes two men to carry one. They also have a type of tree with cottony flowers [probably Malabar silk-cotton tree], like the large mulberry tree of China, three staves [approx. 9.6 m] in height. It flowers twice every year and is both perennial and evergreen. For fruits, they have radishes, 10,000-year jujubes [Persian dates], pomegranates, pearleaf crabapples, big pears, and peaches, all of them large, some as heavy as four or five Chinese pounds [approx. 2.36-2.95 kg]. Their domesticated animals, including camels, horses, donkeys, mules, cattle, sheep, cats, dogs, chickens, geese, ducks, and pigeons are also plentiful. Some of their chickens and ducks weigh over 10 pounds [approx. 5.9 kg]. The products of the land include rosewater, ambergris, *qilin* [giraffes], lions, ostrich, antelope, "fly over the grass" [probably caracal], as well as every color of gemstone, pearls, coral, amber, and other such objects.

The king of the country casts a coin in gold, called the *tangga* [tanka], that is in general circulation. Each one has a diameter of seven-tenths of a Chinese inch [approx. 2.38 cm]. It is equal in weight to one hundredth of an ounce [3.69 g] on our official scales, but compared to Chinese gold, it amounts to only 20 percent as pure.

If you again journey west for a day, you arrive at a city called Modena [Medina]. The tomb of the patriarch Mahama [Muhammad] lies within the city walls. Up until the present day, a ray of light shines from atop the grave and into the clouds. Behind the tomb is a water well. The water is clear and sweet to the taste. It is named *abicancan* [Well of Zamzam; actually located near the Kaaba in Mecca]. Those who travel to foreign lands take this water and store it along the sides of their vessels. If at sea they encounter violent storms, they take some of this water and sprinkle it on the ship, and the wind and waves shall abate.

In the fifth year of the Xuande reign [1430], an imperial order was received from the court of the present dynasty, dispatching the principal envoy, the palace eunuch Zheng He, and others, to go to each foreign country and publicly read out the emperor's edict and bestow rewards. When a flotilla that had separated from the larger fleet arrived at Guli [Calicut], the palace eunuch Hong Bao saw that this country often dispatched envoys there [to Mecca], so he selected and dispatched an interpreter [Ma Huan] and others, seven men in all, carrying a load of musk, porcelain, and other items. They relied on a Guli ship to travel there [to Mecca]. The round trip there and back [to China] took an entire year. They bought all sorts of unusual commodities and rare treasures, along with giraffes, lions, ostriches, and other animals, as well as an accurate painting they had made of the Mosque of Heavenly Paradise [Kaaba] and returned to the capital. The king of the country of Mojia also dispatched envoys carrying local products, accompanied by the interpreter [Ma Huan] and the other Chinese envoys that had originally gone there, and presented these items in court.

Recounted by Ma Huan, September 9, 1451.[1]

INTRODUCTION TO GALEOTE PEREIRA

More than seven decades after the fleets of Zheng He had visited Calicut for the last time, the Portuguese captain Vasco da Gama (1469–1524) arrived on the west coast of India and visited the same Indian kingdom (May–August 1498). According to the diary of one of his sailors, Vasco da Gama's paltry gifts and inept diplomacy failed to impress the local ruler. But court records at Calicut or local oral tradition preserved the memory of earlier, more awe-inspiring, visitors. In a letter, the Italian merchant Girolamo Sernigi (who reported on Vasco da Gama's expedition), writes in July of 1499:

> It is now about 80 years since there arrived in this city of Chalicut [Calicut] certain vessels of white Christians, who wore their hair long like Germans, and had no beards except around the mouth, such as are worn at Constantinople by cavaliers and courtiers. They landed, wearing a cuirass, helmet, and vizor, and carrying a certain weapon [halberd] attached to a spear. Their vessels are armed with bombards, shorter than those in use with us. Once every two years they return with 20 or 25 vessels.[2]

This was an apt description of Zheng He's mighty fleets and the appearance of his Chinese sailors, but the Portuguese had no idea that these flotillas came from the Ming empire in China, which had retreated from sailing the Indian Ocean waters.

After gaining a foothold in India and garrisoning Goa, the Portuguese captured the sultanate of Malacca by force in 1511, gaining access to the main chokepoint of the spice trade. Some Portuguese sailors then began to make brief forays to the Chinese southeastern coast, mostly on native junks, to engage in small-scale trading with the spices they brought from India.

Formal Portuguese diplomatic relations with Ming China got off to a disastrous start. In 1517, a Portuguese squadron of ships was sent from Malacca to Canton with Tomé Pires (ca. 1468–1524) serving as Portuguese ambassador to China, carrying a letter from King Manuel I. After long delays in Canton, Pires was forwarded to Peking in early 1520 to meet with the Zhengde Emperor (r. 1505–1521), but he lacked understanding of the Chinese tributary system. As would recur centuries later with the ill-fated British Macartney Embassy (see chapter 11), misunderstandings and problems with translated letters also doomed the embassy. While waiting for the imperial audience, reports arrived in Peking of the outrageous behavior of one Portuguese captain at Canton and their earlier violent takeover of Malacca. In rapid order, the Ming emperor died before meeting Pires, and the Chinese authorities sent him back to Canton. After rumors surfaced that Portuguese were eating Chinese children and building a military fort, Pires and his men were imprisoned in Canton and tortured. Letters written in 1524 and smuggled out of prison by a man in Pires's retinue, Christovão Vieira, and another imprisoned Portuguese sailor, Vasco Calvo, revealed some new information about China. After this diplomatic catastrophe, the Portuguese were legally excluded from all trade with China, and their ships forcibly turned away. However, they continued to engage in smuggling for the next several decades along the southeast coast, since the illicit trade was so profitable.

The Portuguese soldier and trader Galeote Pereira, who initially sailed to India in 1534, joined a smuggling expedition off the Chinese coast during 1548–1549. His ship was captured by the Chinese coastal guard in March of 1549. Some of his fellow sailors were summarily executed, and the remainder were imprisoned in Fuzhou. A court investigation later ruled that the Chinese viceroy had overstepped his authority, so Pereira and the remaining Portuguese survivors were transferred to Guangxi and other locations for permanent internal exile.

Pereira bribed his way out of exile and was smuggled out of China by 1553, eventually returning to India. He recounted his China experience to Goa seminary pupils in 1561 in an essay titled "Some Things Known About China" (Alguas cousas sabidas da China), which was included in the annual Jesuit mission report to Rome. It was soon translated in an abridged version into Italian (1565) and English (1577). It was also taken up by the Dominican friar Gaspar da Cruz and incorporated into his book on China (1569), the first printed book in Europe exclusively dedicated to China. Da Cruz's work subsequently formed a key source for Juan Gonzáles de Mendoza's highly influential survey of China (1585), which was very popular and translated into every European language. Mendoza's book was considered the definitive work on China for two decades, until the publication of Matteo Ricci's edited journals in 1616 (chapter 10). The Portuguese accounts from the 1520s through 1560s and Spanish reports like that of Martín de Rada (1575) were the first detailed accounts of Chinese government, society, and customs to reach a Western audience since the accounts of Marco Polo and William of Rubruck two centuries earlier.

Pereira gives a general description of the provinces of the Ming empire, a sketch of regional administration, and the topography of the land, the architecture and marketplaces of cities, and the institutions and customs of the people, remarking on such things as human waste collection, public charity, eating habits, and religion, even making a correct distinction between Confucian shrines and Buddhist temples. Pereira provides the first detailed description in the West of the important Chinese civil service exam system, the dignity and privileges of the scholar-officials it produced (called by him *loutea*; Chinese: *laodie* 老爹 or *laoye* 老爺), and the awe that they inspired. He describes from first-hand experience the Chinese judicial system, its tortures and punishments like the cangue, and its hellish prisons. His vivid descriptions would color European views of the cruelty of Chinese punishments for centuries to come. Ironically, Pereira praises the Chinese court system, because suspected criminals there were interrogated openly in public audience, unlike the dungeon tortures of contemporary European courts, which he saw as more corrupt and prone to miscarriages of justice. He was amazed that a foreigner like himself was given a fair hearing at all. Some of these remarks that praised China and disparaged Europe were censored by authorities in Lisbon and Rome when Pereira's account was published in Italian and English translations, but I have restored them from his original Portuguese text.

During the 16th century, Portuguese and Spanish intentions toward China extended beyond peaceful trade and diplomacy. The Portuguese, buoyed by their rapid conquests in India (1510) and Malacca (1511), and the Spanish, filled with hubris from their destruction of the mighty Aztec (1521) and Inca (1532–1572) Empires, believed they could conquer the Middle Kingdom. Several early Iberian correspondents suggested that the populous empire of China could fall to their troops just as easily. The letters from prison by Cristóvão Vieira and Vasco Calvo (1524) gave the first extended eyewitness Portuguese accounts of China but also exposed weaknesses in Canton's defenses and urged the Portuguese crown to send a military expedition of just 10 to 15 ships and 2,000–3,000 men to conquer all of Guangdong and Fujian provinces. Both men remarked that even though the Chinese population was numerous, the people were weak and unarmed and, greatly resenting their own officials, would welcome a Portuguese invasion. Furthermore, they considered the Chinese military a joke, composed of cowards armed with bows and a few puny iron cannons. Lord Macartney would make similar disparaging remarks in 1793 (chapter 11). Pereira's slightly later account was not quite as aggressive as his compatriots Vieira and Calvo, but he did suggest that the Chinese could easily be converted to Christianity, since they didn't even take their own religion seriously and that their weak disposition might make them susceptible to conquest as well.

It perplexed Pereira that no one in "China" responded to that name for their empire. The Portuguese had picked up this nomenclature from their informants in India, where the name China (Sanskrit: Cīna, Mahācīna) had been in use for over a thousand years. It was ultimately derived from the dynastic name of Qin (221–207 BCE), which was also the root of the Greco-Roman word "Thina," employed to refer to a Far Eastern realm in the *Navigating Around the Red Sea* text (chapter 3). The men Pereira spoke to called their country Da Ming 大明 (Great Ming dynasty). The Portuguese were also confused that they found no signs of the "Cathay" of Marco Polo, but the Spanish Augustinian friar Martín da Rada (1533–1578) later clarified that Cathay and Ming China were one and the same, only that the realm was ruled by a new dynasty which had driven out the Mongols.

Primary Portuguese Source, English Translations, and Selected Studies

Boxer, C. R. "A Portuguese Account of South China in 1549-1552." *Archivum Historicum Societatis Iesu* 22 (1953): 57–92.

Boxer, C. R. *South China in the Sixteenth Century.* Hakluyt, 1953.

Ferguson, Donald. *Letters from Portuguese Captives in Canton, Written in 1534 & 1536.* Bombay, 1902.

Lach, Donald E. *Asia in the Making of Europe.* Vol. 1, *The Century of Discovery.* University of Chicago Press, 1965.

Ravenstein, E. G. *A Journal of the First Voyage of Vasco Da Gama, 1497–1499.* Hakluyt, 1898.

Willes, Richard. *History of Travayle in the West and East Indies.* Richard Iugge, 1577.

Excerpts from "Some Things Known About China," by Galeote Pereira (1561)

Administrative Divisions of China

This land of China is parted into 13 shires, the which sometimes were each one a kingdom by itself, but these many years they have been all subject unto one king. Fuquien [Fujian 福建] is made by the Portuguese the first shire, because there their troubles began, and they had occasion thereby to know the rest. In this shire be eight cities, but one principally more famous than others called Fuquieo [Fuzhou 福州]; the other seven are reasonably great, the best known whereof unto the Portuguese is Cinceo [Quanzhou 泉州], in respect of a certain haven joining thereunto [Xiamen Bay], whither in time past they were wont for merchandise to resort.

Cantão [Guangdong 廣東] is the second shire, not so great in quantity, as well accounted of, both by the king thereof, and also by the Portuguese, for that it lieth nearer unto Malacca than any other part of China, and was first descried by the Portuguese before any other shire in that province; this shire hath in it seven cities. . . .

In each one of these shires be set *ponchasis* [*buzhengshi* 布政使; provincial administration commissioners] and *anchasis* [*anchashi* 按察使; surveillance commissioners], before whom are handled the matters of the other cities. There is also placed in each one a *tutão* [*dutang* 都堂; executive censor], as you would say, a governor, and a *chaçim* [*chayuan* 察院, abbreviation of *duchayaun* 都察院; investigating censors], that is a visitor, as it were, whose office is to go in circuit and to see justice exactly done. By these means so uprightly things are ordered there that it may be worthily accounted one of the best governed lands in all the world. . . .

Agriculture and Animal Husbandry
The country is so well inhabited that not one foot of ground is left untilled; only a small number of cattle have we seen this way—we saw only certain oxen that the countrymen use to plow their ground. One ox draws the plough alone not only in this shire but in other places also, wherein is greater store of cattle. These countrymen by art do that in tillage, which we are constrained to do by force. Here be sold the voidings of close stools [human feces], although there is no lack of the dung of beasts, and the excrements of man are good merchandise throughout all China. The dung-farmers seek in every street by exchange to buy this

dirty ware for herbs [i.e., vegetables] and firewood. The custom is very good for keeping the city clean.

There is great abundance of hens, geese, ducks, swine, and goats, but rams have they none; the hens are sold by weight, and so are all other things. Two pounds of hen's flesh, goose, or duck is worth two *foi* [Chinese *fen* 分] of their money, that is, one penny sterling. Swine's flesh is sold at a penny the pound. Beef bears the same price, for the scarcity thereof; however, northward from Fuquieo [Fuzhou] and farther off from the seacoast, there is beef more plenty and sold better cheap. . . .

It just so happens that the Chinese are the greatest eaters in all the world; they do feed upon all things, especially on pork, which, the fattier it is, the less loathsome to them. The highest price of these things aforesaid, I have set down, better cheap shall you sometimes buy them for the great plenty thereof in this country. Frogs are sold at the same price that is made of hens and are good meat among them, as also dogs, cats, rats, snakes, and all other unclean meats.

Chinese Cities
The cities be very gallant, especially near unto the gates, the which are marvelously great and covered with iron. The gatehouses are built on high with towers, and the lower part thereof is made of brick and stone, proportionally with the walls, from the walls upward the building is of timber, and many stories in it one above the other. The strength of their towns is in the mighty walls and ditches; artillery have they none.

The streets in Chincheo [Quanzhou], and in all the rest of the cities we have seen are very fair, so large and so straight, that it is wonderful to behold. Their houses are built with timber, the foundations only excepted, the which are laid with stone; in each side of the streets are awnings or continual porches for the merchants to walk under. The breadth of the streets is nevertheless such that in them 15 men may ride commodiously side by side. As they ride, they must needs pass under many high arches of triumph that cross over the streets made of timber, and carved diversely, covered with tiles of fine clay; under these arches the merchants do offer their smaller wares, and such as want to stand there are defended from rain and the heat of the sun. The greater gentlemen have these arches at their doors; although some of them be not so mightily built as the rest.

Chinese Scholar Officials

I shall have occasion to speak of a certain order of gentlemen that are called *loutea* [*laodie* 老爹 or *laoye* 老爺; "sire," indicating a scholar-official]. I will first therefore expound what this word signifies. *Loutea* is as much to say in our language as "sir," and when any of them calls for his servant, he answers, "*Loutea*," as we would in Portuguese say, "Senhor"; and as we do say, that the king hath made some "gentlemen," so they say that there is made a "*loutea*." And for that among them the degrees are diverse both in name and office, I will tell you only of some principals, being not able to advertise you of all.

The manner how gentlemen are created *louteas*, and do come to that honor and title, is by the giving of a broad girdle, not like to the rest, and a cap, at the commandment of the king. Moreover, the name *loutea* is more general and common, but everybody agrees that this does not signify an equality of honor. Such *louteas* as do serve their prince in weighty matters for justice are created after trial made of their learning [by civil service exam]; but the others that serve in smaller affairs, as captains, constables, sergeants by land and sea, tax collectors and such, whereof there be in every city, as also in this, very many, are appointed by favor. The chief *louteas* are served while kneeling. . . .

In all cities, not only chief in each shire but in the rest also, are means found to make *louteas*. Many of them do study at the expense of the prince, wherefore at the years end they resort unto the head cities, whither the *chaçims* [investigating censors] do come, as it has been said before, as well to give these degrees, as to sit in judgment over the prisoners.

The *chaçims* go in circuit every year, but such as are to be chosen to the greatest offices meet not but from three years to three years, and that in certain large halls appointed for them to be examined in. Many things are asked them, whereunto if they do answer accordingly, and be found sufficient to take their degree, the *chaçims* by and by grants it to them; but the cap and girdle, whereby they are known to be *louteas*, they wear not before that they be confirmed by the king. Their examination done, and trial made of them, such as have taken their degree usually to be given them with all ceremonies, use to banquet and feast many days together—as the Chinese fashion is to end all their pleasures with eating and drinking—and so remain chosen to do the king service in matters of learning. The other examinates found insufficient to proceed are sent back to their study again. Whose ignorance is perceived to come of negligence and default, such a one is whipped, and sometimes sent to prison, where we, lying that year [in our prison] when this kind of act [of examination] occurred, we found many thus punished, and demanding the cause thereof, they said it was for that they knew not how to answer unto certain things asked them.

It is a wonder to see how these *louteas* are served and feared, in such a way that in public assemblies at one shout they give, all the judicial servants tremble thereat. At their being in these places, when they desire to move, be it but even to the gate, these servitors do take them up and carry them in seats of beaten gold. After this sort are they borne when they go in the city, either for their business abroad, or to see each other at home. For the dignity they have, and office they do bear, they be all accompanied; the very meanest of them all that goes around in these seats is ushered by two men at the least that cry unto the people to give place; however, they need it not, for that reverence the common people have unto them. They have also in their company certain sergeants with their maces either silvered or altogether silver, some two, some four, others six, others eight, conveniently for each one his degree. The more principal and chief *louteas* have going orderly before these sergeants, many others with staves, and a great many bailiffs with rods of Indian canes [bamboo] dragged on the ground, so that the streets being paved, you may hear far off as well the noise of the rods as the voice of the criers. These enforcers serve also to apprehend others, and the better to be known they wear liver-red girdles, and in their caps peacock feathers.

Behind these *louteas* come such as do bear certain tablets hung from the end of staves, wherein is written in silver letters, the name, degree, and office of that *loutea* whom they follow. In like manner, they have borne after them *sombreiros* ["hats"; i.e., sunshades] in agreement with their titles; if the *loutea* be mean [low-ranking], then hath he brought after him but one sunshade, and that may not be yellow; but if he be of the better sort, then may he have two, three, or four [sunshades]. The principal and chief *louteas* may have all their sunshades yellow, the which among them is accounted a great honor. The *loutea* for wars, although he might be mean, may notwithstanding have yellow sunshades. The *tutão* [executive censor] and *chaçims* [investigating censors], when they go abroad, have besides all this before them led three or four horses with their guard in armor.

Furthermore, the *louteas*, and indeed all the people of China, typically eat their meat sitting on stools at high tables as we do, and that very cleanly, although they use neither tablecloths nor napkins. Whatsoever is set down upon the board is first carved before it is brought in; they feed with two sticks, refraining from touching their meat with their hands, even as we do with forks, and for this reason they less do need any tablecloths. . . .

Chinese Religion and Temples

The inhabitants of China be very great idolaters, all generally do worship the heavens; and, as we are accustomed to say, "God knoweth it," so say they at every word, *Tien tautee* [*Tian xiaode* 天曉得], that is to say, "The heavens do know it." Some do worship the sun, and some the moon, as they think good, for none are bound more to one then to another. In their temples, the which they do call *meãos* [*miao* 廟], they have a great altar in the same place as we have, true it is that one may go round about it. There set they up the image of a certain *loutea* of that country [likely,

Confucius], whom they have in great reverence for certain notable things
he did. At the right hand stands the devil [probably a deva-king guardian
figure] much more ugly painted then we do use to set him out, whereunto
great homage is done by such as come into the temple to ask counsel, or
to draw lots; this opinion they have of him, that he is malicious and able
to do evil. If you ask them what they do think of the souls departed, they
will answer that they be immortal, and that as soon as anyone departs
this life, he becomes a devil if he has lived well in this world, if otherwise,
that the same devil changes him into a buffalo, ox, or dog. Wherefor to
this devil they do much honor, to him do they sacrifice, praying him that
he will make them like unto himself, and not like other beasts.

They have moreover another sort of temple, wherein both upon the
altars and also on the walls do stand many idols well proportioned,
but bare headed; these bear the name Omithofom [Amituofo 阿彌陀佛;
Amitabha Buddha], accounted by them as spirits, but such as in heaven
do neither good nor evil, thought to be such men and women as have
chastely lived in this world in abstinence from fish and flesh, fed only
with rice and vegetables. Of that devil they make some account; for these
spirits they care little or nothing at all.

Again, they hold opinion that if a man does well in this life, the heavens
will give him many temporal blessings, but if he does evil, then shall he
have infirmities, diseases, troubles, and penury, and all this without any
knowledge of God. Finally, these people know no other thing than to live
and die, yet because they are reasonable creatures, all seemed good unto
them that we spoke in our language, though it were not very sufficient.
Our manner of praying especially pleased them, and truly they are well
enough disposed to receive the knowledge of the truth. Our Lord grant
for his mercy all things to so be ordered, that it may sometime be brought
to pass that so great a nation as this is, perish not for want of help.

Prospects for Conversion

Our manner of praying was so well liked by them that in prison, they persistently beseeched us to write for them something concerning heaven, which we did to their contentment with such reasons as we knew, however not very cunningly. As they do their idolatry, they laugh at themselves. If at any time this country might be joined in league with the kingdom of Portugal, in such a way that free access were had to deal with the people there, they might all be soon converted. The greatest fault we do find in them is sodomy, a vice very common among the lower classes and not very unusual among the best. If they were able to get rid of this sin, in all other things so well disposed that they are, then a good interpreter in a short space of time might do there great good, if, as I said, the country were joined in league with us. . . .

Chinese Justice

Now I will speak of the manner which the Chinese do observe in doing justice, that it be known how far these gentiles do herein exceed many Christians, even though as Christians we are more bound than they to deal justly and in truth. Because the king of China makes his abode continually in the city of Pachym [Peking], his kingdom is so great, and the shires so many, as was said before. In it therefore the governors and rulers, much like our sheriffs, be appointed so suddenly and speedily discharged again, that they have no time to grow anything [hatch plots]. Furthermore, to keep the state more secure, the *louteas* that govern one shire are chosen out of some other shire distant far off, where they must leave their wives, children and goods, carrying nothing with them but themselves. True it is, that at their coming there they do find in readiness all things necessary, their house, furniture, servants, and all other things in such perfection and plenty that they want nothing. Thus, the king is well served without all fear of treason.

In the principal cities of the shires are four chief *louteas*, before whom are brought all matters of the inferior towns, throughout the whole realm. Diverse other *louteas* have the managing of justice, and receiving of rents, bound to yield an account of these to the greater officers. Others do see that there be no evil rule kept in the city; each one as it is required of him. Generally, all these do imprison evildoers, cause them to be whipped and racked, a thing very usual there, and accounted no shame.

* * * * * * * * * * * * * * * **BRIEF EXCERPT** * * * * * * * * * * * * * * * *

These *louteas* do use great diligence in the apprehending of thieves, so that it is a wonder to see a thief escape away in any city, town, or village. Upon the sea near unto the shore many are taken, and look even as they are taken, so be they first whipped, and afterward laid in prison, where shortly after they all die for hunger and cold. At that time when we were in prison, there died of them more than 70. If happily anyone, having the means to get food, do escape, he is set with the condemned persons, and provided for as they be by the king, in such a way as hereafter it shall be said.

Their whips are certain pieces of canes [bamboo], cleft in the middle, in such sort that they seem rather blunt than sharp. He that is to be whipped lies groveling on the ground; upon his thighs the hangman lays on blows mightily with these canes that the bystanders tremble at their cruelty. Ten stripes draw a great deal of blood; 20 or 30 spoil the flesh altogether; 50 or 60 will require a long time to be healed, and if they come to the number of 100, then are they incurable. And these are given to whoever has nothing with which to bribe the executioner.

The *louteas* observe moreover this: when any man is brought before them to be examined, they ask him openly, within the hearing of as many as be present, even if the case be ever so great. Thus did they also behave themselves with us. Because of this, among them can here be no false witness, as daily among us it so happens, for which causes the lives, farms, and honor of men to be at risk, because it is placed in the hands of a dishonest clerk. This good cometh thereof, that many being always about the judge to hear the evidence, and bear witness, the process cannot be falsified, as it happens sometimes with us. The Moors, gentiles, and Jews have all their sundry oaths, the Moors do swear by their Mushaf [the Qur'an], the Brahmans by their sacred thread, the Jews by their Torah, the rest likewise by the things they do worship. The Chinese, though they often swear by heaven, by the moon, by the sun, and by all their idols, in court, nevertheless, they swear no oaths at all. If for some offence an oath be used of anyone, by and by with the least evidence he is tormented, so also are the witnesses that he brings, if they tell not the truth, or do in any point disagree, except they be men of worship and credit [scholar-officials], who are believed without any further matter, the rest are made to confess the truth by force of torments and whips.

And as for the questioning of witnesses in public, in addition to not trusting one man's oath with the life and death of another, it does another good thing that, as there are always people in these audience chambers who can hear what the witnesses say, nothing can be written other than the truth, and in this way, there is no way to falsify the process, as is done among us, where, as what the witnesses say, only the examiner and the notary know, so great is the power of money, and so on. But in this land, besides this order observed of them in examinations, they do fear so much their king, and he where he keeps his residence keeps them so low, that they dare not once stir, so that these men are unique in carrying out their justice, more than the Romans were or any other kind of people.

Again, these *louteas* as great as they be, notwithstanding the multitude of clerks they have, not trusting any others, do write all great processes and matters of importance themselves. Moreover, one virtue they have worthy of great praise, and that is, being men so well regarded and accounted as though they were princes, yet they be patient above measure in giving audience. We poor strangers brought before them might say what we would, as all to be lies and fallacies that they did write, neither did we stand before them with the usual ceremonies of that country, yet did they bear with us so patiently, that they caused us to wonder, knowing specially how little any advocate or judge is inclined in our country to bear with us. . . .

And as for their being heathens, I do not know a better proof of praising their righteousness than the fact that they respected ours, we being prisoners and foreigners. For wheresoever in any town in Christendom should be accused unknown men as we were, I know not what end the very innocents' cause would have; but we in a heathen country, having our great enemies two of the chief men in a whole town, wanting an interpreter, ignorant of that country's language, did in the end see our great adversaries cast into prison for our sake and deprived of their offices and honor for not doing justice, yea not to escape death; for as the rumor goes, they shall be beheaded—now see if they do justice or not?

Somewhat is now to be said of the laws that I have been able to know in this country, and first, no theft or murder is at any time pardoned; adulterers are put in prison, and the fact once proved, are condemned to die, the woman's husband must accuse them. This order is kept with men and women found in that fault, but thieves and murderers are imprisoned as I have said, where they shortly die for hunger and cold.

* *

Chinese Prisons and Punishment

If anyone happens to escape by bribing the jailer to give him meat, his process goes further and comes to the court where he is condemned to die. Sentence being given, the prisoner is brought in public with a terrible band of men that lay him in irons hand and foot, with a board at his neck one handful broad, in length reaching down to his knees, cleft in two parts, and with a hole one handful downward in the table fit for his neck, the which they enclose up therein, nailing the board fast together; one handful of the board stands up behind the neck. The sentence and cause wherefore the felon was condemned to die is written in that part of the tablet that stands before [see figure 7 in appendix to chapter 11].

This ceremony ended, he is laid in a great prison in the company of some other condemned persons, who are provided for by the king as long as they do live. The board aforesaid so torments the prisoners very much, keeping them both from sleeping and eating, for their hands are manacled in irons under that board, so that there is no remedy but death. . . .

Thieves being taken are caried to prison from one place to another in a chest upon men's shoulders, hired for this purpose by the king. The chest is six handfuls high, the prisoner sits therein upon a bench, the cover of the chest is two boards, amid them both a pillory-like hole for the prisoner's neck; there he sits with his head outside the chest and the rest of his body within, not able to move or turn his head this way or that way nor to pluck it in; the necessities of nature he voids through a hole in the bottom of the chest; the meat he eats is put into his mouth by others. There he stays day and night during his whole journey; if his porters happen to stumble or the chest do jostle or be set down carelessly, it gives great pains to him that sits inside, all such motions being like hanging to him. Thus were our companions carried to this city of Funcheo [Fuzhou], seven days journey, never taking any rest as afterward they told us, and their greatest grief was to stay by the way. As soon as they came, being taken out of the chests, they were not able to stand on their feet, and since they were so badly treated, two of them died a few days later.

. . .

And even though the cities are so large as I say, yet the people are so weak, even though they are not few in number, that with very little one could in this country in a very short time do much service to God and to our Lord the King [of Portugal]. . . .

The Name "China"

We usually call this country "China" and the people "Chinese," but as long as we were prisoners, not hearing among them at any time that name, I determined to learn how they were called and asked sometimes by them thereof, for that they understood us not when we called them "Chinese." I answered them, that all the inhabitants of India called them Chinese, wherefore I prayed them that they would tell me for what reason they are so called, whether perhaps any city of theirs bears that name. Hereunto they always answered me that they have no such name, nor ever had. Then did I ask them what name the whole country bears, and what they would answer being asked of other nations what countrymen they were. It was told to me that in ancient times, this country had been many kingdoms, and though presently it were all under one, each kingdom nevertheless enjoyed that name it first had; these kingdoms are the provinces I spoke of before. In conclusion, they said that the whole country is called Tamen [Da Ming 大明; "Great Ming dynasty"], and the inhabitants Tamenjins [Da Ming ren 大明人; "people of the Great Ming"], so that this name China or Chinese, is not heard of in that country. . . .

Hospitals and Charity Houses

They have moreover one thing very good, and that which made us all to marvel at them being gentiles; namely, that there be hospitals in their cities, always full of people; we never saw any poor body beg. We therefore asked the cause of this, and answered it was that in every city, there is a great circuit, wherein be many houses for poor people, for blind, lame, old folk, not able to travel for age, nor having any other means to live. These folk have in the aforesaid houses, ever plenty of rice during their lives, but nothing else. . . . Besides this, they keep in these places swine and hens, whereby the poor be relieved without going a begging.[3]

APPENDIX TO CHAPTER 9

Wubeizhi Chart of Zheng He's Voyages (1628)

Sometime after the seventh and final voyage of Zheng He, nearly all the logs and charts associated with the enterprise were intentionally suppressed or destroyed, following a conservative shift at court against seaborne trade and diplomacy. At least one nautical chart survived this purge and was handed down by a family of famous military officers and published by Mao Yuanyi 茅元儀 in 1628 in his compendium *Record of Military Preparations* (Wubeizhi 武備志).

The original sailing chart was probably a 5.6-meter-long scroll, but when it was published in book format, it was cut into 40 segments for printing. I have reconstructed the original in three long strips (map 9). The chart was probably compiled by an official cartography office for the seventh voyage (1431–1433 and based on information gathered on previous trips. It was the type of navigation scroll given to individual ship commanders. Containing sensitive military information, it would not have been widely disseminated.

The chart commences at the right in the capital of Nanjing and ends along the east coast of Africa. Some 577 place names and legends are inscribed on the chart (see inset legend). For example, Hormuz Island in Iran is written in Chinese transliteration as Hulumosi 忽魯謨斯. There are also dozens of sailing directions from point to point, using compass navigation for the heading and stellar elevations for latitude calculations. The chart renders the coast as a nearly continuous line from southeastern China, through Southeast Asia, India, the Persian Gulf, and down the coast of Africa. Thus, there is no regular north on the chart, nor is it drawn to a consistent scale.

HYPERLINKED ONLINE EDITION AND SELECTED STUDIES

Ma Huan, *Ying-Yai Sheng-lan: The Overall Survey of the Ocean's Shores*, translated and edited by J. V. G. Mills. Cambridge University Press for the Hakluyt Society, 1970.

Xiang Da 向達. *Zheng He hang hai tu* 鄭和航海圖. Beijing: Zhonghua shuju, 1961.

Barbieri-Low, Anthony J., "Interactive Map of 15th Century Indian Ocean World." https://barbierilow.faculty.history.ucsb.edu/Research/ZhengHeMapZoomify/ZhengHe.htm.

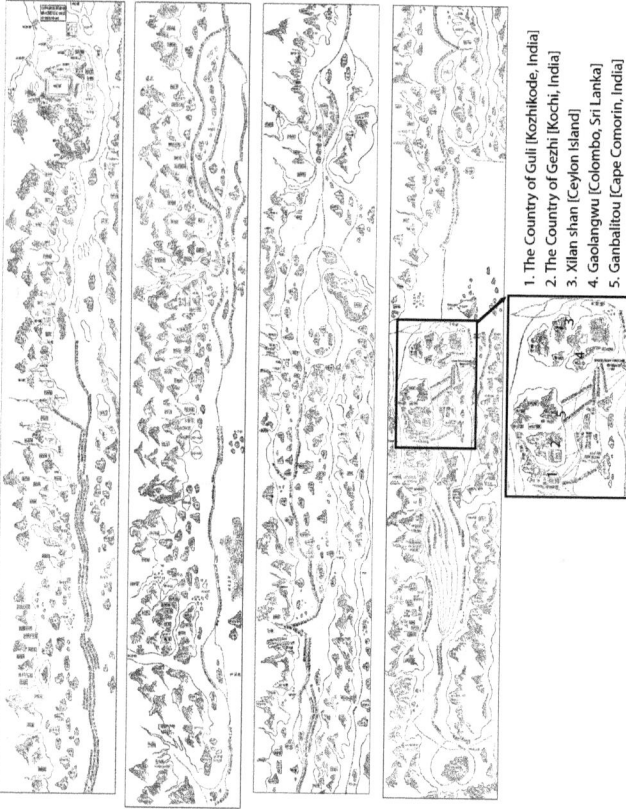

Map 9. The *Wubeizhi* chart of Zheng He's voyages.

1. The Country of Guli [Kozhikode, India]
2. The Country of Gezhi [Kochi, India]
3. Xilan shan [Ceylon Island]
4. Gaolangwu [Colombo, Sri Lanka]
5. Ganbalitou [Cape Comorin, India]

Source: Composite created by the author.

NOTES

1. Feng Chengjun 馮承鈞, *Yingyai shenglan jiaozhu* 瀛涯勝覽校注 (Shangwu yinshuguan, 1935), 1–2, 42–50, 68–72. Translation by the author. Headings from the original.
2. E. G. Ravenstein, *A Journal of the First Voyage of Vasco Da Gama, 1497–1499* (Hakluyt Society, 1898), 131.
3. From Richard Willes, *History of Travayle in the West and East Indies* (Richard Iugge, 1577), 237–243, 245, 246, with updates to modernize the spelling and diction and some redacted passages restored from the oldest Portuguese manuscripts, transcribed in C. R Boxer, "A Portuguese Account of South China in 1549–1552," *Archivum Historicum Societatis Iesu* 22 (1953): 57–92. Headings added by the author.

CHAPTER 10

GATHERING EMPIRICAL KNOWLEDGE FOR THE EARLY MODERN WORLD (1610–1820)

In contrast to the illuminating but superficial accounts of foreign lands from the 15th and 16th centuries by men like Ma Huan and Galeote Pereira (chapter 9), the travelers in this chapter lived among foreigners for decades, were fluent in their spoken dialects and written idioms, and mingled among elites and commoners. Jesuit missionaries like Matteo Ricci produced voluminous journals and corresponded regularly with audiences in Europe, providing detailed empirical knowledge about China and its culture and governance. To support their mission, the Jesuits presented a positive image of China, which encouraged a period of idealization of China in European minds that persisted until the end of the 18th century. Some Chinese also sojourned for prolonged periods in Europe during this period, but one of the only accounts to survive is from a Cantonese sailor named Xie Qinggao, who sailed the world with the Portuguese and visited several countries in Europe.

INTRODUCTION TO MATTEO RICCI

Matteo Ricci (1552–1610), born in Macerata, Italy, became a Jesuit novice in 1571. He sailed first to Goa in 1578 then transferred to Macau in 1582. He aided the pioneer of the Jesuit mission in China, Michele Ruggieri (1543–1607) in his early efforts in Macau and Zhaoqing, in Guangdong, and later took over full responsibilities when Ruggieri was summoned back to Rome in 1588. Ruggieri had employed a method of cultural accommodation wherein the Jesuits dressed like Buddhist monks from India and used some Buddhist terminology in their earliest Christian tracts in Chinese. This led to their being mistaken for Buddhist monks, who were not well respected by Chinese elite society.

Ricci was overtly hostile to Buddhism, so at the suggestion of his disciple Qu Rukui 瞿汝夔, he modified Ruggieri's methodology, growing his hair and beard long and dressing like a Confucian scholar. This new look, and his mastery of the Chinese spoken and written languages, enabled his entrance into elite Ming society, but it was the attraction of Western mathematics, memory methods, and cartography (see appendix to chapter 10) that really lured the Chinese literati to Ricci's door. Ricci hoped that once he attracted these Chinese scholars with Western science, he could slowly introduce them to Christianity, or at least a redacted version of it. While the Virgin Mary found easy acceptance (since she was conflated with the Buddhist goddess of mercy, Guanyin), the dying Christ on the cross was a shocking image that once got Ricci thrown in jail by a eunuch who suspected Ricci was using voodoo to curse the emperor. Ricci was the author of several widely read tracts in Chinese on friendship, Christianity, and memory methods and was eventually accorded the signal honor by the emperor of a burial in Beijing.

While earlier Portuguese and Spanish authors had remarked on such subjects before, Matteo Ricci provided the first real deep dive into issues like the nature of the Chinese spoken and written languages, woodblock printing, handicrafts, filial piety, mourning regulations, foot binding, infanticide, the civil-service exam system, and the structure of the

Ming government. Though Ricci criticized some social practices, like homosexuality and infanticide, as evil, he was careful not to denounce Chinese ancestor worship or sacrifices to Confucius as pagan religious practices, considering them to be secular rituals of veneration. This was an accommodation he made to enable Christian conversion, but this eventually led to the Chinese Rites Controversy that embroiled the Jesuit order. Ricci also claimed in his memoirs that China was not an aggressive empire that sought to conquer neighboring countries. He frames this as an explicit criticism of imperialist countries in Europe.

When Matteo Ricci died of illness and exhaustion in Peking in 1610, his colleagues found in his desk a long manuscript of his memoirs on the history of the mission and other observations on China. This was taken back to Europe by fellow Jesuit Nicolas Trigault, who edited and translated them into Latin. When this was published in 1615 as *De Christiana expeditione apud Sina* (The Christian expedition to China), it became the definitive work on China for an educated audience for over a century.

PRIMARY SOURCE AND SELECTED STUDIES

Hsia, R. Po-chia. *A Jesuit in the Forbidden City: Matteo Ricci, 1552–1610.* Oxford University Press, 2010.

Hsia, R. Po-chia. *Matteo Ricci and the Catholic Mission to China, 1583–1610.* Hackett, 2016.

Spence, Jonathan D. *The Memory Palace of Matteo Ricci.* Viking Penguin, 1984.

Ricci, Matteo. *Storia dell'introduzione del Cristianesimo in Cina.* Edited by Pasquale M. d' Elia. 3 vols. Rome: La Libreria dello Stato, 1942–1949.

Ricci, Matteo. *Lettere: 1580–1609.* Edited by Piero Corradini and Francesco D'Arelli. Quodlibet, 2001.

Trigault, Nicolas. *China in the Sixteenth Century: The Journals of Matthew Ricci, 1583–1610.* Random House, 1953.

EXCERPTS FROM *THE JOURNALS OF MATTHEW RICCI* (1615)

We have been living here in China for well-nigh thirty years and have traveled through its most important provinces. Moreover, we have lived in friendly intercourse with the nobles, the supreme magistrates, and the most distinguished men of letters in the kingdom. We speak the native language of the country, have set ourselves to the study of their customs and laws and finally, what is of the highest importance, we have devoted ourselves day and night to the perusal of their literature. These advantages were, of course, entirely lacking to writers who never at any time penetrated into this alien world. Consequently, they were writing about China, not as eyewitnesses but from hearsay and depending upon the trustworthiness of others. . . .

4. Concerning the Mechanical Arts among the Chinese

It is a matter of common knowledge, borne out by our own experience, that the Chinese are a most industrious people. . . It should be noted that because these people are accustomed to live sparingly, the Chinese craftsman does not strive to reach a perfection of workmanship in the object he creates, with a view to obtaining a higher price for it. His labor is guided rather by the demand of the purchaser who is usually satisfied with a less finished object. Consequently, they frequently sacrifice quality in their productions, and rest content with a superficial finish intended to catch the eye of the purchaser. . . .

Chinese architecture is in every way inferior to that of Europe with respect to the style and the durability of their buildings. In fact, it is dubious just which of these two qualities is the weaker. When they set about building, they seem to gauge things by the span of human life, building for themselves rather than for posterity. Whereas Europeans in accordance with the urge of their civilization seem to strive for the eternal. . . .

The art of printing was practiced in China at a date somewhat earlier than that assigned to the beginning of printing in Europe, which was about 1405. It is quite certain that the Chinese knew the art of printing at least five centuries ago, and some of them assert that printing was known to their people before the beginning of the Christian era, about 50 BCE. Their method of printing differs widely from that employed in Europe, and our method would be quite impracticable for them because of the exceedingly large number of Chinese characters and symbols. At present they cut their characters in a reverse position and in a simplified form, on a comparatively small tablet made for the most part from the wood of the pear tree or the apple tree, although at times the wood of the jujube tree is also used for this purpose.

Their method of making printed books is quite ingenious. The text is written in ink, with a brush made of very fine hair, on a sheet of paper which is inverted and pasted on a wooden tablet. When the paper has become thoroughly dry, its surface is scraped off quickly and with great skill, until nothing but a fine tissue bearing the characters remains on the wooden tablet. Then, with a steel graver, the workman cuts away the surface following the outlines of the characters until these alone stand out in low relief. From such a block a skilled printer can make copies with incredible speed, turning out as many as fifteen hundred copies in a single day. Chinese printers are so skilled in engraving these blocks, that no more time is consumed in making one of them than would be required by one of our printers in setting up a form of type and making the necessary corrections.

This scheme of engraving wooden blocks is well adapted for the large and complex nature of the Chinese characters, but I do not think it would lend itself very aptly to our European type which could hardly be engraved upon wood because of its small dimensions.

Their method of printing has one decided advantage, namely, that once these tablets are made, they can be preserved and used for making changes in the text as often as one wishes. Additions and subtractions can also be made as the tablets can be readily patched. Again, with this method, the printer and the author are not obliged to produce here and now an excessively large edition of a book but are able to print a book in smaller or larger lots sufficient to meet the demand at the time... The simplicity of Chinese printing is what accounts for the exceedingly large numbers of books in circulation here and the ridiculously low prices at which they are sold. Such facts as these would scarcely be believed by one who had not witnessed them. . . .

5. Concerning the Liberal Arts, the Sciences, and the Use of Academic Degrees Among the Chinese

. . .

First, a few words about Chinese writing in general, in which they employ ideographs resembling the hieroglyphic figures of the ancient Egyptians. In style and composition their written language differs widely from the language used in ordinary conversation, and no book is ever written in the colloquial idiom. A writer who would approach very close to the colloquial style in a book would be considered as placing himself and his book on a level with the ordinary people. Strange to say, however, that in spite of the difference that exists between the elegant language which is employed in writing and the ordinary idiom used in everyday life, the words employed are common to both languages. The difference between the two forms is therefore entirely a matter of composition and of style. All Chinese words, without exception, are monosyllabic. I have never encountered a dissyllabic or a polysyllabic word, although a number of words may have two or even three vowel sounds, some of which may be diphthongs.

When I speak of diphthongs I have in mind our European nomenclature. The Chinese are not accustomed to speak of vowels and consonants because every word, just as every object, is represented by its own ideograph, or symbol, used to represent a thought. The number of ideographs is, therefore, equal to the number of words, and the unit of diction is not the word but the syllable. . . .

Although every object has its own appropriate symbol, the symbols do not number more than seventy or eighty thousand in all, because of the manner in which many of them are compounded. When one has acquired a knowledge of about ten thousand of these symbols, he has reached the point in his education where he is ready to begin to write. This is about the least number required for intelligent writing. There probably is no one in the entire kingdom who has mastered all the symbols or has what might be styled a complete ideographic knowledge of the Chinese language. Many of the symbols have the same sound in pronunciation, though they may differ much in written form and also in their signification. . . .

One could not write a book in Chinese from dictation, nor could an audience understand the contents of a book being read to them unless each listener had a copy of the book before his eyes. The meanings of different written symbols having the same sound cannot be determined by the ear, but the forms of the symbols and consequently their meaning can be differentiated by the eye. In fact, it happens not infrequently that those who are conversing together do not fully and accurately understand one another's ideas even though they enunciate very clearly and concisely. At times they have to repeat what they have said, and more than once, or even to write it. If no writing material is at hand, they will trace the symbol on something with water, or perhaps write it with the finger in the air or even on the palm of the listener's hand. . . .

* * * * * * * * * * * * * * * * **BRIEF EXCERPT** * * * * * * * * * * * * * * * *

6. The Administration of the Chinese Commonwealth

. . . From time immemorial the monarchial government was the only one approved by the Chinese people. Aristocracy, democracy, plutarchy, or any other such form was not even known by name.

. . .

The extent of their kingdom is so vast, its borders so distant, and their utter lack of knowledge of a trans-maritime world is so complete that the Chinese imagine the whole world as included in their kingdom. Even now, as from time beyond recording, they call their Emperor, Thiencu [Tianzi 天子], the Son of Heaven, and because they worship Heaven as the Supreme Being, the Son of Heaven and the Son of God are one and the same. In ordinary speech, he is referred to as Hoamsi [Huangdi 皇帝], meaning supreme ruler or monarch, while other and subordinate rulers are called by the much inferior title of Guam [Wang 王, "king"] . . .

Though we have already stated that the Chinese form of government is monarchical, it must be evident from what has been said, and it will be made clearer by what is to come, that it is to some extent an aristocracy. Although all legal statutes inaugurated by magistrates must be confirmed by the King in writing on the written petition presented to him, the King himself makes no final decision in important matters of state without consulting the magistrates or considering their advice. . . . I can assert the following as certain because I have made a thorough investigation of it, namely: that the King has no power to increase a monetary grant to anyone, or to confer a magistracy upon anyone, or to increase the power thereof, except on request of one of the magistrates. One should not conclude from this, however, that the King of his own authority cannot make awards to those who are connected with his household. . . .

To begin with, it seems to be quite remarkable when we stop to consider it, that in a kingdom of almost limitless expanse and innumerable population, and abounding in copious supplies of every description, though they have a well-equipped army and navy that could easily conquer the neighboring nations, neither the King nor his people ever think of waging a war of aggression. They are quite content with what they have and are not ambitious of conquest. In this respect they are much different from the people of Europe, who are frequently discontent with their own governments and covetous of what others enjoy. While the nations of the West seem to be entirely consumed with the idea of supreme domination, they cannot even preserve what their ancestors have bequeathed them, as the Chinese have done through a period of some thousands of years.

. . .

The Chinese will not permit a foreigner to live at large within the confines of the kingdom if he has any intention of ever leaving it or if he has any communication with the outside world. Under no conditions will they permit a stranger to penetrate to the interior of the country. I have never heard of a law to this effect, but it seems quite clear that this custom has developed through the ages from an innate fear and distrust of outside nations. This suspicion exists not only of people who live overseas or at a great distance and are practically unknown to the Chinese. They are also suspicious of friendly as well as of enemy aliens, and even of those with whom they trade, such as the neighboring Koreans, who make use of Chinese laws. . . . If a foreigner should get into China secretly, he would not be put to death or kept in slavery, but he would be prevented from leaving China, lest he should stir up excitement outside to the detriment of the Chinese Government. Hence, the severest punishments are meted out to those who deal with outsiders without the direct consent of the sovereign.

* *

7. Concerning Certain Chinese Customs
. . .

A few words about Chinese banquets, which are both frequent and very ceremonious. With some, in fact, they are of almost daily occurrence, because the Chinese accompany nearly every function, social or religious, with a dinner and consider a banquet as the highest expression of friendship. After the fashion of the Greeks, they speak of drink meetings rather than of banquets and not without reason, too, because, although their cups do not hold more wine than a nutshell, the frequency with which they fill them makes up for their moderate content. They do not use forks or spoons or knives for eating, but rather polished sticks, about a palm and a half long, with which they are very adept in lifting any kind of food to their mouths, without touching it with their fingers. The food is brought to the table already cut into small pieces, unless it be something that is soft, such as cooked eggs or fish and the like, which can be easily separated with the sticks. Their drinks, which may be wine or water or the drink called Cia [*cha* 茶, tea], are always served warm, and this is so even in the hot summer. The idea behind this custom seems to be that it is more beneficial for the stomach and, generally speaking, the Chinese are longer-lived than Europeans and preserve their physical powers up to seventy or even eighty years of age. The custom might also account for the fact that they never suffer from gallstones, so common among the people of the West who are fond of cold drinks. . . .

Chinese books on morals are full of instructions relative to the respect that children should pay to parents and elders. Certainly, if we look to an external display of filial piety, there is no people in the whole world who can compare with the Chinese, as witness the following illustrations. It is a solemn custom, and very strictly observed, that when children sit in the presence of their elders, they are seated to one side and somewhat to the rear, and so for pupils in presence of their teachers. Children are always taught to be respectful in conversation. Even the very poor will work hard to support their parents in plenty to the end of their days. There really is nothing in which these people are more religiously scrupulous than in their devotion to the details of parental funeral rites, in wearing mourning garments, which are white rather than black, and in furnishing a casket or a funeral bier of costly material. In general one would say that their obsequies are too pompous and frequently surpass their means. . . . Three years of mourning, from the day of the death of a father or mother, is the inviolable custom. The reason assigned for this, as one reads in their books, is to repay the three years of their children's infancy, when the parents carried them in their arms and worked so hard to rear them. For the death of other relatives the length of time of mourning is determined by the degree of consanguinity, varying from a year to three months. . . .

It is a custom also, in imitation of their minor sacrifices to the idols, to burn pieces of paper or of white silk. By doing this they imagine that they are offering a robe to the departed in memory of his kindness and generosity. Their caskets can be rendered absolutely airtight by sealing the seams with a certain kind of shiny pitch, and at times the Chinese have been known to keep the body of a parent in the house for three or four years. As long as the remains are kept they place food and drink before the coffin just as they might serve a living person, and during that time the children sit on low stools covered with white cloth rather than on regular chairs, and straw mats are placed upon the floor, beside the casket, to be used in place of beds. They eat neither meat nor seasoned food during the period of mourning, use very little wine, take no baths, attend no festivities, refrain from marital rights, and for some months they do not even appear in public. . . .

8. Concerning Dress and Other Customs and Peculiarities

The Chinese people are almost white, though some of them in the southern provinces are quite dark because of their proximity to the torrid zone. The men's beards are thin and meager and at times they have none at all. Their hair is rough and straight and the moustache late in showing, so that a man of thirty would compare with one of ours of twenty, in that respect. The beard and the head hair are universally black. Red hair they dislike. Their narrow, elliptical-shaped eyes are noticeably black. The nose is small and flat, and their ears are of medium size. . . .

The women are all small in stature, and the smallness of their feet is considered to be a mark of beauty. In order to produce this effect their feet are tightly wrapped in bandages from infancy to prevent growth, and when they walk one would think that their feet had been partly cut off. These foot bandages are worn throughout the whole course of life. They look upon it as unbecoming for women to walk about the streets. Probably one of their sages hit upon this idea to keep them in the house. . . .

9. Concerning Certain Rites, Superstitious and Otherwise

. . .

We shall add here a few shocking practices which the Chinese look upon with indifference and which, God forbid, they even seem to consider as quite morally correct. . . . Many of them, not being able to forgo the company of women, sell themselves to wealthy patrons, so as to find a wife among his women servants, and in so doing, subject their children to perpetual slavery. Others buy a wife when they can save money enough to do so, and when their family becomes too numerous to be supported, they sell their children into slavery for about the same price that one would pay for a pig or a cheap little donkey—about one crown or maybe one and a half. Sometimes this is done when there is really no necessity, and children are separated from their parents forever, becoming slaves to the purchaser, to be used for whatever purpose he pleases. The result of this practice is that the whole country is virtually filled with slaves; not such as are captured in war or brought in from abroad, but slaves born in the country and even in the same city or village in which they live. Many of them are also taken out of the country as slaves by the Portuguese and the Spaniards. These few at least have an opportunity of becoming Christian and of thus escaping the slavery of Satan. . . .

A far more serious evil here is the practice in some provinces of disposing of female infants by drowning them. The reason assigned for this is that their parents despair of being able to support them. At times this is also done by people who are not abjectly poor, for fear the time might come when they would not be able to care for these children and they would be forced to sell them to unknown or to cruel slave masters. Thus, they become cruel in an effort to be considerate. This barbarism is probably rendered less atrocious by their belief in metempsychosis, or the transmigration of souls. Believing that souls are transferred from one body that ceases to exist into another that begins to exist, they cover up their frightful cruelty with a pretext of piety, thinking that they are doing the child a benefit by murdering it. According to their way of thinking, they are releasing the child from the poverty of the family into which it was born, so that it may be reborn into a family of better means. So it happens that this slaughter of the innocents is carried on not in secret but in the open and with general public knowledge. . . .

Yet another barbarity common in the northern provinces is that of castrating a great number of male children, so they may act as servants or as slaves to the King. This condition is demanded for service in the royal palace, so much so, indeed, that the King will have no others nor will he consult with or even speak to any other. Almost the whole administration of the entire kingdom is in the hands of this class of semi-men, who number nearly ten thousand in the service of the royal palace alone. They are a meager-looking class, uneducated and brought up in perpetual slavery, a dull and stolid lot, as incapable of understanding an important order as they are inefficient in carrying it out.

10. Religious Sects Among the Chinese

Of all the pagan sects known to Europe, I know of no people who fell into fewer errors in the early ages of their antiquity than did the Chinese. From the very beginning of their history, it is recorded in their writings that they recognized and worshipped one supreme being whom they called the King of Heaven, or designated by some other name indicating his rule over heaven and earth. It would appear that the ancient Chinese considered heaven and earth to be animated things and that their common soul was worshipped as a supreme deity. As subject to this spirit, they also worshipped the different spirits of the mountains and rivers, and of the four corners of the earth. They also taught that the light of reason came from heaven and that the dictates of reason should be harkened to in every human action. Nowhere do we read that the Chinese created monsters of vice out of this supreme being or from his ministering deities, such as the Romans, the Greeks, and the Egyptians evolved into gods or patrons of the vices. . . .

The most common ceremony practiced by all the Literati, from the King down to the very lowest of them, is that of the annual funeral rites, which we have already described. As they themselves say, they consider this ceremony as an honor bestowed upon their departed ancestors, just as they might honor them if they were living. They do not really believe that the dead actually need the victuals which are placed upon their graves, but they say that they observe the custom of placing them there because it seems to be the best way of testifying their love for their dear departed. . . . This practice of placing food upon the graves of the dead seems to be beyond any charge of sacrilege and perhaps also free from any taint of superstition, because they do not in any respect consider their ancestors to be gods, nor do they petition them for anything or hope for anything from them. However, for those who have accepted the teachings of Christianity, it would seem much better to replace this custom with alms for the poor and for the salvation of souls.[1]

INTRODUCTION TO XIE QINGGAO

Since the Portuguese accounts of the 16th century (chapter 9), there had been a steady stream of increasingly detailed and largely reliable European accounts of China written by missionaries, merchants, soldiers, and diplomats. But there had been no surviving account of Western Europe written by a Chinese-born traveler since the testimony of Rabban Sauma in the 13th century (chapter 7). Jesuit missionaries had brought many Chinese converts back to Europe during the 17th and 18th centuries to study theology or philosophy or to train as missionaries, but unfortunately, almost none of these men has left us a detailed account of their lives or travels. An exception is Li Zibiao 李自標 (Jacobus Li) who served as interpreter to Lord Macartney on his embassy to the court of the Qianlong Emperor in 1793 (chapter 11). Li had written letters to his friends and colleagues in Naples during his travels, but none to his family in China.

It was also during the late 18th century that a new Chinese window onto the Western world was opened by the account of the observant Cantonese sailor, Xie Qinggao 謝清高 (1765–1821). Born in Guangdong, Xie had acquired some Confucian education, but not being very proficient, he left at age 14 to pursue maritime commerce. When his vessel capsized in the South China Sea, he was rescued by a European ship (probably Portuguese) and decided to join its crew. He spent the years between 1779 and 1792 traversing the world aboard trading vessels, visiting the principal ports of Southeast Asia, South Asia, Europe, and the Americas. He began to lose his eyesight so returned to China and settled in the Portuguese colony of Macau, where he supported himself as an interpreter. Archival documents confirm his financial dealings with Europeans in Macau, referring to him as "blind Qing."

More than two decades later, in 1820, a man from his hometown named Yang Bingnan 楊炳南 came to visit Xie. Yang was fascinated by Xie's tales of the various foreign lands he had visited, so he wrote down these accounts, further editing and embellishing them with information from other books, to compile the work known to us as the *Maritime Records* (Hailu 海錄). The account languished in relative obscurity until shortly after the First Opium War (1839–1842), when Commissioner Lin Zexu 林則徐 (1785–1850) promoted it for learning about foreign countries, which he viewed as crucial for Chinese self-strengthening and national preservation.

Xie Qinggao's *Maritime Records* resides firmly in the traditional cataloging genre of Chinese accounts of foreign lands. For each country, Xie provides a record of its location and geography, climate, customs, religion, dress, and natural products. Occasionally, he will talk more extensively about local customs if they are strikingly different from those of China, such as the monogamy, kissing of relatives, or disposal of skeletons in ossuaries that he notices in Portugal. Xie deviates from the tradition, however, when he writes of Western technical knowledge, for he marveled at the potential of mechanical systems like the freshwater mains of London.

As with earlier Chinese geographic works, Xie begins his cataloging account with the countries in Southeast Asia then ventures farther afield to South Asia and other parts of the Indian Ocean world. The final part of the book gives an account of the Great Western Ocean (the Atlantic World), from where the following excerpts are drawn. Xie clearly spent considerable time in Portugal and knew the language, and he sojourned some time in London, but his account of the newly founded United States of America is quite meager and was probably based on second-hand information.

Even though Xie did admire some aspects of Western technology and governance, his *Maritime Records* still presents a Sinocentric view of the world. That view would be shattered by defeats in the two Opium Wars, which would lead to the strikingly different accounts of Europe by Chinese diplomats like Guo Songtao (chapter 11) and scholars like Wang Tao (chapter 12), who sought to learn the secrets of Western power and wanted to reform China.

PRIMARY SOURCE, ALTERNATE ENGLISH TRANSLATIONS, AND SELECTED STUDIES

Ch'en, Kenneth. "Hai-lu 海錄: Fore-Runner of Chinese Travel Accounts of Western Countries." *Monumenta Serica* 7, no. 1/2 (1942): 208–226.

Caltonhill, Mark, trans. "Selections from *Jottings of Sea Voyages*." *Renditions* 53–54 (Spring and Autumn, 2000): 159–163.

Feng Chengjun 馮承鈞. *Hailu zhu* 海錄注. Shangwu yinshuguan, 1938.

Po, Ronald C. "Writing the Waves: Chinese Maritime Writings in the Long Eighteenth Century." *American Journal of Chinese Studies* 22, no. 2 (2015): 343–362.

Xie Qinggao 謝清高 and Yang Bingnan 楊炳南, *Hailu jiaoshi* 海錄校釋. Shangwu yinshuguan, 2002.

Excerpts from *Maritime Records*, by Xie Qinggao (1820)

Portugal

The Kingdom of the Great Western Ocean is also called Boulouhgēisih [Português; Portugal]. The climate is much colder than that of Fujian or Guangdong. From Saan Dēlēi [St. Helena Island, in the south Atlantic], sailing due north for about 20 days, one arrives at its border. Her great southern-facing port is protected by two imposing bastions [probably São Julião da Barra Fort and São Lourenço do Bugio Fort], which are called the "interlocking forts," outfitted with 400 or 500 large bronze cannon and garrisoned by 2,000 soldiers. Whenever any Portuguese oceangoing vessels return home or when ships from other countries arrive here, the authorities first send someone to inspect the crew for any signs of smallpox. If any are present, the ship is not allowed to enter the harbor and must wait until all evidence of the disease has cleared up before it can dock. There are seven major towns in Portugal, analogous to seven prefectural seats in China. After passing by the interlocking forts and into the harbor, one sails a distance of several tens of Chinese miles and arrives at Yuhjaiwōa [Lisboa; Lisbon]. This is a large city, and the king has established his capital here. The city is guarded by bastions and citadels, but there are no encircling walls. From here you travel into the interior and arrive at Gāmbahla [Coimbra], which is also a major town. Many of the [Jesuit missionaries] who enter China to serve as directors of the Bureau of Astronomy, as well as those who arrive in Macau to act as high-ranking monks, are from this region. Further into the interior, there is Wōdaaht [either Évora or Guarda]. Then there is Wàihdīu [Viseu]. The others include Loihlōu [either Faro or Leiria], Alaga [Brage], and Jābéi [Chaves]. These are all large towns that are densely inhabited, and boats and carriages converge on them likes spokes of a wheel. Each houses a heavily armed garrison.

The inhabitants have a white complexion and are fond of cleanliness. They all live in multistoried dwellings. Their vessels and utensils are exquisitely well made and are traditionally white in color as well. All walls and buildings are typically whitewashed, and when they become just a little faded, they plaster them again. Their womenfolk also consider that those with the whitest skin are the most noble.

The king of the country is titled *lēi* [*rei*], and the heir-apparent is titled *làihfāandē* [*infante*], while the king's other sons are called *péilàhmsāibéi* [*principes*] and his daughters *péilàhmsōsí* [*princesas*]. Their prime minister is called *gōndē* [*conde*; i.e., count] and the general-in-chief is titled *malagēija* [*marechal*; i.e., marshal]. . . .

Each year, Portugal sends out one civil and one military official to the overseas possessions it controls to aid in administration; for the larger territories, they send three or four men. If there is a major problem, these six men [the two sent from Portugal and the four local officials] consult together on a solution. If those officials sent from the home country have not yet brought their dependent family members, then they must wait for the mayor and the other three local officials to discuss the matter, and only if it accords with local sentiments and customs will it be allowed. The delegated officials [from Portugal] are not able to act arbitrarily. If all the delegated officials already have family members with them, then people will obey the leader of the delegated officials, and the local officials will usually not dispute this. They will just consider this their shared problem.

Men wear short upper clothes and trousers down below, and both are extremely tight fitting, so that they really restrict the body. On special occasions, they don an extra piece of clothing over the shirt that is short in the front but long in the rear that looks like the wings of a cicada. Chief officials have something attached to each shoulder [epaulets] which resembles a bottle gourd; the gold ones are the most valued, with the silver ones coming second. Their hats are round with vertical sides and a flat top and a brim all around.

Women also wear short, tight-fitting clothing on their upper body but do not wear trousers. Instead, they wear flouncy skirts with as many as eight or nine layers. Poor women have clothing made of cotton, while the wealthier ones wear silk, but in each case, they consider the thinnest and lightest cloth to be the best. Younger women expose their breasts, while older women cover them up. When women go out of the house, they always wear a broad veil that covers the head and hangs down to the knees [in the back]. Wealthy women furthermore cover their faces with a veil of black silk gauze, so fine and delicate that from a distance, it looks like a whisp of smoke. Some of these are worth as much as 20 gold pieces. Many people play with strings of beads they hold in their hands [rosaries], and for the rich these are often made of pearls or diamonds. Men and women both wear leather shoes.

From the monarch down to commoners, no one has two wives. Only when a wife dies can the husband marry again. And when a husband dies, the wife can also remarry. When those families with daughters are selecting a prospective son-in-law, the man's family first inspects the size of the girl's trousseau, and only if they are satisfied will they approve the match. Parents consider it very shameful if their adult daughters remain unmarried, so even if they have to exhaust the family's resources [to provide a dowry], they will not hesitate to do so. And yet people hardly regard it as important whether a man has a wife or not. Marriage is not forbidden to people with the same surname; only full brothers and sisters can't get married. Widows who want to get remarried are even allowed to wed an uncle or nephew. All close relatives who want to get married must first go to a priest and seek permission to wed. Only when the priest has approved the match can a wedding commence. The Catholic priest is like the abbot of a Buddhist temple.

The custom of the people is to uphold the Catholic faith, and in every location, they have established numerous churches. Every seven days, women go together to the church to pray. When marrying a wife, the prospective bride and groom will both arrive at the church to listen to the sermon of the priest and then return home together. For those grooms who are becoming a son-in-law [to a family of higher status], they will return home to live in the bride's natal household. When men and women are discussing marriage, the parents and a matchmaker must first report this to a priest, who will then make a general announcement to let the entire community know. If the potential bride and groom have already made a private pact, this will still be approved if they have made a public declaration of their love. For those couples who have made a public announcement, even if this is based upon a private pact, no one can dispute the potential marriage, even the parents.

Women who have committed adultery or other crimes and who desire to mend their ways will go to the church and make a confession to the priest. During confession, the priest sits in a little niche chamber that has a window on one side. The woman kneels below the window and whispers to the priest, telling him the true facts of the sin. The priest then makes a formulaic statement and tells her that her sins have been forgiven. If the priest were to tell anyone in the community the matters he had heard in confession, he would be considered to have committed a crime and could be hanged. Those men and women who have broken the law or who fear being punished by their fathers can go to the church and make a plea to the priest. If the priest consents and absolves the person of guilt, reporting this in writing to the patriarch of the family, then even if he were extremely angry, the father would not dare to bring up the crime again.

When people die, they are all buried in the church. When there are new arrivals to be buried, they select from the corpses of those who had died previously, remove their skeletons, and discard them in a corner of the church, allowing the new arrival to be buried in their place. Both births and deaths are reported to the church, and a priest makes a record of the family genealogy, although people rarely know their ancestors back more than three generations.

When a new king ascends the throne, they do not change the reign title [and restart the years] but record the year according to the Christian calendar. The start of each year occurs on the seventh day after the winter solstice. They divide each year into 12 months but don't care whether or not the first day of the month falls on a new moon. Some months even have 31 days. They consider that the moon only shines by borrowing the light of the sun. This calendrical method is really inadequate.

Around 50 days or so after the winter solstice, men and women of the country refrain from eating meat, calling it the "vegetarian fast" [Lent], and only end it after 49 days. Three days before the close of the fast [on Good Friday], women everywhere pray at churches, calling it "searching for the ancestor." Three days later [on Easter Sunday], the priest will take all the carved figures of Christ stored in the church and place them in the main nave or even set them up on the side of the streets. Those who are the first to see [the figures of Christ] will announce to all around that they have "found the resurrected one." The following day, the monks and all the soldiers and commoners of the town will send the figures back to the various churches for storage. The bishop will come out to greet the statues, wearing a long vestment that reaches the ground, employing four junior priests to hold up the corners of his train. The bishop stands under a cloth canopy about one Chinese stave [approx. 3.2 meters] in length and five or six Chinese feet [approx. 1.6-1.9 meters] in breadth, held up by four poles at the corners. Four men are selected from wealthy households, each holding one of the poles of the canopy. When the bishop is under this canopy, he grasps a circular mirror in one hand that carries the sign of the cross. Surrounded by an honor guard, he proceeds down the street. Those who witness the procession would kneel at the side of the road, and only when the bishop has passed would they arise.

Among the womenfolk, there are also those who leave their families to become nuns and who reside separately in a temple [a convent] whose gates are locked to the world. Food and clothing all enter the convent through an aperture in the wall, and for the rest of their lives, these nuns will never again leave the convent. In some cases, there have been nuns whose family members were in the service of the king. If her father or mother were guilty of some crime, the nun would write a request for clemency to the throne, and the parents would always receive a pardon, regardless of the severity of the offense.

When a soldier or commoner has an audience with the king or another high official, he first removes his hat outside the door. Then he hastens inside and clasps the boot [of the king or minister] in his hands and kisses it. After this, with his hands hanging at his side, he bows deeply with one foot extended, retreats several paces, then stands upright and speaks without kneeling. When a son greets his father, one who has not seen him for a long time will also remove his hat outside the door then rush inside and clasp his arms around his father's waist. The father will pat the son on the back with both hands, and the two will kiss each other a few times. The son will then bow deeply with one foot extended, retreat several paces, then stand up and speak. If the son had not yet attained manhood, he would not hug the father's waist but will hasten inside, clasp his father's hand, and kiss it; the rest of the ritual is the same. When one visits his mother, it is the mother who will hug the child and kiss him several times on the mouth. The child will then place his hands behind him, bow deeply with foot extended, and retreat as in the other cases. If he sees her frequently, he will nevertheless still place his hands at his back and bow deeply with extended foot like before. When a child is very young, he will greet his parents in the morning and evening, kissing their hands; the rest being like I described before. Greeting a grandfather is like greeting one's father, and greeting a grandmother is like greeting one's mother. Brothers and near relatives who are very close to one another will hug each other after having not seen the other for a long time, then perform a bow. The ritual for greeting someone of an older generation is similar to that for one's father, but they do not kiss each other. If an elder greets someone of a similar age, they will also hug one another, but the one of lower social status will genuflect slightly. When a daughter is greeting her parents or grandparents, if she is a child, then it is like the ritual for a boy; if she is grown, then she will hastily enter, clasp one of their hands and kiss it. After she backs away, she will grasp her skirt with both hands and drop a few curtsies. When she greets her parents-in-law, the ritual will be similar. When close male and female relatives greet one another, the man will make a low bow with one foot extended, while the woman will grasp her skirt with both hands and curtsy a few times, and then they will sit down. When two women greet one another, they will stand facing the other, grasp their petticoats, and make several curtsies, alternating between bending the left or right knee, and then be seated. When [male] friends and relatives meet on the road, each will remove his hat. Men will often venture out accompanied by their wife and children and then return home. When relatives call on one another, the womenfolk will usually sit and keep each other company and talk. When a woman goes out sightseeing, she is usually chaperoned by her husband, father, or some male relative who holds her hand while they walk together. Sometimes, one man will hold hands with two women while they walk all together. This is a broad outline of their rituals and customs.

It is their custom to honor the wealthy and despise the poor. When one's family has great wealth and high social standing, those who are poor, even if they are brothers, uncles, or nephews, would not have the audacity to enter their home and would not presume themselves worthy enough to eat or speak with them.

Their local products include gold, silver, copper, iron, galvanized iron, sal ammoniac, snuff tobacco, timber, fish, grape wine, soap, coarse woolens, camlet cloth, serge, and clocks. The peasants sow mostly wheat, and no rice, and their plows are harnessed to horses.

* * * * * * * * * * * * * * * **BRIEF EXCERPT** * * * * * * * * * * * * * * * *

England

The country of Yīnggātleih [England], the land of the red-haired barbarians, faces across the sea southwest of Fahtlòhnggēi [the Franks, i.e., France]. Sailing from St. Helena Island and traveling north by northwest for about two months, passing the lands of Portugal, Léuihsung ["Greater" Luzon; i.e., Spain], and France, one finally arrives at England. A solitary island in the midst of the ocean, the territory is several thousand Chinese miles in circumference.

The land is sparsely populated, but there are a large number of wealthy families who all reside in multistory buildings. The people are eager for success and greatly esteem profit, making their livelihood from maritime commerce. If there are regions overseas with a potential for profit, these people will all contend for it. Their traders blanket all the oceans, and they consider Mìhngala [Bengal], Maahndaahtlasaat [Madras], and Maahngmáaih [Mumbai] as their overseas possessions. Male commoners 15 and over can be conscripted to provide military service for the king, but those over 60 are no longer liable. Moreover, they maintain a larger army of foreign mercenaries. Consequently, even though the country is small, they can still field a powerful army of over 100,000 soldiers, so many other foreign countries fear them. The main ocean port is called Láahnlèuhn [London]. After you enter the estuary [of the Thames], you sail upstream more than 50 Chinese miles, and you arrive at the place called London, which is one of the largest cities in the country. Various houses and pavilions extend as far as the eye can see. Her forests are lush and verdant, and the residents are populous and affluent. It also serves as the country's capital and is controlled by a major official [the lord mayor of London].

The water [of the Thames] is exceedingly clear and sweet. There are three major bridges over the river, referred to as "the three decorated bridges" [probably Blackfriars Bridge (opened 1769), Westminster Bridge (opened 1750), and Old London Bridge (1209–1831)].

Near each bridge is a system of waterwheels that raise river water and send it flowing through a series of large tin-pipe water mains to all the streets in the city. Persons who desire water don't have to be hassled with transporting it in buckets on shoulder poles, for they need only connect their home to the tin water main in their street with a smaller copper pipe, buried under his house's outer wall. Others use their own smaller water wheels to procure water and cause it to flow into storage vessels. The king collects a water tax, based on the size of the household. The three bridges divide the city into three water districts, and each day, they rotate the water supply between districts. They send persons to travel everywhere to announce to the residents of that district that they are each ordered to get water now. Each family then opens their copper pipe or operates their small water wheel pump, and the water starts to flow automatically into storage containers. Once they have enough supply for three days, they shut off the pipe. That district then shuts down its waterwheel, and the water supply immediately dries up. The next day, they rotate to another district, so in three days, all the districts are supplied. Then the whole rotation begins again.

England's laws are very harsh, and no one dares to commit robbery. This is quite a rare occurrence for a foreign land. The country has many prostitutes. Even bastards born from adultery will be raised to adulthood, and no one would dare to kill or harm them.

Men and women both normally wear white, but during mourning, they garb themselves in black. Their soldiers' uniforms are all bright red. Women wear long dresses that trail on the ground, with a tight-fitting bodice on top and loose petticoats below. The waist is cinched with a belt, for they desire to appear slender. The end of the belt is decorated with a gold clasp called a *bokgūlóuhsíh* [buckle]. Between their two shoulders they wear silk cords that form a net-like decorative pattern, stitched like embroidery atop their dress. When there is some festive occasion where guests are invited to a banquet, [the host] directs beautiful young girls wearing splendid attire to dance and sing songs to accompany it. Such sweet voices and nimble movements; they call this a "dance battle." All the daughters of wealthy and noble families practice these arts from a young age, for this is what the customs of the country consider delightful.

Under their military code, five soldiers form a section, and each section has a leader [a corporal]. Twenty soldiers form a platoon. Military commands are strictly enforced, and no one would dare cower in battle. The soldiers' main weapon is a rifle that can be fired repeatedly; they have no other special skills.

When their vessels are on the high seas for the purpose of trade and they come across a capsized ship, they always deploy a sampan [a dinghy] to rescue the crew. When they have got the men, they provide them with food and drink and sufficient traveling money, enabling each sailor to return to his native country. If a ship captain does not render this assistance, he is liable for punishment. This is an indication of their good governance. Their remaining social customs are largely similar to those of the other countries of the Western Ocean.

The local products include gold, silver, copper, tin, lead, iron, galvanized iron, rattan, coarse woolens, serge, camlet cloth, clocks, glass, and Dutch-style grain alcohol. They don't have any tigers, leopards, or David's deer. . . .

* *

America

The country of Mēléihgōn [America] is located west of England. From the island of St. Helena, you travel west-northwest for about two months; you then sail due west from England and can arrive in America in about 10 days or so. It is also an isolated island in the midst of the ocean. The territory [of the recently created United States] is rather narrow. It used to be an enfeoffed territory [a colony] of England, but now it is an independent country. The customs of America are the same as those in England. This is called the land of the "flowery flag" [the Star-Spangled Banner] by the Cantonese.

For transportation in and out of the country, they often use fire-powered [steam] boats. These ships are outfitted, internally or externally, with rotating wheels on an axle. In the middle of the ship, they have installed a furnace. The fire excites water to rush at the wheels, which then rotate and churn the river water. Without employing any human labor, the ship is propelled forward speedily on its own power. The system is so clever and ingenious, but it almost impossible for people to get a peek at its inner workings. The various countries of the Indian Ocean have sought to emulate this technology.

The countries from Portugal to America are generally called the Great Western Ocean. The people esteem technical skills and craftiness, making their livelihood through maritime commerce. From the king down to commoners, everyone practices monogamy. The hills abound with strange birds and beasts that no one has yet given names to, but they don't have tigers, leopards, or David's deer.[2]

APPENDIX TO CHAPTER 10

Complete Diagram of the Myriad Countries of the Earth (1602)

Matteo Ricci's famous world map, *Complete Diagram of the Myriad Countries of the Earth* (Kunyu wanguo quantu 坤輿萬國全圖), was the first to combine the Western cartographic tradition, new information from the Voyages of Discovery, and traditional Chinese geographic information. It revealed vast new information to learned Chinese, solidifying Ricci's reputation as a great scholar.

This enormous printing was Matteo Ricci's third version of his world map and the only one to survive. His first version was made in Zhaoqing, Guandong, in 1584, not long after Ricci arrived. The local magistrate had seen a world map hanging on Ricci's wall and wanted a Chinese version of it. The production was rudimentary, with limited captions. The second edition was produced in 1600 in Nanjing, responding to a direct request from a middle-level Chinese official, and incorporated many more Chinese annotations as well as meridian lines. The surviving third edition, reproduced here, was printed in Beijing in August of 1602 on six enormous wooden printing blocks, with the collaboration and sponsorship of Ricci's friend, the geographer Li Zhizao 李之藻. Because of its popularity, a fourth edition was published on eight panels in 1603, with further revisions by Ricci. The fame of Ricci's map even reached the Wanli Emperor (r. 1572–1620), who requested a newly printed copy and a virtual audience in 1608.

Ricci's map and its preface informed the Chinese not only the shape and locations of the five continents but that the world was a sphere and could be divided into meridians of latitude and longitude. Some coordinates for Chinese places on his map were based on Ricci's own observations. Being a devout Jesuit, however, Ricci shunned the insights of Copernicus (1473–1543) and still represented the cosmos as being geocentric, following the classical precedent of Claudius Ptolemy. Ricci's base map was copied from Abraham Ortelius's renowned *Theatrum Orbis Terrarum* atlas (1570). Ricci coined Chinese transliterations of many of the names for regions and countries, some of which are still in use today, such as Ouluoba 歐羅巴 ("Europe"). Ricci also moved the prime meridian from the Atlantic to the Pacific, which allowed China to be situated roughly at the center of the map. He notes in his journal that he made this change "out of deference to their ideas." It is noteworthy that some legendary places from the Chinese geographic tradition appear on Ricci's map, including the Country of Women (Nüren Guo 女人國), which Ricci placed east of the Black Sea, in Georgia, corresponding to the Amazon myth. Though Ricci's maps created a splash during the late Ming period, they do not seem to have had a lasting impact on pre-19th century Chinese cartography, which was rather conservative.

Scholarly Publication, Online Editions, and Selected Studies

Asian Art Museum of San Francisco. Interactive Matteo Ricci Map. https://asianart.org/webapps/riccimap/.

Cams, Mario and Elke Papelitzky, eds. *Remapping the World in East Asia: Toward a Global History of the "Ricci Maps."* University of Hawai'i Press, 2024.

Ch'en, Kenneth. "Matteo Ricci's Contribution to, and Influence on, Geographical Knowledge in China." *JAOS* 59, no. 3 (1939): 325–359.

d' Elia, Pasquale M. *Il mappamondo cinese del p. Matteo Ricci, S.I.* 3rd ed. Pechino, 1602. Conservato presso la Biblioteca Vaticana. Vatican Apostolic Library, 1938.

Hostetler, Laura, ed. *Reimagining the Globe and Cultural Exchange: The East Asian Legacies of Matteo Ricci's World Map.* Brill, 2024.

Ricci, Mateo. "Kunyu wanguo quantu (Map of the Ten Thousand Countries of the Earth)." 1602. University of Minnesota Libraries, James Ford Bell Library. https://umedia.lib.umn.edu/item/p16022coll251:8823.

Yee, Cordell. "Traditional Chinese Cartography and the Myth of Westernization." In *The History of Cartography,* edited by David Woodward. University of Chicago Press, 1994.

Map 10a. *Complete Diagram of the Myriad Countries of the Earth* (Kunyu wanguo quantu 坤輿萬國全圖).

1) Europe has over thirty countries all of which are monarchies and do not embrace any other religion but Catholicism. Officials are divided into three classes; the first tends to matters of religion; the second, temporal affairs; and the third, matters pertaining to the army. All kinds of grains, fruits, and metals are produced. Wine is made from grapes. The workers are skillful and clever, while the people are well versed in astronomy and philosophy. In their daily activities they are solid and honest, and have high regard for the five relationships. Modes of production are plentiful. Princes and ministers are prosperous and healthy. Communications are kept up with foreign countries at all times, while her merchants roam over the entire earth. She is about 80,000 Chinese miles away from China. No relations existed between the two regions until about seventy years ago.

Source: Matteo Ricci, with collaboration of the printer Zhang Wentao and the scholar Li Zhizao (August 17, 1602). Height 1.19 m, width 4.14 m. Reprinted with permission of James Ford Bell Library, University of Minnesota. Caption translation by Kenneth Ch'en, with modifications by the author.

Map 10b. *Complete Diagram of the Myriad Countries of the Earth* (Kunyu wanguo quantu 坤輿萬國全圖).

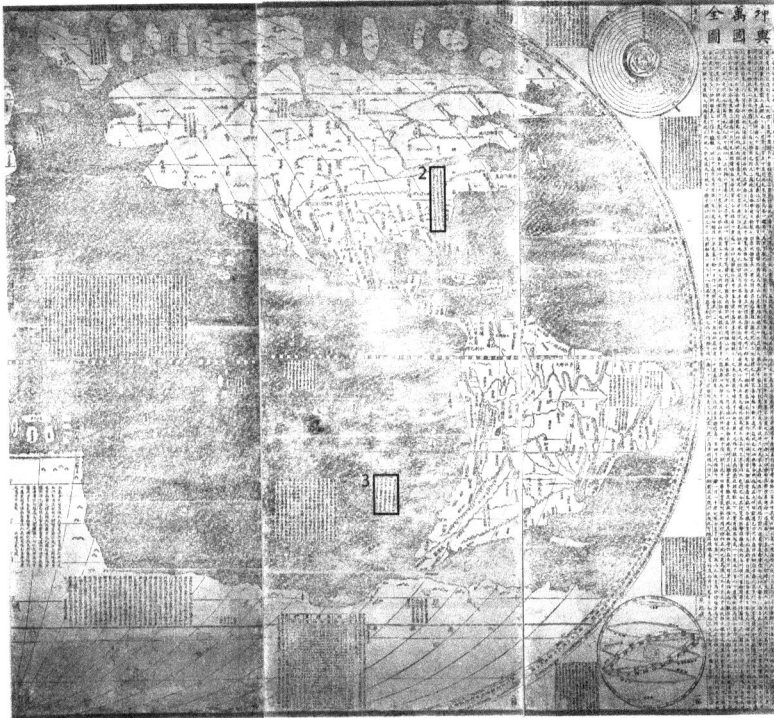

2) The general name for the land stretching from Labrador to Florida is Canada. However, each region has a separate name. The people are kind and hospitable to strangers from other lands. They use hides and furs for clothing and depend upon fishing for their livelihood. In the mountainous regions are natives who are constantly fighting each other and live on snakes, ants, spiders, and other insects.

3) In the past no one knew about North and South America and Magellanica. However, about a hundred years ago, Europeans sailed the high seas and arrived at their shores, and thus got to know of the land. The countries are vast and the inhabitants barbarous; consequently very little is known of the interior regions and their peoples.

NOTES

1. Excerpted from Nicolas Trigault, *China in the Sixteenth Century: The Journals of Matthew Ricci, 1583–1610* (Random House, 1953), 4–5, 19–23, 26–27, 41–45, 64–65, 72, 73, 75, 77, 79, 81, 85–87, 93, 96. Reprinted with permission of Penguin Random House. Headings original to Ricci's Italian manuscript.

2. Xie Qinggao 謝清高 and Yang Bingnan 楊炳南, *Hailu jiaoshi* 海錄校釋 (Shangwu yinshuguan, 2002), 200–203, 250–251, 264. Translation by the author. Headings from the original.

CHAPTER 11

POLITICAL AMBASSADORS (1793–1878)

Since the earliest documented diplomatic encounters during the Near Eastern Bronze Age, the objectives of diplomacy have been to seek out allies and deter adversaries, to detect a rival's strengths and weaknesses, and to negotiate territory and resources, all while maximizing gains and minimizing losses. But diplomacy's overarching purpose has always been to avert or postpone war. Chapter 2 documented the activities of the ancient diplomatic envoys Megasthenes and Zhang Qian. This chapter places Lord George Macartney, the first British envoy to the Qing court, and Guo Songtao, the first Chinese ambassador to Great Britain, into a parallel conversation.

INTRODUCTION TO LORD GEORGE MACARTNEY

There is a misconception that premodern Chinese relations with foreign powers were invariably regulated through a tributary system. Under this framework, China regarded itself as the civilized center of the world, while all other countries were subordinate tributary powers that offered submission and local products in exchange for generous gifts or trade privileges. Such a system may have prevailed when China was objectively paramount in the Eurasian world, like during the early Tang or early Ming periods (see chapter 9). However, when confronted with militarily comparable or superior adversaries, Chinese polities engaged in concerted peer-polity diplomacy, deviating from this idealized model. The Song appeasement treaties concluded with the Liao (1005 CE) and Jin (1141 CE) are notable examples.

During the Qing dynasty (1644–1912), the ruling Manchus managed foreigners through several institutions, overlaying the traditional Chinese tributary worldview with their own practice of overlordship. The Court of Colonial Affairs handled Mongols, Zunghars, and Russians, concluding marriage alliances and arranging trade caravans. The Imperial Household Department managed resident Jesuits who arrived after Matteo Ricci (chapter 10), while the Ministry of Rites handled tributary relations and trade with Korea, Burma, Thailand, Vietnam, and the Ryūkyū. European merchants like the Portuguese, Dutch, and English were managed through the Canton system, which had evolved between 1700 and 1760, and required all foreign trade to be conducted through a monopoly of approved merchants in Guangzhou (the *cohong* 公行), who also collected transit dues. All foreigners were required to return to Macau after the trading season, and grievances could be submitted only to the *cohong*, not to Chinese officials.

The East India Company and the British government were unhappy with this restrictive system and wanted more commercial transparency, diplomatic equality, and a greater market for their goods. They were accustomed to the dominance and prestige they enjoyed in India. After the failed attempt at negotiations by James Flint (1759) and the aborted embassy of Charles Cathcart (1787–1788), the East India Company and the prime minister selected Lord George Macartney (1737–1806) for a new embassy. Macartney was a Northern Ireland peer and former ambassador to Catherine the Great of Russia and had also served as governor of

Madras in India. Macartney left London in September of 1792 and arrived in China in June of 1793. He sailed on a menacing 64-gun warship, accompanied by an entourage of 95 men, including scientists, botanists, metallurgists, interpreters, and draftsmen, bearing gifts of the best of English manufactures and scientific instruments to impress the Chinese. Macartney kept a detailed journal of his voyage, excerpted below.

The British circumvented the usual procedures and traveled directly to Tianjin in the north, claiming that they wanted to salute the Qianlong Emperor's 80th birthday at Rehe (present-day Chengde). They insisted that their presents would be damaged traveling overland or upriver from Canton. The Qing court insisted that Macartney comply with proper guest ritual and perform a kowtow of submission before the emperor, which consisted of three separate full prostrations, lightly knocking one's forehead on the ground three times each. In Macartney's thinking, this ritual was similar to the *proskynesis* required of subjects before the Persian emperor in the ancient world, which no self-respecting free Greek male was willing to do, except before the gods. Macartney offered various compromises, but steadfastly and politely refused to prostrate himself.

During the imperial audience on September 14, 1793, after dropping to one knee and bowing his head (see figure 3), Macartney presented a letter from George III asking for a permanent ambassador in Beijing, new ports for trade, and fixed and equitable tariffs. The Qing emperor denied all these requests in an edict drafted before Macartney had even arrived in the capital. Macartney was disappointed, but he still considered the embassy a measured success, since he had gained some commitments for minor reforms in the Canton system, had gathered critical intelligence, and was told that Britain could send another embassy.

However, politicians in Britain viewed the embassy as a failure and an insult. Historians later saw it as a pivotal moment in modern history, where two irreconcilable worldviews inevitably clashed: a backward and irrational China and a modernizing and rational Britain. The greatest fault, they believed, lay in China's inability to perceive the challenge of the West and its failure to modernize. Later historical approaches to this encounter interpret it more as a geopolitical conflict between two expanding empires, both of which were aware of the political, strategic, and economic stakes.

Macartney's journal reveals his understanding that Britain could easily get its way in these matters by force, given the outdated and thinly deployed Qing military forces, which he viewed as "slovenly," "unwarlike," and "effeminate." When frustrated by Qing intransigence, he ominously declares that "a couple of English frigates would be an overmatch for the whole naval force of their empire." But when he calculated that this would lead to slaughtering millions of Chinese people, destroying the prospects for trade, and causing conflicts with jealous European powers, he concluded that he much preferred the "gentle" approach of diplomacy.

Four decades later, after another botched embassy in 1816 by Lord Amherst (who also refused to kowtow) and growing tensions in Canton, Britain abandoned diplomacy entirely and went with the "hasty" approach advised against by Lord Macartney, provoking the Opium Wars (1839–1841, 1856–1860) that would gain England treaty ports, legal extraterritoriality, and all the privileges Macartney had sought in 1793, but at enormous human cost.

Primary Sources and Selected Studies

Cranmer-Byng, J. L. "Lord Macartney's Embassy to Peking in 1793, from Official Chinese Documents." *Journal of Oriental Studies* 4, no. 1–2 (1957–1958): 117–186.

Fairbank, John King. "A Preliminary Framework." In *The Chinese World Order; Traditional China's Foreign Relations*, edited by John King Fairbank and Ta-tuan Ch'en. Harvard University Press, 1968.

Harrison, Henrietta. *The Perils of Interpreting: The Extraordinary Lives of Two Translators between Qing China and the British Empire.* Princeton, NJ: Princeton University Press, 2021.

Hevia, James L. *Cherishing Men from Afar: Qing Guest Ritual and the Macartney Embassy of 1793.* Duke University Press, 1995.

Lien-Sheng, Yang. "Some Historical Notes on the Chinese World Order." In *The Chinese World Order; Traditional China's Foreign Relations,* edited by John King Fairbank and Ta-tuan Ch'en. Harvard University Press, 1968.

Macartney, George. *An Embassy to China; Being the Journal Kept by Lord Macartney during His Embassy to the Emperor Ch'ien-Lung, 1793–1794.* Edited by J. L. Cranmer-Byng. Longmans, 1962. (Reprint, Folio Society, 2004).

Staunton, George Leonard. *An Authentic Account of an Embassy from the King of Great Britain to the Emperor of China.* 2 vols. W. Bulmer, 1797.

EXCERPTS FROM THE JOURNAL OF LORD GEORGE MACARTNEY, 1793–1794

Monday, August 5 [1793]. . . . The river here appeared to be as broad as the Thames at Gravesend. Great numbers of houses on each side, built of mud and thatched, a good deal resembling the cottages near Christchurch in Hampshire, and inhabited by such swarms of people as far exceeded my most extravagant ideas even of Chinese population. Among those who crowded the banks we saw several women, who tripped along with such agility as induced us to imagine their feet had not been crippled in the usual manner of the Chinese. It is said, indeed, that this practice, especially among the lower sort, is now less frequent in the northern provinces than in the others. These women are much weather-beaten, but not ill-featured, and wear their hair, which is universally black and coarse, neatly braided, and fastened on the top of their heads with a bodkin.

The children are very numerous and almost stark naked. The men in general well-looking, well-limbed, robust and muscular. I was so much struck with their appearance that I could scarce refrain from crying out with Shakespeare's Miranda in *The Tempest*:

Oh, wonder!
How many goodly creatures are there here!
How beauteous mankind is!
Oh, brave new world
That has such people in it!

. . .

Thursday, August 15. . . . During the greater part of the passage our conductors, Cheng-jui [Zhengrui 徵瑞, 1734–1815] the Legate, together with Wang [Wang Wenxiong 王文雄, 1749–1800] and Chou [Qiao Renjie 喬人傑, 1740–1804], visited me almost every day, but this morning they came with an appearance of more formality than usual. . . . They then introduced the subject of the Court ceremonies with a degree of art, address, and insinuation that I could not avoid admiring. They began by turning the conversation upon the different modes of dress that prevailed among different nations, and, after pretending to examine ours particularly, seemed to prefer their own, on account of its being loose and free from ligatures, and of its not impeding or obstructing the genuflections and prostrations which were, they said, customary to be made by all persons whenever the Emperor appeared in public. They therefore apprehended much inconvenience to us from our knee-buckles and garters, and hinted to us that it would be better to disencumber ourselves of them before we should go to Court. I told them they need not be uneasy about that circumstance, as I supposed, whatever ceremonies were usual for the Chinese to perform, the Emperor would prefer my paying him the same obeisance which I did to my own Sovereign. They said they supposed the ceremonies in both countries must be nearly alike, that in China the form was to kneel down upon both knees, and make nine prostrations or inclinations of the head to the ground, and that it never had been, and never could be, dispensed with. I told them ours was somewhat different, and that though I had the most earnest desire to do everything that might be agreeable to the Emperor, my first duty must be to do what might be agreeable to my own King; but if they were really in earnest in objecting to my following the etiquette of the English Court, I should deliver to them my reply in writing as soon as I arrived at Peking. . . .

Friday, August 16. This day at half after six p.m. we arrived at the suburbs of Tungchow [Tongzhou 通州], where (our navigation being now ended) we quitted our yachts and went on shore, but before I proceed further I must set down a few particulars which have struck me, lest in the multiplicity of things before me they should slip from my memory. Indeed, observations ought always to be written upon the spot; if made afterwards upon the ground of recollection they are apt to vary their hue considerably. . . .

The most refined politeness and sly good breeding appeared in the behavior of all those Mandarins with whom we had any connection; but although we found an immediate acquiescence in words with everything we seemed to propose, yet, in fact, some ingenious pretense or plausible objection was usually invented to disappoint us. Thus, when we desired to make little excursions from our boats into the towns, or into the country, to visit any object that struck us as we went along, our wishes were seldom gratified. The refusal, or evasion was, however, attended with so much profession, artifice, and compliment that we grew soon reconciled and even amused with it.

We have indeed been very narrowly watched, and all our customs, habits and proceedings, even of the most trivial nature, observed with an inquisitiveness and jealousy which surpassed all that we had read of in the history of China. But we endeavored always to put the best face upon everything, and to preserve a perfect serenity of countenance upon all occasions.

I therefore shut my eyes upon the flags of our yachts, which were inscribed "The English Ambassador bringing tribute to the Emperor of

China," and have made no complaint of it, reserving myself to notice it if a proper opportunity occurs. . . .

Monday, August 19. I went down to the pandals this morning, where I met the Tartar Legate, Wang and Chou, and several other Mandarins, who were assembled there to give orders for the operations of the next day. . . . On our return from the pandals, Wang and Chou walked up with us to our quarters, and told us that the Emperor's answer was come to our request of having a European missionary to attend us, and that we might choose any of the Europeans in the Emperor's service then at Peking. . . .

They then renewed the subject of the ceremonial relative to which they had been perfectly silent for several days. It seems to be a very serious matter with them, and a point which they have set their hearts upon. They pressed me most earnestly to comply with it, said it was a mere trifle; kneeled down on the floor and practiced it of their own accord to show me the manner of it, and begged me to try whether I could not perform it. On my declining it, they applied to my interpreter to do it, who though a Chinese, said he could only act as I directed him; they seem a little disappointed in finding me not so pliant in this point as they wished. As to themselves, they are wonderfully supple, and though generally considered as most respectable characters, are not very scrupulous in regard to veracity, saying and unsaying without hesitation what seems to answer the purpose of the moment. Their ideas of the obligations of truth are certainly very lax, for when we hinted to them any contradictions that occurred, or deviations from their promises in our affairs, they made very light of them, and seemed to think them of trifling consequence. . . .

Saturday, August 24. Sir George Staunton went to Yuan-ming Yuan [Yuanmingyuan 圓明園, "Old Summer Palace"], and took with him Mr. Barrow, Dr. Dinwiddie, Thibault, and Petitpierre, and other artists and workmen, to give them directions about arranging the machinery and disposing in their proper places the planetarium, orrery, globes, clocks, lusters, etc. . . . Some of the Chinese workmen, not accustomed to handle articles of such delicate machinery, were interrupted in their attempts to unpack them by our interpreter, who told them that, till put up and delivered, they must still be considered as under our care; upon which the Legate interposed and said, no, they are *cong-so* [*gongshu* 貢輸] "tributes" to the Emperor, and consequently we had nothing more to do with them. Our interpreter replied that they were not tributes (*cong-so*), but "presents" (*sung-lo*) [*songli* 送禮]. The Grand Secretary put an end to the conversation by saying that the expression of *sung-lo*, or presents, was proper enough. . . .

Thursday, August 29. This day I put up the state canopy and their Majesties' pictures in the presence chamber, and delivered my paper relative to the ceremonial [the kowtow] to be transmitted to Jehol [Rehe 熱河, present-day Chengde, Hebei]. I had a good deal of difficulty in persuading Father Raux to get it translated into Chinese and to put it into the proper diplomatic form, so much is every person here afraid of intermeddling in any state matter without the special authority of Government; and he only consented on condition that neither his writing nor that of his secretary should appear, but that I should get it copied by some other hand. Little Staunton [George Thomas Staunton] was able to supply my wants on this occasion, for having very early in the voyage begun to study Chinese under my two interpreters, he had not only made considerable progress in it, but he had learned to write the characters with great neatness and celerity, so that he was of material use to me on this occasion, as he had been already before in transcribing the catalogue of the presents.

In the paper I expressed the strongest desire to do whatever I thought would be most agreeable to the Emperor, but that, being the representative of the first monarch of the Western world, his dignity must be the measure of my conduct; and that, in order to reconcile it to the customs of the Court of China, I was willing to conform to their etiquette, provided a person of equal rank with mine were appointed to perform the same ceremony before my Sovereign's picture that I should perform before the Emperor himself. The Legate shook his head, but Wang and Chou said it was a good expedient, and offered immediately to go through the ceremony themselves on the spot; but as they had no authority for the purpose, I civilly declined their proposal. . . .

Friday, September 6. . . . This evening our interpreter amused us with an extract from one of the Tientsin [Tianjin 天津] gazettes, which seem to be much on a par with our own newspapers for wit and authenticity. In an account given there of the presents said to be brought for the Emperor from England the following articles are mentioned: several dwarfs or little men not twelve inches high, but in form and intellect as perfect as grenadiers; an elephant not larger than a cat, and a horse the size of a mouse; a singing-bird as big as a hen, that feeds upon charcoal, and devours usually fifty pounds per day; and, lastly, an enchanted pillow, on which whoever lays his head immediately falls asleep, and if he dreams of any distant place, such as Canton, Formosa, or Europe, is instantly transported thither without the fatigue of traveling. This little anecdote, however ridiculous, I thought would not be fair to leave out of my journal. . . .

Tuesday, September 10. This day the Legate, Wang and Chou renewed the conversation of yesterday relative to the ceremony, in the course of which I told them it was not natural to expect that an ambassador should pay greater homage to a foreign prince than to his own liege Sovereign, unless a return were made to him that might warrant him to do more. Upon which they asked me what was the ceremony of presentation to the King of England. I told them it was performed by kneeling upon one knee and kissing His Majesty's hand.

"Why then," cried they, "can't you do so to the Emperor?"

"Most readily," said I, "the same ceremony I perform to my own King I am willing to go through for your Emperor, and I think it a greater compliment than any other I can pay him." I showed them the manner of it, and they retired seemingly well satisfied. In the afternoon Chou came to me alone, and said that he had just seen the Minister, and had a long conference with him upon this business, the result of which was that either the English mode of presentation (which I had shown them in the morning) or the picture ceremony should be adopted, but he had not yet decided which. I said nothing.

Soon after the Legate arrived and declared that it was finally determined to adopt the English ceremony, only that, as it was not the custom in China to kiss the Emperor's hand, he proposed I should kneel upon both knees instead of it. I told him I had already given my answer, which was to kneel upon one knee only on those occasions when it was usual for the Chinese to prostrate themselves.

"Well then," said they, "the ceremony of kissing the Emperor's hand must be omitted.

To this I assented, saying, "As you please, but remember it is your doing, and according to your proposal, is but half the ceremony, and you see I am willing to perform the whole one." And thus ended this curious negotiation, which has given me a tolerable insight into the character of this Court, and that political address upon which they so much value themselves. . . .

Saturday, September 14. This morning at four o'clock a.m. we set out for the Court under the convoy of Wang and Chou, and reached it in little more than an hour, the distance being about three miles from our hotel. I proceeded in great state with all my train of music, guards, etc. Sir George Staunton and I went in palanquins and the officers and gentlemen of the Embassy on horseback. Over a rich embroidered velvet, I wore the mantle of the Order of the Bath, with the collar, a diamond badge, and a diamond star.

Sir George Staunton was dressed in a rich embroidered velvet also, and being a Doctor of Laws in the University of Oxford, wore the habit of his degree, which is of scarlet silk, full and flowing. I mention these little particulars to show the attention I always paid, where a proper opportunity offered, to oriental customs and ideas. We alighted at the park gate, from whence we walked to the Imperial encampment, and were conducted to a large, handsome tent prepared for us on one side of the Emperor's. After waiting there about an hour his approach was announced by drums and music, on which we quitted our tent and came forward upon the green carpet.

He was seated in an open palanquin, carried by sixteen bearers, attended by numbers of officers bearing flags, standards, and umbrellas, and as he passed we paid him our compliments by kneeling on one knee, whilst all the Chinese made their usual prostrations. As soon as he had ascended his throne I came to the entrance of the tent, and, holding in both my hands a large gold box enriched with diamonds in which was enclosed the King's letter, I walked deliberately up, and ascending the side-steps of the throne, delivered it into the Emperor's own hands, who, having received it, passed it to the Minister, by whom it was placed on the cushion. He then gave me as the first present from him to His Majesty the *ju-eu-jou* or *giou-giou* [*ruyi yu* 如意玉], as the symbol of peace and prosperity, and expressed his hopes that my Sovereign and he should always live in good correspondence and amity. It is a whitish, agate-looking stone about a foot and a half long, curiously carved, and highly prized by the Chinese, but to me it does not appear in itself to be of any great value [see figure 3].

Figure 3. "Lord Macartney in Audience with the Qianlong Emperor."

Source: William Alexander, British Library, London, UK. By permission of the British Library archive/Bridgeman Images.

The Emperor then presented me with a *ju-eu-jou* of a greenish-colored stone of the same emblematic character; at the same time he very graciously received from me a pair of beautiful enameled watches set with diamonds, which I had prepared in consequence of the information given me, and which, having looked at, he passed to the Minister. Sir George Staunton, whom, as he had been appointed Minister Plenipotentiary to act in case of my death or departure, I introduced to him as such, now came forward, and after kneeling upon one knee in the same manner which I had done, presented to him two elegant air-guns, and received from him a *ju-eu-jou* of greenish stone nearly similar to mine. Other presents were sent at the same time to all the gentlemen of my train. We then descended from the steps of the throne, and sat down upon cushions at one of the tables on the Emperor's left hand; and at other tables, according to their different ranks, the chief Tartar Princes and the

Mandarins of the Court at the same time took their places, all dressed in the proper robes of their respective ranks. These tables were then uncovered and exhibited a sumptuous banquet. The Emperor sent us several dishes from his own table, together with some liquors, which the Chinese call wine, not, however, expressed from the grape, but distilled or extracted from rice, herbs, and honey. In about half an hour he sent for Sir George Staunton and me to come to him, and gave to each of us, with his own hands, a cup of warm wine, which we immediately drank in his presence, and found it very pleasant and comfortable, the morning being cold and raw.

Amongst other things, he asked me the age of my King, and being informed of it, said he hoped he might live as many years as himself, which are eighty-three. His manner is dignified, but affable, and condescending, and his reception of us has been very gracious and satisfactory. He is a very fine old gentleman, still healthy and vigorous, not having the appearance of a man of more than sixty.

The order and regularity in serving and removing the dinner was wonderfully exact, and every function of the ceremony performed with such silence and solemnity as in some measure to resemble the celebration of a religious mystery. The Emperor's tent or pavilion, which is circular, I should calculate to be about twenty-four or twenty-five yards in diameter, and is supported by a number of pillars, either gilded, painted, or varnished, according to their distance and position. In the front was an opening of six yards, and from this opening a yellow fly-tent projected so as to lengthen considerably the space between the entrance and the throne.

The materials and distribution of the furniture within at once displayed grandeur and elegance. The tapestry, the curtains, the carpets, the lanterns, the fringes, the tassels were disposed with such harmony, the colors so artfully varied, and the light and shades so judiciously managed, that the whole assemblage filled the eye with delight, and diffused over the mind a pleasing serenity and repose undisturbed by glitter or affected embellishments. The commanding feature of the ceremony was that calm dignity, that sober pomp of Asiatic greatness, which European refinements have not yet attained.

I forgot to mention that there were present on this occasion three ambassadors from Tatze or Pegu [in Myanmar] and six Mohammedan ambassadors from the Kalmucks [Kalmyks] of the southwest, but their appearance was not very splendid. Neither must I omit that, during the ceremony, which lasted five hours, various entertainments of wrestling, tumbling, wire-dancing, together with dramatic representations, were exhibited opposite to the tent, but at a considerable distance from it.

Thus, then, have I seen "King Solomon in all his glory." I use this expression, as the scene recalled perfectly to my memory a puppet show of that name which I recollect to have seen in my childhood, and which made so strong an impression on my mind that I then thought it a true representation of the highest pitch of human greatness and felicity. . . .

Friday, October 25. Sung-yun [Songyun 松筠] told me that the Emperor was very much pleased with the accounts he had been enabled to give him of our prosperous journey, and had sent me a testimony of his benevolence (a cheese and some sweetmeats) with a gracious repetition of kindness and regard. We had a good deal of desultory conversation upon the general subjects of our last meeting, during which he took occasion to say that we should find it an easy matter to set everything to rights with the new Viceroy of Canton, who was so reasonable and so just that I might depend upon it he never would countenance the most trifling oppression. He again declared that greater indulgence and favor were intended to be shown to the English than they had ever experienced before, and seemed anxious to impress this opinion upon me. If the Court of Peking is not really sincere can they possibly expect to feed us long with promises? Can they be ignorant that a couple of English frigates would be an overmatch for the whole naval force of their empire, that in half a summer they could totally destroy all the navigation of their coasts and reduce the inhabitants of the maritime provinces, who subsist chiefly on fish, to absolute famine? . . .

Wednesday, November 6th, 1793. At daybreak we fell into the Yangtze River, commonly called the Kiang-ho [Jianghe 江河], which was about a mile and a half wide at the place where we crossed it. On the southern shore stands the town of Tchin-chien [Zhenjiang 鎮江], which is large, well situated, well built, and well inhabited, but the walls seem to be much out of repair and going fast to decay. A garrison of at least two thousand men all turned out to show themselves, with colors and music, and appointed as if going to be reviewed. They consisted of different corps, differently dressed and armed, according to their respective services, some with matchlocks, some with bows and arrows, and some with halberds, lances, swords and targets. Many of them wore steel helmets, as they are supposed to be, though I suspect they are only of burnished leather or glittering pasteboard. The uniforms which are very showy and of different colors, red, white, blue, buff and yellow, must be very expensive, but after all these troops have a slovenly, unmilitary air, and their quilted boots and long petticoats make them look heavy, inactive and effeminate. . . .

Wednesday, December 4th, 1793. . . . Having occasion to light his pipe, and his attendants being absent, I took out of my pocket a small phosphoric bottle, and instantly kindled a match at it. The singularity of a man's carrying fire in his fob without damage startled him a good deal. I therefore explained to him the phenomenon and made him a present of it. This little incident led to a conversation upon other curious subjects, from which it appeared to us how far the Chinese (although they excel in some branches of mechanics) are yet behind other nations in medical or surgical skill and philosophical knowledge. Having often observed numbers of blind persons, but never having met a wooden leg or a deformed limb here, I concluded that good oculists were very rare and that death was the usual consequence of a fracture. The Viceroy told me I was right in my conjecture. But when I told him of many things in England, and which I had brought people to instruct the Chinese in, if it had been allowed, such as the reanimating drowned persons by a mechanical operation, restoring sight to the blind by the extraction or depression of the glaucoma, and repairing or amputating limbs by manual dexterity, both he and his companions seemed as if awakened out of a dream, and could not conceal their regret for the Court's coldness and indifference to our discoveries.

From the manner of these gentlemen's inquiries, the remarks which they made, and the impressions which they seemed to feel, I have conceived a much higher opinion of their liberality and understanding. Whether in these two respects the Minister be really inferior to them, or whether he acts upon a certain public system, which often supersedes private conviction, I know not. But certain it is that in a conversation with him at Jehol, when I mentioned to him some recent inventions of European ingenuity, particularly that of the air balloon, and that I had taken care to provide one at Peking with a person to go up in it, he not only discouraged that experiment, but most of the others, which from a perusal of all the printed accounts of this country we had calculated and prepared for the meridian of China. Whatever taste the Emperor K'ang-hsi [Kangxi] might have shown for the sciences, as related by the Jesuits in his day, his successors have not inherited it with his other great qualities and possessions. For it would now seem that the policy and vanity of the Court equally concurred in endeavoring to keep out of sight whatever can manifest our pre-eminence, which they undoubtedly feel, but have not yet learned to make the proper use of. It is, however, in vain to attempt arresting the progress of human knowledge. The human mind is of a soaring nature and having once gained the lower steps of the ascent, struggles incessantly against every difficulty to reach the highest. I am indeed very much mistaken if all the authority and all the address of the Tartar Government will be able much longer to stifle the energies of their Chinese subjects. Scarcely a year now passes without an insurrection in some of the provinces. It is true they are usually soon suppressed, but their frequency is a strong symptom of the fever within. The paroxysm is repelled, but the disease is not cured.

Wednesday, December 18th, 1793. . . . I can't omit that in the course of our navigating from Nan-chou-fu [Nanchangfu 南昌府, in Jiangxi], we have had an uncommon profusion of military honors lavished upon us everywhere as we passed along, which I attribute to the Viceroy's having given particular directions for the purpose as he preceded us. As the Chinese consider the province of Canton [Guangdong] to be the most obnoxious to invasion from the sea, the military posts in it are very numerous. There seemed to be an affected reiteration of salutes wherever we appeared, in order, I presume, to impress us with an idea of the vigilance and alertness of the troops, and to show that they were not unprepared against an enemy. Nevertheless, as they are totally ignorant of our discipline, cumbersomely clothed, armed only with matchlocks, bows and arrows, and heavy swords, awkward in the management of them, of an unwarlike character and disposition, I imagine they would make but a feeble resistance to a well-conducted attack. The circumstance of greatest embarrassment to an invader would be their immense numbers, not on account of the mischief they could do to him, but that he would find no end of doing mischief to them. The slaughter of millions would scarcely be perceived, and unless the people themselves soon voluntarily submitted, the victor might indeed reap the vanity of destruction, but not the glory or use of dominion.

January 2nd–7th, 1794. . . . Now I am very much mistaken if, by a proper management, we might not gradually and in some years be able to mold the China trade (as we seem to have done the trade everywhere else) to the shape that will best suit us. But it would certainly require in us great skill, caution, temper and perseverance, much greater perhaps than it is reasonable to expect. I dare say there are many hasty spirits disposed to go a shorter way to work, but no shorter way will do it.

If, indeed, the Chinese were provoked to interdict us their commerce, or do us any material injury, we certainly have the means easy enough of revenging ourselves, for a few frigates could in a few weeks destroy all their coast navigation and intercourse from the island of Hainan to the Gulf of Pei-chihli [Beizhili 北直隸] and if I were to indulge the speculations of an ambitious or vindictive politician, I doubt not but we might vulnerate them as sensibly in many other quarters. We might probably be able from Bengal to excite the most serious disturbances on their Tibet frontier by means of their neighbors there, who appear to require only a little encouragement and assistance to begin. The Koreans, if they once saw ships in the Yellow Sea acting as enemies to China might be induced to attempt the recovery of their independence. The thread of connection between this Empire and Formosa [Taiwan] is so slender that it must soon break of itself, but a breath of foreign interference would instantly snap it asunder. . . .

Such might be the consequence to this Empire if we had a serious quarrel with it. On the other hand, let us see what might be the consequence to ourselves. It is possible that other nations, now trading or expecting to trade with China, would not behold our success with indifference, and thus we might be involved with much more formidable enemies than Chinese. But I leave that consideration aside and proceed to others.

Our settlements in India would suffer most severely by any interruption of their China traffic which is infinitely valuable to them, whether considered singly as a market for cotton and opium, or as connected with their adventures to the Philippines and Malaya.

To Great Britain the blow would be immediate and heavy. Our great woollen manufacture, the ancient staple of England, would feel such a sudden convulsion as scarcely any vigilance or vigor in Government could for a long time remedy or alleviate. The demand from Canton for our woollens alone can't now be less than £500,000 to £600,000 per annum, and there is good reason to believe that with proper care it may in some years be stretched to a million. We should lose the other growing branches of export to China of tin, lead, copper, hardware, and of clocks and watches, and similar articles of ingenious mechanism. We should lose the import from China not only of its raw silk, an indispensable ingredient in our silk fabrics, but of another indispensable luxury, or rather an absolute necessary of life: tea. We should also in some measure lose an excellent school of nautical knowledge, a strong limb of marine power, and a prolific source of public revenue.

These evils, it would seem, must infallibly follow from a breach with China. Whether in time other markets might not be found or created to make us amends, I am not yet sufficiently acquainted with this part of the world (and still less with the disposition of the Court of Spain) to hazard a decision; but it is not impossible that, though prodigious inconveniences and mischiefs would certainly be felt at the moment from a rupture, means might be discovered to reverse or repair them. But all these inconveniences and mischiefs which I have stated as objects of apprehension may happen in the common course of things without any quarrel or interference on our part.

The Empire of China is an old, crazy, First rate man-of-war, which a fortunate succession of able and vigilant officers has contrived to keep afloat for these one hundred and fifty years past, and to overawe their neighbors merely by her bulk and appearance, but whenever an insufficient man happens to have the command upon deck, adieu to the discipline and safety of the ship. She may perhaps not sink outright; she may drift some time as a wreck, and will then be dashed to pieces on the shore; but she can never be rebuilt on the old bottom.

The breaking-up of the power of China (no very improbable event) would occasion a complete subversion of the commerce, not only of Asia, but a very sensible change in the other quarters of the world. The industry and ingenuity of the Chinese would be checked and enfeebled, but they would not be annihilated. Her ports could no longer be barricaded; they would be attempted by all the adventurers of all trading nations, who would search every channel, creek, and cranny of China for a market, and for some time be the cause of much rivalry and disorder. Nevertheless, as Great Britain, from the weight of her riches and the genius and spirit of her people, is become the first political, marine and commercial power on the globe, it is reasonable to think that she would prove the greatest gainer by such a revolution as I have alluded to, and rise superior over every competitor.

But to take things solely as they now are, and to bound our views by the visible horizon of our situation, without speculating upon probable events (which seldom take place according to our speculations), our present interests, our reason, and our humanity equally forbid the thoughts of any offensive measures with regard to the Chinese, whilst a ray of hope remains for succeeding by gentle ones. Nothing could be urged in favor of an hostile conduct, but an irresistible conviction of failure by forbearance. . . .

* *

Observations on China

. . .

When Marco Polo, the Venetian, visited China in the thirteenth century, it was about the time of the conquest of China by the western or Mongol Tartars, with Kublai Khan, a grandson of Genghis Khan, at their head. A little before that period the Chinese had reached their highest pitch of civilization, and no doubt they were then a very civilized people in comparison of their Tartar conquerors, and their European contemporaries, but not having improved and advanced forward, or having rather gone back, at least for these one hundred and fifty years past, since the last conquest by the northern or Manchu Tartars; whilst we have been every day rising in arts and sciences, they are actually become a semi-barbarous people in comparison with the present nations of Europe. Hence it is that they retain the vanity, conceit, and pretensions that are usually the concomitants of half-knowledge, and that, though during their intercourse with the embassy they perceived many of the advantages we had over them, they seemed rather surprised than mortified, and sometimes affected not to see what they could not avoid feeling. . . .

The common people of China are a strong hardy race, patient, industrious, and much given to traffic and all the arts of gain; cheerful and loquacious under the severest labor, and by no means that sedate, tranquil people they have been represented. In their joint efforts and exertions, they work with incessant vociferation, often angrily scold one another, and seem ready to proceed to blows, but scarcely ever come to that extremity. The inevitable severity of the law restrains them, for the loss of a life is always punished by the death of the offender, even though he acted merely in self-defense, and without any malice prepense.

Superstitious and suspicious in their temper they at first appeared shy and apprehensive of us, being full of prejudices against strangers, of whose cunning and ferocity a thousand ridiculous tales had been propagated, and perhaps industriously encouraged by the government, whose political system seems to be to endeavor to persuade the people that they are themselves already perfect and can therefore learn nothing from others; but it is to little purpose. A nation that does not advance must retrograde, and finally fall back to barbarism and misery. . . .

Though much circumscribed in the course of our travels we had opportunities of observation seldom afforded to others, and not neglected by us. The genuine character of the inhabitants, and the effects resulting from the refined polity and principles of the government, which are meant to restrain and direct them, naturally claimed my particular attention and inquiry. In my researches, I often perceived the ground to be hollow under a vast superstructure, and in trees of the most stately and flourishing appearance I discovered symptoms of speedy decay, whilst humbler plants were held by vigorous roots, and mean edifices rested on steady foundations. The Chinese are now recovering from the blows that had stunned them; they are awaking from the political stupor they had been thrown into by the Tartar impression, and begin to feel their native energies revive. A slight collision might elicit fire from the flint, and spread flames of revolt from one extremity of China to the other. In fact, the volume of the empire is now grown too ponderous and disproportionate to be easily grasped by a single hand, be it ever so capacious and strong. It is possible, notwithstanding, that the momentum impressed on the machine by the vigor and wisdom of the present Emperor may keep it steady and entire in its orbit for a considerable time longer; but I should not be surprised if its dislocation or dismemberment were to take place before my own dissolution. Whenever such an event happens, it will probably be attended with all the horrors and atrocities from which they were delivered by the Tartar domination; but men are apt to lose the memory of former evils under the pressure of immediate suffering; and what can be expected from those who are corrupted by servitude, exasperated by despotism and maddened by despair? Their condition, however, might then become still worse than it can be at present. Like the slave who fled into the desert from his chains and was devoured by the lion, they may draw down upon themselves oppression and destruction by their very effort to avoid them, may be poisoned by their own remedies and be buried themselves in the graves which they dug for others. A sudden transition from slavery to freedom, from dependence to authority, can seldom be borne with moderation or discretion.

Every change in the state of man ought to be gentle and gradual, otherwise it is commonly dangerous to himself and intolerable to others. A due preparation may be as necessary for liberty as for inoculation of the smallpox which, like liberty, is future health but without due preparation is almost certain destruction. Thus, then the Chinese, if not led to emancipation by degrees, but let loose on a burst of enthusiasm would probably fall into all the excesses of folly, suffer all the paroxysm of madness, and be found as unfit for the enjoyment of freedom as the French and the negroes. . . .[1]

INTRODUCTION TO GUO SONGTAO

Since Macartney's embassy, the Qing court had resisted calls to send permanent ambassadors to Europe or America, but their hand was forced in 1875 after the Margary Affair, in which a junior British diplomat was murdered traveling from Burma to Yunnan. The resulting unequal treaty, the Chefoo Convention (August 21, 1876), required China to send an envoy of apology to Great Britain, which was later converted into a permanent diplomatic posting. Guo Songtao 郭嵩燾 (1818–1891) was sent by the Qing court to be this first ambassador to Great Britain (in residence, January 1877--August 1878) and later was named concurrent ambassador to France (April–August 1878).

Guo was a promising Confucian scholar from Hunan (*jinshi,* 1847). Along with his friend and fellow Hunanese, Zeng Guofan 曾國藩 (1811–1872), he helped to put down the disastrous Taiping Rebellion, personally leading some battles. He then held several important positions at court and in the provinces. But his reputation and career were damaged by his minority viewpoint that encroaching Western powers could only be dealt with diplomatically, for he thought it impossible to defeat them militarily. Along with Senior Grand Secretary Li Hongzhang 李鴻章 (1823–1901), Guo advocated that reform was the only way to strengthen China against European imperialism. Li selected Guo for the posting in Great Britain, since it was in line with his thinking on diplomacy and reform, but Guo realized it was a thankless job and represented a humiliating exile.

Guo traveled by steamship to London, arriving in January, 1877. He served until censured and recalled in August 1878. The diary of his initial journey, *Record of an Envoy's Journey to the West* (Shixi jicheng 使西紀程), was published immediately in China. It contained statements that shocked court conservatives, such as one suggesting that European power and prosperity were not just the results of superior weapons but were built upon principles of justice and order, comparable to those of China. This implied that Europeans were not just unethical barbarians. To compete with Europe, Guo argued, China needed to reform its core institutions, not just copy Western technology. The backlash against Guo's observations destroyed what was left of his career. Conservatives at court banned his book and destroyed the printing blocks. His countryman Wang Tao 王韜 had traveled to Great Britain almost a decade earlier and made similar observations, but his memoir would not be published until 1890 (chapter 12) so did not generate the same level of outcry.

Guo discharged his official duties in London admirably, maintaining an active social calendar and diplomatic activity. However, his vice ambassador, Liu Xihong 劉錫鴻 (d. 1891), who was placed by the conservative faction at court to spy on Guo, constantly undermined him. Liu even fed embarrassing stories about Guo to the British press to humiliate him and sabotage the mission.

This chapter includes a slight abridgement of a letter that Guo wrote in 1877 from London to Li Hongzhang, the most powerful leader of the reform camp and his personal protector at court. It contains numerous admiring observations about the benefits of European technologies, financial institutions, and parliamentary governance. Guo was particularly impressed by the efficiency and impact of railroads and telegraph lines. He suggested that they could help China move information more quickly, shrink the enormous gulf between officials and the people, and even squelch future rebellions before they could grow. He also debunked Chinese critics' views that rail and telegraph lines would disturb local fengshui.

Guo was ashamed of China's lack of progress compared to Japan, which had committed to reform and sent hundreds of advisors and students to Great Britain to learn law, financial administration, technical skills, and military training. The letter also reveals Guo's thoughts on important issues of Chinese domestic policy, like banning opium smoking, as well as his views on current international crises, particularly those regarding Kashgar and Ili in northwestern China.

PRIMARY SOURCES, ALTERNATE ENGLISH TRANSLATION, AND SELECTED STUDIES

Day, Jenny Huangfu. "The Scholar." In *Qing Travelers to the Far West: Diplomacy and the Information Order in Late Imperial China*. Cambridge University Press, 2018.

Frodsham, J. D. *The First Chinese Embassy to the West; The Journals of Kuo-Sung-t'ao, Liu Hsi-hung and Chang Te-yi*. Clarendon, 1974.

Guo Songtao 郭嵩燾. *Guo Songtao riji* 郭嵩燾日記. 4 vols. Hunan renmin chubanshe, 1980–1983.

Guo Songtao 郭嵩燾. *Guo Songtao shi wen ji* 郭嵩燾詩文集. Edited by Yang Jian 楊堅. Yuelu shushe, 1984.

A LETTER FROM LONDON TO LI HONGZHANG, BY GUO SONGTAO (MARCH 22, 1877)

Observations on British Progress

Here, the general atmosphere of politics, education, and social customs is constantly renewed and progressing. Inquiring into the origins and development of the country, in the beginning, the king and the common people engaged in a struggle for political power and slaughtered one another. The chaos and civil war lasted for dozens of years, even up to a century. Only after it had come to that point did things later reach a settlement. It was not the case that from the outset, they had a long-accumulated tradition of perfect virtue and splendid teachings. For the last 100 years, government officials and the people have worked collaboratively, paying special attention to the governance of the country. Matters are communicated to the sovereign, and the policies are implemented. With every passing day, this system has flourished until it has attained a superior level of governing. To this day, her monarch [Queen Victoria] is acclaimed for being wise and capable. Public sentiment and lifestyles advance and improve. And yet if we assess this undertaking of building up their wealth and power, it really only began after the

Qianlong reign period [1736–1795]. The steamboat was first invented during the Qianlong period [by John Fitch of Connecticut, 1787], but at first it was not very beneficial. Not until the sixth year of the Jiaqing reign period [1801] did they begin to employ them to travel on the seas. They followed the same method with the invention of the locomotive, which began in the 18th year of the Jiaqing reign [1813]. After that, they paid increasing attention to the study of electricity, using electromagnetic machines to transmit messages. By the 18th year of the Daoguang reign period [1838], they had established telegraph service in their national capital. They gradually expanded its reach until by the fourth year of the Tongzhi reign period [1865] it had reached India. It was in the 20th year of the Daoguang reign period [1840] when England was at war with China that steamships arrived off the coast of Guangdong. When they were at war again in the 10th year of the Xianfeng reign period [1860], telegraph service had already been extended from India to reach Shanghai. England only launched this initiative a few decades ago, but taking advantage of China's decline, they have arrived from over 70,000 Chinese miles away in what seems like the blink of an eye. However, this situation also allows one to see that once the energy of heaven and earth is unleashed, there is no checking it. Chinese scholar-bureaucrats continue to just rely on themselves in trying to suppress the natural operation of the universe. None of them have been successful at doing this so far.

Railways
Since arriving here several months ago, I have really begun to see the convenience of trains. A round trip of 300 or 400 Chinese miles [approx. 150–200 km] takes only half a day. The English gentry have strongly encouraged China to proactively develop their own railways, saying that the wealth and power of England was built upon this. At the outset, there were those in England who also had doubts about railroads and tried to block their construction. For example, to sustain vehicle transportation between the port of Southampton and London previously required the use of 30,000 horses, so people worried that the railroad would be detrimental

to their livelihoods. But after the railways were opened up, the number of horses employed has increased to sixty or seventy thousand. This is because the route was made more convenient, so trade increased daily. Since the train only operates along one line, those people within a few dozen miles of the railroad have been employing more and more horses to come there to take the train.

Last winter, I went to Shanghai and saw a railway map held by the Gezhi Academy. It showed a line running directly from India to Yunnan, a branch going out from Lin'an [south of Kunming, Yunnan] and hastening east toward Canton, one going north from Chuxiong [west of Kunming] through Sichuan and proceeding to Hankou, and one going from Canton, skirting the mountains, exiting from Hunan, and joining up with the other line at Hankou. Another went from Nanjing to Zhenjiang [in Jiangsu] and emerged in the east in Shanghai. A further route went east from Nanjing to emerge in Ningbo, and a northern branch went to Tianjin, continuing on to reach the capital in Beijing. When I saw this map, I thought it was a little bit strange, for Yunnan had only just recently been opened to commerce, and already, preparations were underway for a railway line. When I arrived in London, I got a hold of this same map and realized that it had been drawn more than 10 years ago. In general, wherever the English are determined to go, they will eventually get there. The India railway currently only reaches to Assam [far northeastern India]. . . .

Japanese Commitment to Reform

The Japanese consul in Great Britain [Ueno Kagenori, 1845–1888], upon meeting me, said, "Westerners can exploit the natural resources of the world. They have done the hard part; we just have to do the easy part. How is it fitting for us to sit idly by and neglect this? China's vast lands and enormous population are the envy of every other country. But they think it is extremely pitiful when they hear that up until now, we have still not done anything to exert ourselves in this regard." I became shamefaced and was unable to answer.

Reflections on China's Historical Foreign Relations

Last year, when I entered the capital [of Beijing], my original intention was to inquire into ancient and modern matters and distinguish between similarities and differences, successes and failures. Trade relations between China and the West have been conducted since the Sui [581–618 CE] and Tang [618–907 CE] dynasties, for more than 1,300 years, but we incurred their enmity because of the ban on opium smoking, which was followed by an obligation to open an increasing number of treaty ports, including internal ones along the Yangzi River. Conditions are becoming daily more oppressive, and the troubles are getting deeper. It is advisable to investigate the whole course of events and to lay out, item by item, the reality behind how they became so rich and powerful and to discover their motives. Then, we can understand the main points of how China should deal with its own situation and how it should handle its relations with others. We should plan on writing this up in a book, submit it to the Zongli Yamen 總理衙門 [Office in Charge of Affairs Concerning All Nations], and distribute it to all the schools in the empire to clear up the confusion of our scholar-bureaucrats.

How the imperial court deals with the intentions of people from distant lands still has some far-reaching and positive qualities to it, and we should make sure that the officials and commoners are informed and aware of it. Once the principles behind this are clarified, we can assuredly determine the long-term foundation of our country, for a hundred million generations. I had already laid all this out to you when I passed through Tianjin, but when I made it to Beijing, I was unable to get the words out of my mouth, because I was turned back by all the clamorous discussions I encountered.

Chinese Reaction to Opium Smoking and to Various Foreign Technologies

I would personally say that there are some aspects to the Chinese mind that are absolutely impossible to comprehend. Nothing from the West has done more severe harm than opium smoking. The British gentry also feel ashamed over the harm it has caused to people and how it was used as a tool for provoking a conflict with China. They have strenuously tried to ban and eradicate it. But China's scholar-bureaucrats willingly sink into this vice, indifferently and without regret. What has been our country's national shame for the last several decades has exhausted our financial resources and has poisoned our people, and yet there is not a single man who takes this as deeply troubling. Every household possesses Western clocks and toys, and the use of foreign woolen textiles has become widespread in poor and remote areas. It has even become the custom in Jiangsu and Zhejiang to eschew our national currency and exclusively use foreign money—which further adds to inflation—totally indifferent and ignorant as to how wrong this is. But as soon as some people hear about the construction of a rail line or a telegraph, they feel bitter hatred and will form a mob to cause difficulties. Some even consider the appearance of any Western machine to be a cause for public outrage. When Jiegang [a.k.a. Zeng Jize 曾紀澤; 1839–1890] once took a small steamboat from Nanjing to Changsha to return home to mourn his parent, many officials and gentry put up a huge clamor that went on for years without cease. Such people willingly allow others to harm us through opium and enable them to exploit our "fat and grease" [the fruits of the people's labor], while employing all their energy to block the potential sources of revenue for our country. I have no idea what their motives are!

Public Sentiment in China

After 30 years of dealing with foreign affairs, our senior provincial officials still have no understanding and hold the court hostage to their ideas, calling it "public opinion." As a result, the imperial court also praises such views, calling them public opinion. Alas! Genuine public sentiment in the empire has been pent up and repressed, and the people have had no way to express their complaints to higher authorities for some time. And yet this faction uses brash and ignorant rhetoric to agitate and incite vagrants [against foreigners and missionaries] to accomplish their ends. Government officials have participated in this and have even taken a guiding role.

Guo Songtao's Experience as a Reformer

The weakening of the Song dynasty [960–1279] and the demise of the Ming [1368–1644] were both caused by this type of arrogant and ignorant persons. I, Songtao, am a native of Chu [Hunan Province], and I grew up in the ignorant and backward countryside. Furthermore, I was unfamiliar with commerce and was never near any foreigners. Nevertheless, I gained some knowledge through the reading of books, contemplating the principles of things, and chronologically investigated the course of events of the past and present. Amid the laughter and ridicule heaped upon me by the whole world, I still sought the guiding principles through which we could protect the nation and govern the country so that China could stand on its own and not be destroyed. I spoke confidently of this, without any misgivings. But from beginning to end, we never experienced any mutual understanding, so I was forced to go into hiding 70,000 Chinese miles from my country [to Great Britain]. Within the first two months of coming here, after being impeached on two occasions before the emperor, I suddenly and completely regretted my original aspiration and never dared to again bring this up for discussion. But still, there are things I have seen and heard here that I certainly must relate to you, Grand Secretary.

Japanese Studying Law and Finance in Great Britain

There are more than 200 Japanese in Great Britain studying special skills. Every port has some of them, and in London alone there are 90. I personally met more than 20 of them, and they all can speak English. There is one named Nagaoka Moriyoshi 長岡護美 [1842–1906], a former feudal lord who administered an entire domain [Kumamoto]. Now he has been downgraded to a regular hereditary nobleman and is currently studying law here. Their former secretary to the minister of finance, Inoue Kaoru 井上馨 [1835–1915], arrived here as an envoy to concentrate on financial administration. He plans to completely imitate the practices here and implement them in Japan. The telegraph bureaus that they established in Japan were also set up and managed after a period of successful study in London. Very few of the Japanese here are actually studying military skills. The military is of peripheral importance, whereas the creation of each type of governmental institution is fundamental to state building. You, Li Hongzhang, are soon to manage all the armed forces, thus you are especially interested in having men study the arts of war. In my humble opinion, the military structure of each province lacks any organizational principle behind it, and we can't rely on the mustering of temporary recruits into local militias as a permanent practice.

British Military Training

For the last few decades in the West, there hasn't been much worry about a major war, so they can just make decisions based on the strategic situation. Looking into the recruitment and training of soldiers in London, they first make them read books, so they are informed about the arts of war, and only then are they selected for service. They send a doctor to examine them to determine if their circulatory system is robust and their muscles and bones are strong. Later, they are instructed in how to leap and systematically and thoroughly taught the technical skills of using guns and cannons. They are then placed into army units. The foundational roots of their military practice are very thick indeed. How could China possibly have the capability to implement something like this! One person's skill cannot go very far. Right now, we are considering wasting so much money learning to "slay dragons," but after completing this skilled training, we still may not be able to employ it.

Recommendations for Study Abroad Students

I would like to have the government-sponsored study-abroad students brought by Li Danya switch to studying iron and coal surveying, smelting methods, building railways, and electrical science, things with practical application. At the same time, I recommend that we order the governors-general of each province to select several talented young people, provide for their expenses, and send them first to the Technical Bureaus in Tianjin, Shanghai, and Fujian to learn the proper technical models, teach them foreign languages, and then send them abroad, where each would take different courses of study according to their natural gifts. Each Technical Bureau should also add two or three additional instructors to handle the influx of students. It is only necessary right now to get this started, to give people some expectation of where we are headed.

* * * * * * * * * * * * * * * **BRIEF EXCERPT** * * * * * * * * * * * * * * * * *

Building Railways and Telegraph Lines

There is a man here named Sidiwensen [Sir Rowland Macdonald Stephenson, 1808–1895] who claims he has built many railways in several different countries. He is particularly earnest in this regard and urges China to do so with haste. I respectfully submit a rough summary of his plan for your inspection.

In my opinion, needing foreigners to do everything for us is certainly not a sustainable policy. We should first send Chinese people to understand their methods. When Egypt, which is in Africa, built railroads, they first sent men to England to study and train, and later they imitated those methods at home. This is the best method for us to emulate. I humbly request your instructions, and I will rely on these when I discuss the matter with Li Danya.

In my opinion, among the essentials to properly governing our country today, knowing that the things that need to be implemented are numerous, none is more pressing than urgently planning these domestic matters for establishing the foundation for prosperity and strength. With these two things [railways and telegraphs], we can build a state which will endure for a thousand years without decline. Their great and far-reaching benefits hardly need to be discussed. Among these, there are two advantages that are quite obvious: China covers an area of over ten thousand square miles, and postal communication from the most distant places takes several dozen days to arrive. We often suffer from being cut off from information. But with these two things, traveling 10,000 Chinese miles would become like moving between the threshold and courtyard of one's own house. If there were suddenly a flood, drought, or bandit outbreak, an incident that occurred in the morning could be known by evening. So we would not have to worry about treacherous people who might exploit such situations to secretly launch a rebellion. This is the first major advantage.

In China, the gap between the government and the common people is far too wide, and this is advantageous to either party who might want to shield the ears and eyes of the court to benefit their own selfish interests. Because of this, public sentiment often gets pent up and is unable to reach the higher authorities. When we implement railways and telegraphs, all our wealthier subjects could voluntarily contribute to the expenses of the country, and many would be overjoyed to do so. And these railway and telegraph lines that pass through the country would be like the natural circulation of blood through the human body, so obviously, neither the good nor the evils of government could be concealed. So we would not have to be concerned with greedy officials who restrain public sentiment for their own illicit benefit. This is the other major advantage. . . .

Domestic critics simply claim that in whatever location foreign machinery arrives, it will harm the local fengshui. Their statements are preposterous. To construct rail and telegraph lines, these must certainly be placed along our current postal routes. They are all constructed on flat surfaces, without any boring or destruction. As for mechanized coal mining, one must pump out the water to seek deeper seams, for the quality of the coal is better the deeper one excavates. Chinese coal mining concentrates on opening broad surface trenches, whereas Western coal mining focuses on deep penetration, but they are both mining operations, whether they are shallow or deep. What reason is there to obstruct [the importation of Western coal mining techniques]?. . .

As for the clamorous discussions around this issue, one doesn't have to involve himself in uncovering the fallacy of the other side's argument. Why not approach the problem by basing one's argument on the real facts of the situation to prove one's point? When our Chinese people build these enterprises ourselves, we shall reap the profit. There is no reason for groups to rise up and thwart these activities. Several decades from now, the foreigners will have gradually built up such infrastructure in whatever places they arrive, and the momentum of their power and influence will be sufficient to curb any competition, and the profit they derive from this will be enough to entice scoundrels and troublemakers to work for them towards their own purposes, resulting in all the power and wealth being vested with the foreigners and China being unable to stand on its own. Mencius said, "Heaven's plan in the production of mankind is this: that they who are first informed should instruct those who are later in being informed, and they who first apprehend principles should instruct those who are slower to do so." The responsibility for having foresight and forethought certainly rests upon our senior ministers at court. Hence, once the political principles of this are clarified, then the debates among the scholar-bureaucrats will cease on their own. It just requires the imperial court to make a firm decision and implement it, and that is all.

* *

Banning Opium Smoking

As for the fundamental, major policy plans of the country, when we evaluate the present moment and assess our current strength, it would be difficult to implement even one or two of them. But there are numerous pressing issues, which aren't even related to the major policy plans, but the various political measures necessarily must follow upon resolving these. If you don't concentrate on these first, then even if you have good measures and fine intentions and try to achieve a successful outcome, it will ultimately fail.

I am referring first to a ban on opium smoking. Opium smoking was originally prohibited during the Yongzheng reign [1722–1735], when it was just an ingredient in medical prescriptions. Due to the orderly and enlightened nature of governance and moral education back then, officials upheld the law, and no one among the common people dared to smoke it. Coming down to the Daoguang reign [1821–1850], this habit started to become more prominent, though when I was young, I had never heard about it. At that time, material resources were abundant, every family was well fed and clothed, and the people were law abiding and circumspect. Subsequently, the harm caused by opium smoking grew, and societal morals worsened day by day. Floods, droughts, and banditry ensued, feeding off one another. The intensity of the "foreign disaster" we are facing today actually started with a ban on opium smoking [1839]. The leaders of the Jintian rebels [the Taiping], gathered together the coastal militiamen dispersed [by the Opium War, 1839–1842] into the remote valleys, and this gradually developed into a major rebellion [Taiping Rebellion, 1850–1864]. So this opium smoking not only undermines people's livelihoods and depletes their financial resources; it was also a major cause of the great rebellion. Nowadays, the foreigners consider smuggling opium into China to have been a major mistake, but somehow, China has grown accustomed to this habit and made peace with it! I personally believe that as long as the harm from opium smoking is not eradicated, nothing can be accomplished. As for the method of prohibiting it, there is a very simple and easy way to do it. The method consists in dealing with the officials first and then the people, starting with the scholar-officials and then extending their example to the commoners. On the one hand, use the art of persuasion, while on the other, use punishment to accomplish a thorough eradication. This method is also intended to inspire people's sense of integrity and encourage self-reliance. Within a period of 20 years, we can ensure that it is completely eliminated, and there will be not even the slightest lingering trouble. I have already explained this in detail earlier, and I have nothing to add. This is the first matter.

Cultivating Wastelands

Secondly, we must speak about cultivating wastelands in Jiangsu and Zhejiang. At first, I heard that foreigners were focused only on garnering profit for their merchants and didn't do much business in agriculture. Then, I finally started to realize that this was not the case. Their trade relations concentrate on enriching the people. They levy taxes on only a few things like tea, alcohol, and tobacco, with every other commodity remaining untaxed. Each year, an account is drawn up that shows the revenue from trade and the profit that the merchants have garnered, from which the government only takes about 1.25 percent. Houses and other material assets also incur a property tax, roughly like our household tax in China. Those whose annual household income is below £300 are not taxed at all. There is a tax on honorary medals one has acquired and other engraved items of commemoration; those animals beyond just dogs raised as pets are also taxed. Alongside such customary taxes, the bulk of the government-managed revenue comes from a land tax, which shows how diligent they are when it comes to exploiting the strength of the land. . . .

Right after Jiangsu and Zhejiang experienced the Taiping Rebellion, 12 or 13 years ago, there were many uncultivated fields. The common people were afraid of expending effort in clearing fields and plowing and were conserving their energy. The prefectures and counties found it too troublesome to send in reports of cash and grain revenue, and moreover, they obscured the names of landholders. Wastelands were allowed to become overgrown with weeds, while land under cultivation was concealed from the tax rolls. This should be managed by the Board of Revenue and encouraged and guided by the governors-general and provincial governors. This is the second matter.

The Issue of Kashgar

Thirdly, I must say that the territory of Kashgar should be ceded to Yakub Beg [1820–1877]. The Di and Qiang barbarians revolted several times in Jiuquan Commandery [in present-day Gansu], but Emperor Guangwu of the Han [r. 25–57 CE] remained their sovereign by conferring on them seals of office. The Tuyuhun confederacy had moved their people to the area of the Datong River [west of present-day Lanzhou], so the Tang dynasty official Guo Yuanzhen [656–713 CE] petitioned the throne that they be officially installed in that place. To draw the important frontier areas closer, it is best to adopt a "loose rein" policy, and then one won't have anything to worry about. As for the Han dynasty establishing the offices of the Wu and Ji Colonels [48 BCE] and the Tang dynasty setting up the office of Protector General to Pacify the West (640 CE), the farthest garrisons were more than 10,000 Chinese miles distant from the capital and the closest were still several thousand miles away, so when surrendered Xiongnu or other barbarians occasionally revolted, their leaders were enfeoffed with a title as a matter of convention. Thus, these events are not even worthy of comparison. Kashgar is close to Andijan [in Uzbekistan], and our power is insufficient to use the Pamir Mountains as another Great Wall to block their people from coming in. The various tribes of Kokand [in the Ferghana Valley] are attached to Russia. The remnants of the Muslims [Yakub Beg] took advantage of the chaos in Kashgar to attack and occupy that territory. Still, overawed by China's might, he longs to be entrusted as our vassal, and last year, Thomas Wade [minister plenipotentiary and chief superintendent of British trade in China] made such a request on their behalf. I say that we should comply with their thoughts on this matter and conclude an oath with Yakub Beg, ordering him to hand back each of the cities. If in return we are able to station a senior official in Urumqi, his good faith and prestige would be enough to keep them submissive and would guarantee that there would be no trouble for the next century. If we were to rely exclusively on military might to capture those cities, it would be a very protracted and endlessly expensive campaign. If we were fortunate enough to be victorious, the remaining Muslims would certainly defect and ally themselves with Russia and engage in looting and military disturbances year-round. This will only allow the Russians to take advantage of the situation and sit idly by enjoying its benefits, while China just suffers its aftermath, without knowing how to ever bring about a positive outcome.

Those who govern a country must concentrate on long-term planning. Military leaders know only how to go on the attack. Hence, a proposal to cede this territory would never come from our generals. We can rely only on the imperial court to weigh the priorities, formulate the appropriate plans, and make these guidelines known. Then the general's might can be extended and the imperial court's kindness can penetrate deeply into the hearts of people far away, causing them to bow their heads and become obedient. Therefore, I say that Thomas Wade's intercession for Yakub Beg is an opportunity that cannot be missed. This is the third matter.

Treaty with Russia over Ili Valley

Fourthly, I must speak about the matter of the cities in the lower Ili River Valley [present-day northwest Xinjiang]. We should negotiate a long-term treaty with the Russians over this. The national power of Great Britain and Russia is sufficient to make them adversaries, but British people concentrate on expanding their territory to stimulate profit, whereas the Russians raid territories just to expand their borders. Without any reason, they took advantage of the chaos [caused by Yakub Beg in Kashgar] to attack and occupy Ili [1871], and their aim was just to plunder its territory and nothing else. I feel that settling this Xinjiang matter and causing the Russians to hand Ili back to us is certainly going to require a great deal of difficult discussion. There is no article under Western international law that permits a country to take advantage of disorder to occupy someone else's territory. Since the Russians are in the more powerful strategic position, they will require a war indemnity before we can redeem the land, and nothing less than a huge amount will satisfy them. Thus, nothing would be better than to turn the tables on them. Instead of requiring them to lower the redemption fee that we would pay to get the territory back; rather, we should ask them to pay that same amount to purchase it from us. If they want to completely occupy a deserted Japanese island like Sakhalin [1875], then it is obvious that they won't be willing to easily return Ili to us. Instead of vaguely leaving the issue unsettled, which could give rise to a war, there is no better course of action than to make a clear treaty with them and draw the borders, ensuring peace for several decades. If it is absolutely unavoidable, we could ask the Russians to cede to us the territory west of the Amur River in exchange for Ili and still be able to justify this. This is the fourth matter.

Abolishing Internal Tariffs

Fifthly, I want to talk about abolishing the *likin* 釐金 (internal tariff). I am very knowledgeable about raising *likin* to provision militias [against the Taiping], and I was very effective in carrying it out. When Hunan started to implement the collection of the *likin* tariff, I did my best to endorse and promote it. When I was in Guangdong [as inspector general, 1863–1866], I also submitted a report on the situation of the *likin*, citing ancient precedents to support the present needs. So I feel that I can grasp the main points. Nevertheless, I feel that, in general, raising special taxes to fund the military is not how the country was originally governed. It has been more than 10 years since those military affairs were completed, so to not even discuss abolishing the temporary tariff is wrong. When such taxation methods persist for too long, it gives rise to fraud and misuse. The provinces have no urgent needs and just consider the *likin* revenue to be spare money, not paying much attention to it. Hence, the longer the practice endures, the revenue continues to diminish daily. If there were suddenly a military emergency and we needed to make plans to provision an army, but people just abided by the old rules, and customary practices became permanent conventions, then there would be no additional pressure we could exert in this matter. This would be a path leading doubly to ruin. Last year, following the case in Yunnan [Margary Affair, 1875–1876], we had to abolish the *likin* in the foreign concessions of the treaty ports. This exclusively absolved foreigners from paying this duty. But how is this different from chasing a fish into deeper waters or driving a sparrow into a denser thicket? Not only did this lose us the support of our domestic merchants, but we also did considerable harm to our national prestige.

My meaning is that I consider it appropriate to immediately abolish collecting the *likin* in every province, and the exemption clause for the foreign concession should naturally be removed as well. At the same time, we should establish a treaty which states that raising funds for a military emergency shall be an exception to this rule. However, we should collect a local tax on tea from Fujian, on silk from Zhejiang, and other regional commodities, to be used for regional expenses, so taxes on these items shall not be completely exempted along with the *likin*. We can win out in this situation by following a reasonable course of action. In the old days, there was a levy of eight taels of silver [approx. 300 g] on every box of tea. Since the opening of the first five treaty ports, this has dropped sharply to two and half taels. I once spoke to the inspector general of the Imperial Maritime Customs Service, Sir Robert Hart [1835–1911], and his major point was that we can find a way so that the tax on profit from commodities for each province can be negotiated separately, and there should be no reason why such a regional tax would not be permitted. The control of China's own internal revenue should come entirely from the imperial court. I am fairly ignorant about this issue, but I am still very unsettled by it. This is the fifth matter.

Final Reflections on Foreign Relations

When I was first sent to be an overseas envoy, I witnessed some things about China's relations with foreign powers that, following precedent, should be communicated to the court. I cherished the desire to explain these for some time. However, I thought only about the slander I had encountered in the capital, which placed me in an untenable position in the world, so that no matter where I went, I could find no acceptance. The court had no choice but to accept the words of these people, which increased the general contempt for me. When I had just been sent abroad, I was repeatedly censured back at court, so I did not dare to bring these issues up for discussion, for fear of inviting further blame upon myself. However, I consider that Your Excellency is one of the most important ministers in the country and that you are concerned with the success or failure and advantages or disadvantages in China's foreign relations, so it is fitting that I submit them for your careful consideration. So what I have humbly set forth above, after realistically sizing up the situation, is a brief mention of the most important ones. These measures can be adopted and implemented, for they are very simple and easy. They are not just some lofty discussions which are difficult to implement, made just to please myself. . . .

Nevertheless, my study of antiquity enables me to prove my points about current events and allows me to comprehend the essentials of historical foreign affairs. From the Han through the Tang dynasties, they promoted an "appeasement of the foreigners" strategy that had been used to govern the country during the Three Dynasties. I have a deep understanding of the similarities and differences between their situation and today, in terms of what is advantageous or disadvantageous. I consider that this correct principle has been extinct in the Chinese world for over seven hundred years, ever since the Southern Song dynasty. This is the only area where I have complete confidence in my knowledge and am not willing to make concession to others. If only Your Excellency could adopt my suggestions, present them to the court, and encourage it to implement them, then the benefits to our country could be numerous indeed![2]

APPENDIX TO CHAPTER 11

Selections from *Picturesque Representations of the Dress and Manners of the Chinese, Illustrated in Fifty Coloured Engravings with Descriptions,* by William Alexander (John Murray, 1814).

The visual imaginary of the East, originally fueled by the monstrous races and fantastic creatures of Ctesias (chapter 1), was later stimulated by the lavish descriptions of Marco Polo's popular account (see chapter 8). The European perception of how Chinese buildings, people, and landscapes should look was later strongly influenced by designs on imported blue-and-white porcelain, leading to the fanciful visions of the exotic Oriental East seen in the Chinoiserie designs of the late 17th and early 18th centuries.

William Alexander (1767–1816) was one of the first professionally trained artists to travel to China and publish representations of various locations and social classes. Straight out of art school, Alexander was chosen by Lord Macartney as the junior draftsman for the embassy. Alexander was excited at the opportunity to build his reputation and at the prospect of future commercial success from his drawings. Like many men on the embassy, Alexander kept a journal, but he also made numerous sketches of his time in Vietnam and China. He would spend the rest of his life transforming these sketches into finished watercolors and engravings for royal presentation and commercial sale. Alexander complained in his journal that he had limited access to sketch the most important embassy events (like Macartney's audience with the emperor, which he painted after the fact – see fig. 3) and was often confined to a walled residence in Beijing or aboard a ship, not allowed to wander city streets or the countryside looking for subject matter. Though many of his sketches were limited to those characters and vistas he could witness from the deck of his ship on the Grand Canal, he still captured a remarkable cross-section of late 18th–century Chinese society, including peasants, fishermen, beggars, actors, criminals, blacksmiths, musicians, booksellers, Buddhist monks, children, mourners, soldiers, porters, merchants, boatpeople, barbers, puppeteers, and many others.

When he returned from the voyage, he immediately began reworking his sketches into finished watercolors. Though he produced several etchings for the official account of the embassy written by Sir George Staunton (1797) and exhibited more than a dozen of his China paintings at the Royal Academy, Alexander garnered his greatest fame for two collections of tinted engravings. *The Costume of China* was a collection of 48 high-quality aquatint etchings he published in 1805 (see book cover). This was followed in 1814 by *Picturesque Representations of the Dress and*

Manners of the Chinese, a lower-quality collection, from which I have reproduced five captioned plates.

For the illustrations in these two volumes, Alexander would often draw elements from different locations or reuse stock characters and buildings from other drawings. For his landscape and water scenes, he would also add elements people expected from Chinoiserie designs, like pagodas or willows. His Chinese people have disconcertedly Western faces and are usually placed against a stark white background, like some specimen in a natural history catalogue. Alexander had an eye for the details of costumes and the intricacies of ships, but his attempts to render Chinese written characters always resulted in gibberish, and his people are often awkwardly posed, wrongly proportioned, and expressionless.

The descriptions Alexander attached to his plates are more illuminating. He praises only a few aspects of Chinese culture. Overall, his account is quite critical and condescending. Following the trend of negative impressions of China that began in earnest with that of Commodore George Anson (1697–1762), Alexander views China through an imperialist lens, encouraged by increasing English self-confidence. To Englishmen like Macartney and Alexander, China was a backward and decaying monarchy, with quaint but primitive customs that were not worthy of emulation, but were worth studying and recording by naturalists, ethnographers, and draftsmen, who had documented similar "primitive" cultures in India, Africa, or the Americas.

Alexander's captions show that he was critical of practices like foot-binding and the brutality of Chinese punishments. He also denigrates the Chinese work ethic, belittles popular religious practice as "superstitious," dismisses Chinese music as "execrable" and their literature as "inferior," and disparages Chinese shipbuilding as outdated. While he notes that some high-class women could be considered refined and attractive, he refers to female peasants as "ill-featured" and "void of expression." Like his superior Macartney, he reserves his most biting comments for Chinese soldiers, whom he calls "naturally effeminate" and "enervated," depicting them with outdated weapons, flimsy armor, or wearing ridiculous tiger uniforms. He calls their military tactics "ridiculous and absurd."

Alexander's images of China would be reprinted numerous times and would have a tremendous influence on the European visual imagination of China for nearly half a century, until the advent of photography.

ONLINE LINKS AND SELECTED STUDIES

Alexander, William. *The Costume of China, Illustrated in Forty-Eight Coloured Engravings.* William Miller, 1805. https://archive.org/details/costumeofchinail00alex.

Alexander, William. *Picturesque Representations of the Dress and Manners of the Chinese, Illustrated in Fifty Coloured Engravings with Descriptions.* John Murray, 1814. https://archive.org/details/representationsdres00alex.

Chen Yushu. "William Alexander's Image of Qing China." *Monumenta Serica* 67, no. 2 (2019): 397–440.

Legouix, Susan. *Image of China: William Alexander.* Jupiter Books, 1980.

Sloboda, Stacey. "Picturing China: William Alexander and the Visual Language of Chinoiserie." *British Art Journal* 9, no. 2 (2008): 28–36.

Wood, Frances. "Closely Observed China: From William Alexander's Sketches to His Published Work." *British Library Journal* 24, No. 1 (1998): 98–121.

Figure 4. "An Offering in the Temple."

Source: William Alexander, *Picturesque Representations of the Dress and Manners of the Chinese, Illustrated in Fifty Coloured Engravings with Descriptions* (John Murray, 1814), plate 6.

"The figure kneeling before the deities mounted on pedestals is a priest of the sect of Fo (Buddhism). He is burning incense, or rather paper that is covered over with some liquid that resembles gold. Sometimes, in lieu of this, tin foil is burnt before the altars of China, and this is the principal use to which the large quantities of tin sent from this country is applied. On the four-legged stool is the pot containing the sticks of fate, and other paraphernalia belonging to the temple, and behind it is the tripod in which incense is sometimes burned. These superstitious rites are performed several times by the priests every day, but there is no kind of congregational worship in China. The people pay the priests for taking care of their present and future fate."

Figure 5. "An Itinerant Musician."

Source: Alexander, *Picturesque Representations*, plate 19.
"The Chinese have full as great a variety of musical instruments as most other nations, but they are all of them indifferent, and the music, if it may be so called, produced out of them, execrable. The merit of our traveling musician consists in beating a sort of tambourine, or rather a shallow kettle-drum, with a mallet held between the toes of one foot, while he strikes a pair of cymbals with the other, and, at the same time plays upon a sort of guitar accompanied by his voice. It would seem also that he is equally skilled in wind instruments, of which a flute and trumpet make their appearance out of the mouth of his bag; a pair of rattles connected by a piece of riband [ribbon] lie on the ground, and near them a hollow piece of wood, nearly heart-shaped, which, when struck with a mallet, emits a dull disagreeable sound, like the hollow bamboo carried by the watchman, for which this is sometimes substituted. A Chinese band always play in unison, and never in parts: this indeed is an art they have not yet reached, and those few who have heard European harmony pretend to dislike it. A Chinese ear is best gratified with the sounds of noisy instruments, as gongs, kettle-drums, shrill trumpets, jingling bells and cymbals, or with the faint and reedy tones, scarcely audible, of a little bamboo organ, which swell and die away not unlike those of an Eolian harp."

Figure 6. "A Bookseller."

Source: From Alexander, *Picturesque Representations*, plate 23.

"In so arbitrary a government as that of China, it would scarcely be supposed that the press should be free; that is to say, that everyone who chooses it may follow the profession of a printer or a bookseller without any previous license, or without submitting the works he may print or expose for sale to any censor appointed by government; but then he must take his chance to suffer in his person all the consequences that may result from the impression that may be made on the minds of the civil officers as to the tendency of the work. A libel against the government, an immoral or indecent book, would subject both printer and publisher to certain punishment both in his person and purse. The Chinese have not made any great progress in literature, and still less in the sciences: they most excel in the history of their own country, in morality, and in practical jurisprudence. Their dramatic works are constructed on the same model as those of the Greeks, to which it is hardly necessary to add they are infinitely inferior. Their novels and moral tales are better; but the works in most esteem are the four classical books supposed to be written or compiled by Confucius. Their printing is not performed by moveable types, like ours, but by wooden blocks the size of the page; and this mode appears to have been in use long before the Christian era."

Figure 7. "Punishment of the *TCHA*, or Cangue."

Source: Alexander, *Picturesque Representations*, plate 39.
"The punishment of the cangue may be compared to that of our pillory, with this difference, that in China a person convicted of petty crimes or misdemeanors is sometimes
sentenced to carry the wooden clog about his neck for weeks, or even months; sometimes one hand, or even both hands, are inserted through holes, as well as the neck.
The annexed representation is not a common one, and far less painful than the plain
heavy tablet of wood, the whole weight of which must be supported on the shoulders;
whereas in this it is mere confinement, without the person being compelled to carry a
heavy load. The nature of the offence is always described in large characters, either
on the edge of the cangue, or, as in the present instance, on a piece of board attached
to it."

Figure 8. "A Chinese Lady of Rank."

Source: Alexander, *Picturesque Representations*, plate 43.
"If we except the unnatural custom of maiming the feet, which swells and distorts the ankles, and wrapping the latter up in bandages, the dress of Chinese ladies in the upper ranks of life is by no means unbecoming. In the head dress, in particular, they sometimes exhibit great taste, and great variety; and the materials of which their garments are made, and especially those parts of them which consist of their own embroidering, are exceedingly beautiful. Confined by education in their mental acquirements, a great part of their time is employed in works of this kind, in looking after and cultivating plants growing in pots which decorate their apartments and inner courtyards, and in attending to birds, which are either kept for singing, or some particular beauty of form or plumage. The buildings in the background form part of a view of Peking, near one of the western gates."

NOTES

1. George Macartney, *An Embassy to China; Being the Journal Kept by Lord Macartney During His Embassy to the Emperor Ch'ien-Lung, 1793–1794*, ed. J. L. Cranmer-Byng (Longmans, 1962), 74, 84–85, 86–88, 90, 97, 99–100, 114, 119, 122–24, 170, 174, 190–191, 202–203, 210–213, 222, 226, 239–240. Only the original journal text, which is in the public domain, was excerpted, and some spellings have been Americanized. The dates are from the original.

2. Guo Songtao 郭嵩燾, *Guo Songtao shi wen ji* 郭嵩燾詩文集, edited by Yang Jian 楊堅 (Yuelu shushe, 1984), 188–195. Translation and headings by the author.

CHAPTER 12

CULTURAL AMBASSADORS (1845–1870)

The political ambassadors we met in the previous chapter carried great responsibilities in their official functions: representing their country abroad, analyzing shifting geopolitical situations, and engaging in delicate negotiations with monumental consequences. But these men were also keen cultural observers of the foreign environment in which they found themselves.

The paired travel writers in the current chapter, Eliza Jane Bridgman and Wang Tao, held no official titles or functions. Bridgman was the first American female Protestant missionary to teach and preach in China, and Wang was the first classically trained Confucian scholar to reside in the United Kingdom for an extended period. They represent different lenses through which to view the fateful collision between civilizations in the 19th century and are even more illuminating when read in parallel with one another.

Both travelers held reform-minded views on women's education in China. For Bridgman, this was an integral part of her American Protestant upbringing, but for Wang, this view was rare among his contemporaries, though it can be explained by his long-term exposure to missionary ideas. Both travelers were gawked at as curiosities during their sojourns abroad, though Bridgman experienced overt hostility and xenophobia in the charged atmosphere of South China after the Opium War.

Curiously, both cultural ambassadors saw religion as a means to bridge the chasm between Western and Eastern cultures. Bridgman believed that the Gospel could liberate China from ignorance and vice, creating a brotherhood in Christ, whereas in his speech given at Oxford, Wang suggested that behind the superficial differences between Confucianism and Christianity, a "Great Unity" formed the foundation of belief East and West, and cultural exchange could lead to a shared spiritual future.

INTRODUCTION TO ELIZA JANE GILLETT BRIDGMAN

Though some Western women had sojourned or lived in East Asia since at least the 13th century (e.g., William of Rubruck met a woman from Lorraine named Paquette in Karakorum), the American Protestant Eliza Jane Bridgman (née Gillett, 1805–1871) was the first Euro-American woman to live for an extended period in China and write about her observations.

Some intrepid Protestant missionaries like the pioneering Robert Morrison (1782–1834), who had translated the New Testament into Chinese, had worked in South China since 1807, but one of the main terms of the Treaty of Nanjing (1842) after the First Opium War opened five new treaty ports for trade. The subsequent American-negotiated Treaty of Wangxia (July 1844) opened the way to buy land for churches and missionary schools in these ports. It was in this environment that the young Eliza Jane arrived in Hong Kong from America in April 1845. Soon after, she met and married Elijah Coleman Bridgman (1801–1861), an experienced American Protestant missionary.

Eliza Jane Bridgman differs from many of the other Western observers in this anthology because she had access to the cloistered world of Chinese women, at least those in the merchant class and lower rungs of society. She felt great sympathy for these "Daughters of China" and dedicated her career to their education and to lifting them up out of what she considered ignorance and oppression. She complained that Chinese women and girls were "not considered worth the pains, time and money, of being taught to read." As one can read in the excerpts from her memoir, she was also sharply critical of the practices of foot binding and female infanticide, arranged marriages, and concubinage. She founded the first schools for girls in Canton and Shanghai and later founded the Bridgman Academy (1864) in Peking.

Her method of Chinese language acquisition differed greatly from that of Matteo Ricci (chapter 10) or her contemporary male Protestant missionaries. Rather than wear herself out studying thousands of Chinese characters indoors, she advised that one must "mingle with the people; hear them talk; and learn as the little child does." Even communicating the Gospel did not require perfect literacy in Chinese, she argued.

The most dramatic episode in her memoir occurred when Eliza, her husband, and some other missionary companions went on a boating excursion outside Canton. After a pleasant morning spent at a local farmhouse, they were assaulted by angry young men on the riverbank, who hurled large stones at their boat, injuring two boatmen. Such anti-missionary sentiment was common in South China after the Opium War, and this particular episode followed an incident during which several Chinese men had been killed by English traders. She also mentions anti-foreign placards that had been posted in the city. Eliza attributed such anti-foreign sentiment to the work of the devil.

PRIMARY SOURCES AND SELECTED STUDIES

Bridgman, Eliza Jane Gillett. *Daughters of China: Sketches of Domestic Life in the Celestial Empire.* Robert Carter & Brothers, 1853.

Bridgman, Eliza Jane Gillett, ed. *The Life and Labors of Elijah Coleman Bridgman.* Anson Randolph, 1864.

Bridgman, Eliza Jane, and Elijah Bridgman. "Journal of Elijah Bridgman and autobiography of Eliza Jane Bridgman, 1834–1870." Yale Divinity School Archives, call no. RG 8, Series I.

Spence, Jonathan. "Women Observers." In *The Chan's Great Continent: China in Western Minds.* W. W. Norton, 1998.

EXCERPTS FROM *DAUGHTERS OF CHINA,* BY ELIZA JANE GILLETT BRIDGMAN (1853)

The Plight of Women in China

In China, as in other countries, woman's influence is immense. It is so in the family, and in the state, in morals and in religion. But what God has ordained for the best and noblest of purposes is in China exercised for evil, because the Father of Lies has here held almost undisputed dominion. From time immemorial the Chinese Empire has been his grand university, where the most captivating forms of idolatry, have been devised and carried into practice.

In such a state of religious society—where all is set in the wrong direction—nothing but the truth of God—through his blessing—can break up this dreadful incubus, overthrow idolatry, and bring the people to know, to love, and to serve Jehovah their Maker. The Bible can, and will do all this, as soon as it is given to the people, and they are taught to know, and led to obey it.

But who shall teach the woman of China? The missionary—the ordained minister of the Gospel, who goes forth to preach—cannot gain access to the daughters of the land. The usages of society debar these from the public assembly. Woman, in all ordinary cases, is secluded, and cannot come out to hear the preaching of the Gospel.

Shall woman then be there neglected? Can nothing be done to give to her the glorious Gospel, and elevate her, to her proper sphere! . . .

After a voyage of 131 days, a good part occupied in the study of the Chinese language, we arrive at Hong Kong, April 24th [1845]. . . .

Visiting a Chinese Lady

It is the people especially whom we wish to study, and it was the condition of woman in particular that led us to China, so let us pay a visit to some Chinese ladies, whose sphere, according to their own classics, is in the "inner apartments.". . .

June 3rd [1845]—On Friday last we prepared ourselves in our best attire to accompany Dr. and Mrs. Parker, Mr. and Mrs. Carr of Hong Kong, and some English gentlemen, to wait upon Mrs. Pwan, the lady of Pwan Tingkwa, a salt merchant of wealth who is somewhat favorable to foreigners.

The only mode here of traveling any distance is in boats and sedan chairs. In this case, as there were several ladies, we took four sedans, each borne upon the shoulders of two men called "coolies." We were preceded by Dr. Parker, while the other gentlemen walked by the side of the sedans. We left the Hong at one o'clock, having received intelligence through a messenger sent at twelve, that "his excellency was not up."

The weather was extremely warm; but having to pass through several streets in order to avoid the gaze of the Chinese, we had the curtains of the sedans closely drawn. It was almost suffocating, but 15 or 20 minutes brought us to our destination. The front gate was opened by attendants in waiting, and the sedans were lowered to the ground with care in an open court. The gentlemen and ladies were respectively directed to different apartments, with the exception of Dr. Parker, who, acting as interpreter for the ladies, was allowed to accompany us.

The room was full of children and women, from their appearance, I should judge, in subordinate capacities, such as nurses, waiting maids, etc.; also, some old women as supervisors; in all, probably, not less than 30 individuals. These came as much to see the sight as we did and did not hesitate, at once, to examine every part of our dress and pass comments thereon, as well as on the lightness of our complexions, which always attracts the notice of a Chinese lady, and she contrasts it with her own tawny skin.

All this is done without the least thought or intention of being impolite; indeed, they begged us to be seated at small tables accommodating two persons, with a chair at each end. The receiving apartments are furnished in this way. Baskets of flowers suspended from the ceiling, and a view in the open court, or perhaps a garden, give these rooms a more cheerful aspect than the external part of the dwelling without windows would lead one to anticipate.

Pwan Tingkwa is employed in the service of the government; this gives him some distinction. He has 10 wives. The lady of the house, or "number one wife," did not make her appearance until a little time had elapsed. At length, she entered the room, and the others gave place, while she received her visitors and refused to sit herself until every one of her guests was seated.

She was a beautiful young creature, not over 21 years of age. Her hair was arranged in their usual tasteful manner and adorned with flowers, pearls, and other ornaments. She was attired in a simple dress of grasscloth [hemp], tight about the throat, with large sleeves, exposing a beautiful hand and wrist full of bracelets. Underneath her grasscloth tunic she wore an embroidered skirt that nearly concealed her little feet. Her manners were graceful and elegant. To the remarks of the ladies she responded courteously, never allowing herself to sit while any of the ladies were standing.

Tea was served in small cups with covers, but without milk or sugar. Soon after this, we were invited into another apartment. Mrs. Pwan, our lady host, took Mrs. Parker by the hand and led the way, while several other of Mr. Pwan Tingkwa's ladies attended to the rest of us, and we followed; the company of relatives, nurses, servants and children succeeded, all eager to satisfy their curiosity by gazing at us.

A repast was prepared, consisting of jellies, fruits, nuts, etc., which in the East is called *tiffin* [*dim sum* 點心]; the Chinese call it "a bit for the heart." It was easy to distinguish the lady of the house; she moved us to be seated, while she presided, the others standing, and the servants fanning us while we partook of the delicacies. According to Chinese etiquette, Mrs. Pwan passed some fruit or jelly on her fork or with her chopsticks to each lady, and we would return the compliment, she rising very gracefully and receiving it; they even go so far as to put it into your mouth.

Tiffin being finished, we repaired to her private bedroom. It was furnished with a mirror, bureau, bedstead with mattress, the bedclothes neatly laid in folds and put aside in the back part of the bed. We followed Mrs. Pwan, all the attendants accompanying us, through the different apartments of this spacious building, still unfinished. The carving was elegant. The rooms were furnished with divans, center tables, mirrors, and chandeliers. The ceilings were beautifully painted with birds and flowers. A gallery was appropriated to the "sing-song" [theater].

In going down the staircase, we passed the room where "his excellency" and guests (the gentlemen who accompanied us) were regaling themselves with refreshments; they could not help turning their heads to catch a glimpse of the fair Chinese ladies. At length, the time arrived for us to leave; the females of the house, one and all, retired to the inner apartments, and the gentlemen conducted us to our sedans. On returning home, we suffered our faces to be exposed, and gazers were not a few, eagerly striving to get a peep at the "*fau quipo*" [*fan qipo* 番妻婆 "foreign devils' wives"], as they stigmatized us. . . .

My mind would revert to those Chinese ladies, and the sex in general. I could not learn that one in Pwan Tingkwa's household, knew how to read in her own language; and as to their employments, much time is spent at the toilet, embroidery perhaps occupies a part, and then the amusements of the theatre, and others equally frivolous fill up the rest. Not many months after this visit, I heard of the death of Mrs. Pwan and also learned that her husband was never pleased with her. . . .

The Dress, Features, and Temperament of Chinese Women

Female education in China is still in its incipient stage of progress. Though all true Christians will readily admit that it is the steady and holy influence of the Gospel that has given her, in enlightened countries, her true position in society—the position that the great Creator designed she should occupy; yet but few have begun to realize what woman is without the Gospel.

The Chinese lady, in the better classes, is not without attractions. She is generally bland and courteous in her manners; her toilet is often arranged with taste and beauty, though her decorations are usually profuse and gaudy.

Her dress is well adapted to the season. In the heat of summer, her attire is simply grasscloth [hemp]; as the weather becomes cool, this is exchanged for silk and other richly embroidered materials.

The whole Chinese system of ethics requires females to be so secluded that their opportunities of intercourse with foreign ladies are few; when they do meet them, however, their address is singularly confiding and affectionate, and they enter into conversation with sprightliness and vivacity. But what do they talk about? Your age; the number of your children; your ornaments; the style of your dress; and your *large feet!*

Examine the countenance of the Chinese: the features are regular, and though there are peculiarities which mark the race, such as the obliquity of the eyes, flat nose, tawny skin, and when uneducated a certain inane expression common to both sexes, yet when the Chinese lady is favored with an interchange of sympathies with one of her own sex from another country, there is light in her eye and joy in her heart. It is not the flash of a bright and highly cultivated intellect—for, alas! She is not considered worth the pains, time and money, of being taught to read. But the women of China have souls, and there are deep fountains there, sending out, as far as their situation admits, streams of maternal and sisterly affection. And there are fountains of evil too, and the courses that issue from them are broad and deep. Ungovernable temper often spreads discord in the domestic circle, and the strong folds of idolatrous superstition bind her tender offspring by an oath of perpetual fidelity to the altars of false deities. . . .

Learning the Chinese Language

The Chinese written language, without doubt, is very copious. The number of written characters can hardly be ascertained. But does anyone suppose that it is necessary to have at command one-half or one-quarter of these written characters in order to impart a knowledge of the simple truths of the Gospel? If so, allow me to remove that impression.

The number of spoken dialects is also very numerous, and some knowledge of the local phraseology is certainly indispensable. The means of acquiring this are very simple: *mingle with the people, hear them talk, and learn as the little child does.* Indeed, we must follow our Savior's direction and "become as little children," in order to get access to the Chinese mind.

To habits of daily intercourse, it is thought by some who have been long in the field, should be added a few hours of study on the written character. This exercise, if it is not too long at one time, is pleasant and will afford a variety of occupation.

Christian missionaries, of course, differ in their views on this subject and pursue various methods. There are some, whose knowledge of the character is very limited, who are very successful preachers of the Gospel and are readily understood by the Chinese. I have in my mind's eye one well known to the Chinese in all the regions where he lived, who by his constant practice of being familiar with the people, wearing a smile, and greeting his acquaintances in a kind and friendly way, possesses an influence in that neighborhood which will never wear out. Such a one learns to reach the heart of the heathen. The children recognize the feeling, and to them such a missionary is always welcome.

The health of several persons has been seriously injured, and some have lost it entirely, by too close in-door application, to Chinese during the first or second year of a residence in the East, and then too, if a feeling of discouragement takes possession of the mind in the outset, it acts like an incubus—induces sedentary habits, and often the individual disheartened, sinks under the pressure of disease, or returns to his native land. . . .

Another Visit to Some Chinese Ladies
On the third instance, I enjoyed another visit to some Chinese ladies at the house of How-kwa, one of the Hong merchants. We are allowed to pay our respects to them, but alas! Their lords will not permit them to reciprocate this visit in our own dwellings; the higher class at least will not, those in more humble life are not so particular.

Our party consisted of J. D. Sword, Esq., five children, and two nurses; Mr. and Mrs. Delano, child and nurse; Mr. Trott; Dr. and Mrs. Parker; Dr. Bridgman; and myself.

On this occasion boats were in requisition, instead of sedans. The gentlemen went in one boat, and the ladies and children in another.

As we drew near to the landing, the ladies in full Chinese dress made their appearance on the terrace. It seemed crowded, and they were all chatting together. As we were leaving the boat, we were met by a man whose duty it was to conduct us to the female apartments.

Mrs. How-kwa, a lady in middle life, gave us a polite reception and introduced her son, a young man, who kept close by her side. There were a good many ladies present, and our conductor, as we passed along, remarked, "This is Mr. How-kwa's number two *wifoo*," "this number three," "number four," and so on.

These ladies, although handsomely dressed, carry in their manner and bearing a sense of inferiority, which it is rather painful to witness. They regard themselves and act in a subordinate capacity. They are sometimes purchased for a sum of money, which varies I apprehend according to their youth and beauty; and their condition is so humiliating, that in some cases it is a state of servitude, and they may be discharged at the pleasure or caprice of their lords.

Mrs. How-kwa was a more dignified lady than any I had previously seen. She evidently did not consider these "small wives" (as they are sometimes called) on an equality with her, though she could claim all their children as her own. I suppose if a concubine be sent away, she cannot usually have her children to solace her loneliness, in a state worse than widowhood, and yet if it be her master's pleasure he can compel her to take them, if he does not wish the trouble and expense of maintaining them, which is often the case if they be girls.

When we look at woman's condition in China, in all its aspects, we need not wonder that before her female offspring have drawn but few inhalations of a heathen atmosphere, with the prospect placed before the child which the mother knows, and feels in all its force, she quenches the fire of maternal love and closes its existence by suffocation.

This act I imagine is usually committed immediately after birth. They do not wait for the eye to sparkle, and the smile of the expanding infant to work upon the maternal bosom—this would be too much for a mother's heart, even for a heathen Chinese mother. . . .

A Perilous Outing in the Countryside

Up to July 25th, 1846, we had experienced in all our excursions no other than exhibitions of kind and friendly feeling, and we had no fear or hesitancy about going together anywhere among the people in the vicinity of Canton.

But this was not to last, and we must now turn to the dark side of the picture. The seat of the great Enemy is here, and he only waits for suitable opportunities, for a demonstration, through those over whom he reigns, of his hatred and hostility to foreign influence.

It was Saturday, the close of a warm and laborious week, and our health required a change of scene, the springs of life needed resuscitating. Dr. Bridgman proposed a longer excursion than we had previously taken, and there were friends with us, who were happy to join the party.

In the early part of this month, July, there had been a quarrel between some of the merchants and people, recourse was had to arms, the mob would not desist from violence, they were fired upon and several Chinese killed. But as the excitement had passed away at Canton, so we supposed it had in the surrounding villages; otherwise, we should probably have remained at home.

A Hong-boat was procured. These are used generally by the merchants. They are much larger than the *tanka* boats. One accommodates six persons or more, has a comfortable cabin with Venetian blinds, and arrangements for reclining. Some are even large enough for a center-table and chairs and require several boatmen.

Our company consisted of Rev. Mr. Pohlman, who has been before mentioned and who was then staying in Canton, being treated for his eyes, under the care of Dr. Parker; Mr. Bonney, and Sze Ping, to whom allusion has been before made, and who was then Mr. Bonney's teacher, Dr. Bridgman and myself.

To the places where we went on this occasion, my husband, in the early years of his residence at Canton, had gone frequently with confidence and without molestation.

On our way, we neared some junks lying in the river that had come from Siam [Thailand] and Cochin-China [southern Vietnam]; the mast of one of them had been scathed by lightning; and the people were offering sacrifices and giving thanks to the gods for their deliverance from death. The gentlemen went on board and distributed tracts and copies of the Testament. Then we passed on, about four miles down the river and turned into a creek, where was a pleasant landing place. There we walked up a hill and came to a farmhouse. It was a mere shed, but there was a poor woman there who understood the rites of hospitality. She prepared *tiffin* in her humble way, gave us a rough bench, the best she had, and begged us to be seated at the table and take some tea. There was something in her manner particularly attractive, because her politeness seemed to proceed from real kindness of heart. While we were partaking of her good tea, a crowd of people came round to look at us. My costume especially attracted their attention, and I took off my bonnet and allowed them to see the style in which my hair was dressed. I was quite willing to gratify their curiosity, as they seemed very respectful.

We then walked out to some Chinese graves, built with hewn stones, in the form of a semicircle. Here they were threshing rice. The prospect from this hill was charming. I had seen nothing that so much resembled the scenery in our dear native land. As far as the eye could reach were hill and dale clothed with verdure, and the river winding along in the midst. On our return, our hostess, above alluded to, told her son to accompany us and show us the way down the hill, which he did courteously and saw us safely in our boat.

We re-embarked and proceeded on our way, along the creek, passed under a bridge, and came to a Chinese village; the sun was declining; the lengthened shadows of the shrubbery upon the water gave indications of the approach of evening. It being a warm summer's day, I put off my bonnet and went outside on the deck of the boat to enjoy the cool air. I had not stood there but a moment before Sze Ping said to me, "You had better come inside, those are bad people on the shore."

I observed a crowd of boys and others making a noise; and presently heard the sound of pebbles against the sides of the boat. Sze Ping looked alarmed and closed the blinds. There came small stones with more force. The excitement seemed to increase. There was quite a mob, pieces of mud, and heavier stones came. We barricaded the windows with anything we could find.

At this crisis, two boatmen rushed into the little cabin wounded, it was fearful—it seemed as if death was very near; the stones grew large and heavy. Dr. Bridgman went out on the deck to expostulate with them; they would not give heed to anything that he said, but replied, "You lie, you have killed our people, and we will kill you;" and the mud and the stones came thick and heavy. I begged my husband to come inside, his situation was so much exposed. He at length yielded to my entreaties. He was covered with mud and dirt but not wounded.

The two boatmen were sadly cut, and I took my pocket handkerchief and with some cold water tried to staunch the blood—the storm of stones increased; our Venetians were getting broken, and we were expecting every moment to be knocked down by the heavy stones that came in quick succession.

One man with a demon-like expression plunged into the water and filched away an oar, and two or three of the boatmen escaped to the shore with fright.

The tide was against us, and we had to pass under another bridge before leaving the creek. There was one young man, about 17, left in the bow of the boat, who remained firm at his post. The mob gathered on the bridge, and as the boat emerged from under it, threw down a stone large enough to sink the boat or kill any person upon whom it might fall. It struck upon a beam of the boat, cracking the beam, but harming none of us. The young hero of our battered craft took the stone and sat upon it, still rowing with all his might.

This was their last effort. After being a half-hour under this shower of stones, a wall on one side, and a broad sheet of water, on the other, prevented further pursuit. They had done their utmost to destroy the "foreign devils," as they called us, but an arm stronger than theirs foiled their attempt. We soon reached the Macau passage, a branch of the river above Canton, with grateful hearts to our Heavenly Father for his protection in a time of such imminent peril. Our boat was almost a wreck.

It was nearly dark when we arrived at the landing; a favorable hour, for we were in a sad plight. My dress was covered with the blood of the wounded boatmen; the others were covered with dirt, though not a hair of our heads was injured.

The heavy stone, which no doubt was intended as our death blow, was taken home, and weighed nearly 100 pounds.

The wounds of the boatmen were not serious; medical attendance was procured, and they soon recovered. The young hero, who kept his post and performed his duty so well, was suitably rewarded; and we presented our thanksgivings to our Heavenly Father for so great deliverance. . . .

The Confinement of Chinese Women

By long-established custom, woman in China is confined to the inner apartments, her feet are cramped, and she never goes out except on some particular idolatrous days, when it is their special duty to visit the temples and make offerings. She must have no will of her own but be entirely subordinate to her mother-in-law. Not treated as a companion by her husband, untaught in books, what are her resources? Alas! Her mind becomes a prey to unmeaning superstitious rites, her temper often sour and irritable, and her household a scene of jargon and discord. No wonder that in subsequent life, after such a long season of subjection, she becomes herself in turn the tyrant and uses her sway to the best of her ability. No movement of any importance can be made without the consent or acquiescence of paternal grandparents. It is said that the emperor's mother has almost unbounded sway in the imperial household.

After these aged parents or grandparents die, then comes the show of devotion; filial offerings, in abundance, are paid at the tombs, and on the occasion of a marriage, the worship of the ancestral tablets is an important part of the ceremony.

The women of China possess intellect, but it wants cultivation; they have hearts, but they require the Gospel's sanctifying influence; they need also early, judicious training.

The daughter, at her marriage, becomes a part of another family and is entirely given up by her own. The son, at his marriage, remains in his father's house and pays divine honors to him when he is dead; therefore, a father considers it a great calamity to have no son to make offerings when he is gone, and the rich often multiply their wives until they can obtain the desired blessing. In some cases they adopt one of another family.

Women of intellect and observation feel their deprivations, especially when they become acquainted with foreign ladies. Often in calling upon those of the richer class, we ask them to return our visits; they answer, "No, we cannot; we would like to come, but no have this custom, Chinamen not like laugh so much." . . .

* * * * * * * * * * * * * * * **BRIEF EXCERPT** * * * * * * * * * * * * * * * *

Infanticide, Foot Binding, and Female Degradation and Ignorance

Nothing can exceed the ignorance and the degradation of a great proportion of the females in China. Shut up and crippled from their infancy, the higher classes spend their time in the decoration of their persons, the amusements of the theater, and games of chance. A little embroidery, perhaps, may occupy a small portion of their time; but the most beautiful specimens of work are done by men; the women are astonishingly deficient in the use of the needle, and as to being able to read their own language, probably not one in a hundred, even of the better class, receive any instruction from native teachers.

Parents sometimes destroy their female offspring soon after birth, and in cases of want, some of both sexes are left to starvation in the streets. All this seems to be done without compunction of conscience.

Many are maimed, to be made beggars; their eyes are put out, a foot perhaps amputated. Sometimes children are exhibited in the streets, apparently covered with smallpox, to excite pity and extort money. You examine the child, and it is all a deception; something is put upon the face that appears like the disease, by which the passerby is deceived. It is well known that the Chinese place little or no value upon their daughters, and if questioned as to how many children they have, they answer according to the number of their sons, omitting to bring their daughters into the account.

I once asked a tailor, "Why do your people always rejoice at the birth of a son and not at the birth of a daughter?" "Because the girls are so much trouble and expense, they cannot work and get money."

Again, I asked an officer of government, "Why do you not teach your daughters as well as your sons to read?" He replied, "It is of no use." I said, "Will you send your little daughter to me to be taught?" His answer in broken English was, "No can do;" meaning that it would be of no use.

. . .

Very few Chinese women know how to sew as to make their own clothes. We wish to gather female children into schools to give them habits of industry, and that they may acquire skill in various kinds of work. The women here are very passionate. Nothing is more common than domestic broils, which are carried to a fearful extent. Betrothed at an early age, the marriage is founded upon the will of the parents, the parties have no choice in the matter; hence they must be often unsuited to each other.

. . .

This [Chinese] seamstress says, "The ladies who come from other countries all know how to read and write, but the poor Shanghai women do not. They have to work all the time and have no opportunity to learn." As she said this, there was something in the expression of her countenance that told a truth which she seemed to feel: the great difference between the Chinese female and the English or American lady.

. . .

Most of the pupils who entered our family as boarder, had their feet already compressed. To this prevalent custom there were some exceptions; at least their feet had not been long bound, and there were five or six whose parents were persuaded to let them remain without being bandaged.

This is a most barbarous practice, but it has such a strong hold upon the people that it will be a long time before it will be relinquished.

Our ears are often assailed by the cries of children who are suffering from compression. In the higher classes, this process commences very early, at four or five years of age. The elder women are called upon to do it. The four toes on each foot are bent completely under, and then a long fold of bandage is put on, which is tightened every three or four days. The pain is very severe.

This fashion is the mark of a lady and considered indispensable to a suitable betrothal, which also takes place early. A girl whose feet are permitted to grow to the usual size, would not be selected as the wife. She may be bought for a sum of money if she has a pretty face, for the second, third, or fourth wife, but the large feet affect her rank seriously in domestic life, and hence the prevalence of the custom in the middling and lower classes. . . .

* *

Anti-Foreign Placards

August 15th, 1851. During the current month, a vile placard has been published against foreigners, and some of the pupils have been ridiculed and threatened by their relatives for continuing under our instruction.

One of them, on returning from a visit to her friends, mentioned some false and wicked remarks that were made by them about us. She was asked if she believed them; she answered, "No." Have we dealt truly or deceitfully with you? "Always truly," was the reply. Do you think the foreigners wish to kill and destroy the Chinese, as the placard stated? "No; I think the foreigners often help the poor Chinese, when our own people will not."[1]

INTRODUCTION TO WANG TAO

Wang Tao 王韜 (born Wang Libin 王利賓; November 10, 1828–May 24, 1897) was the first classically trained Confucian scholar to spend considerable time living abroad in the United Kingdom and France. He used his observations to argue for reform in China. Scholars consider him a member of the "pioneering generation" of reform-minded thinkers in China, one with decades of experience working with Westerners in Shanghai and Hong Kong, even before he sojourned in Europe and experienced their culture firsthand. He was an intellectual ancestor of reformers like Liang Qichao (chapter 13) and revolutionaries like Sun Yat-sen.

Wang Tao was born into a minor gentry family in Suzhou. After his father's death in 1849, he moved to Shanghai and worked as a Chinese editor for the London Missionary Society press for 13 years, helping with a new translation of the Bible into Chinese. After a secret letter he wrote to the leadership of the Taiping rebels was discovered in 1862, he changed his name to Wang Tao and fled to Hong Kong, where he collaborated with missionary Sinologist James Legge (1815–1897) on translating the Confucian Classics. In 1867, Legge, who had returned to Scotland for his health, invited Wang to join him in his hometown to continue their collaboration.

Wang left Hong Kong for Europe on a steamer on December 15, 1867, and arrived in Marseilles around 40 days later, after stops in

Singapore, Ceylon, Aden, and Suez. He spent a few months in Paris and London, visiting museums and parks, attending the theater, and socializing extensively within Legge's circle. He even lectured on Sino-Western cultural exchange at Oxford University, speaking in Chinese. He spent most of the next two years working on the Confucian Classics translation with Legge in the town of Dollar, Scotland, taking breaks to hike in the highlands or attend festivals. He returned with Legge to Hong Kong in March 1870.

Wang's observations from his sojourn in Europe provided ample material for his subsequent journalistic career (he founded the first wholly Chinese-run newspaper), with his editorials advocating for technological, educational, and political reform. They also fueled his literary career in a series of fictional short stories and novels about the adventures of Chinese men traveling abroad or Western women in China. His travel diaries were compiled and published in 1890, titled *Illustrated Record of My Jottings on Carefree Travel* (Manyou suilu tuji 漫游隨錄圖記) and accompanied by woodblock-print illustrations.

Like ambassador Guo Songtao (chapter 11), Wang wrote with admiration and wonder about innovative Western technologies like trains, suspension bridges, and telegraphs, but he also focused his attention on institutions like the patent office, taxation system, and customs service, which encouraged the innovation and commerce that gave England a competitive advantage.

Like other contemporary Chinese travelers, Wang commented on the "gender inversion" in Western society. But like Liang Qichao (chapter 13), Wang appears to have mistaken Western courtesy and chivalry for women really being valued over men. But unlike most other Chinese travelers, he did not view the assertiveness, emotional openness, and educational status of Western women negatively. In his writings and lectures, Wang envisioned a future of cooperation and even unification with the West, a "great unity" (*datong* 大同), utilizing a Confucian term very familiar to Chinese intellectuals.

Primary Source, Alternate English Translations, and Selected Studies

Chapman, Ian., trans. "Selections from *Jottings of Carefree Travels*." *Renditions* 53–54 (Spring and Autmn, 2000): 164–173.

Cohen, Paul A. *Between Tradition and Modernity: Wang T'ao and Reform in Late Ch'ing China*. Council on East Asian Studies, Harvard University, 1974.

McAleavy, Henry. *Wang T'ao: The Life and Writings of a Displaced Person*. China Society, 1953.

Tian Xiaofei. *Visionary Journeys: Travel Writings from Early Medieval and Nineteenth-Century China*. Harvard University Asia Center, 2012.

Teng, Emma Jinhua, "The West as a 'Kingdom of Women': Woman and Occidentalism in Wang Tao's Tales of Travel." In *Traditions of East Asian Travel*, edited by Joshua Fogel, 97–124. Berghahn, 2006.

Wang Tao 王韜, *Manyou suilu tuji* 漫游隨錄圖記. Ji'nan: Shandong huabao chubanshe, 2004.

Excerpts from *Illustrated Record of My Jottings on Carefree Travel*, by Wang Tao (1890)

Classified Record of Climate and Customs

English climate is seldom balmy and frequently cold. During the year, the sky is more often cloudy than clear. Midsummer is never intensely hot, but nor is midwinter severely cold. Everywhere, forests, shrubs, and flowering plants grow dense and luxurious, and their leaves and petals endure for quite some time without withering. The scenery is quite charming, and it truly seems to be a paradise.

The land is commonly said to be rich and fertile, but actually only 20-30 percent of it is arable, suitable for planting wheat or other cereals. The remaining territory of the flatlands and moors is covered with flourishing grass, which is often used to graze livestock. Between spring and summer, the livestock are not penned in but set loose in the outskirts of towns, very similar to the way animals are put out to pasture in north China. There is no need to guard or tie up the herd, for there have never been any incidents of rustling cattle or sheep, so we can see that the customs of the people are fairly honest and good. . . .

Many of the bridges that have been constructed in London are quite magnificent and ornate. Some are built from stone, while others are fashioned out of iron. However, the most amazing and incomparable thing is the suspension bridge, which spans across the sky. When seen at a distance, it looks like a rainbow encircling the heavens and straddling across a great distance. Such a unique creation is especially rare in China.

England also produces table salt, which is made by boiling the brine from wells, quite similar to the salt ponds and brine wells in Sichuan and Yunnan in China. Their method of making salt is like this: They pour molten iron into a huge shallow depression and excavate the soil underneath to place a furnace. Then, they use a coal fire to evaporate the water. The success of this method is attributed to the broadness of the pan, for it is much quicker than using an iron cauldron. . . .

There are also villas [country clubs], which people jointly construct as a place for tranquility and leave from work, where they can take rest and amuse themselves. Monthly dues are collected from each household to provide for expenses. The rooms are all tidy and exquisite, with antique vases and classically elegant furnishings, and the beds, mattresses, and bed-curtains are all quite resplendent. One can be relieved from the heat of summer and avoid the cold of winter there. Being able to read the latest newspaper gossip or play chess there is enough to dissolve life's worries and break out of one's solitude, and these are also considered refined pastimes.

As for hotel restaurants where guests can take their meals in the morning or evening, these are also extremely luxurious. Numerous meat dishes and wine are served, making it quite an extravagant way to provide for life's necessities. For the wealthiest people, the monthly expenses for meals can amount to hundreds of gold coins, and apartment rent is correspondingly lofty. Sojourning abroad is certainly not an easy thing to do. There are also restaurants where the most skilled chefs are from France. It goes without saying that their food and drink are wonderful and clean. Some of their cooking cauldrons are made of cast iron, but coated with porcelain on the inside, which is a newfangled style.

The British attach great importance to literacy. When they are very young, they are entered into private schools, so when they reach adulthood, they can manage business operations in the four quadrants of their empire. Thus, even among lowly workers and crude servants, most of them know how to read and write.

Both women and men alike learn to recite when they are young. In general, calligraphy, painting, calendrics, astrology, cartography, geography, and oceanography have all been thoroughly studied, arriving at their essential principles. There are many Chinese men who have qualms about this type of educated woman. The prevailing custom in England is that women are more honored than men. In marriage, one selects their own partner, and husband and wife live together into old age, without the introduction of concubines. Household servants are mostly young girls and old women, but at the mansions of noble families, and for the job of coachmen, only males are employed. . . .

British customs and mores are straightforward and honest, while their production is prosperous and diversified. A wealthy family will spend extravagantly, but an impoverished household will work very diligently. Every day, one is surprised by some novel and ingenious technology, for this country has very few indolent or unproductive people.

The most admired persons are the ones who are known for being polite, respectfully cautious, and sincere. Throughout the country, interactions between officials and commoners rarely lead to incidents of altercations or high-handed treatment. Those foreigners who are sojourning in England are never deceived or swindled. They are treated with familiarity and affection and almost never with suspicion. It is quite rare for me to witness such social customs in a foreign country, let alone back in China.

A General Overview of Institutions

Outside of London, they have established a customs house which is spacious and stately, quite magnificent in scale. All the merchant ships from the various countries that arrive here carrying goods are subject to strict inspection. According to regulations, all the commodities on the ship must be taken out and exhaustively listed and displayed in the customs house, where they will be assessed according to their weight. The laws are comprehensive, but there is never any abuse through concealment or loopholes, because it is strict and impartial like this.

When I changed my route from Paris and arrived in England, my luggage arrived from Southampton about eight days later. The customs office sent someone to deliver it to me, so I compensated him with a small tip. The customs office has its own transportation company to handle their business, so people need not have anxiety about it. Tea and cigarettes that people carry with them, intended as gifts for friends, will not be charged a duty, and the boxes will not be opened, so one can say that they are lenient in the treatment of strangers from far away. According to British regulations, inspections of imports are stricter than those on exports, and there is no tax levied on exports at all. So one can observe in this their favorable treatment toward merchants. Thus, although paying taxes is always considered onerous, people have no real resentment over it.

Of all the beneficial and rapid systems of the West, nothing surpasses that of vehicular travel. Whether it is traveling to and from London or within the capital itself, everyone relies on the swiftness and convenience of the locomotive. Its structure is roughly like that of a giant cabinet, with doors opening on the left and right to allow entrance and egress. Several dozen passengers can sit within it, and either four or six wheels are emplaced below. When traveling, several cars are linked together with steel hooks, and the front car is equipped with a fire box and boiler. Once the fire is stoked, the mechanism starts to move, turning the wheels rapidly, and all the connected carriages drag each other forward.

Carriages are divided into three classes. The upper-class cabins are more spacious, and their seat cushions and curtains are resplendent and clean, making the passengers feel comfortable. The middle-class cabins are the next best, but the lower-class carriages have no roof or canvas covering at all, being exposed to the elements. They are only used to carry bulk cargo or for servants to sit in. A locomotive can travel about two or three hundred Chinese miles in a double-hour (50–75 kph).

The train tracks are cast from iron to make a running channel, and the upper part of the rail has a convex profile to accommodate the wheels. The tracks are lined up precisely within norms, being leveled and stabilized to facilitate fast travel, obviating the troubles of bumps, ruts, slopes, or slants. When the train route encounters a rocky mountain, they will bore a great tunnel through it, level and straight as a whetstone.

To the side of the railway track, they have strung a continuous iron wire that extends unbroken for thousands of miles. It uses an electrical machine with a hidden mechanism to transmit language. If there is something you want to say, the electricity moves it along the wire with the swiftness of lightning, traveling thousands of miles in an instant. So it is like conversing face to face with a person in another city, responding to their questions. This technology is so profound and subtle that there are aspects of it which are difficult for me to analyze or describe.

If two trains were to meet head-on, it would be sudden and unavoidable, running the risk of collision and derailment. Therefore, the timing for departures and arrivals of all itineraries are predetermined. At strategic transportation hubs along the railway, they place train stations and postal offices, which are strictly manned by officers in continuous rotation without a moment of leisure.

When I was residing with the English merchant Mr. Spencer and his family, each time we went to the house of the former director the Chinese Maritime Customs, Horatio Nelson Lay [1832–1898] for dinner, our train would travel in an underground subway tunnel, and then after about 15 minutes, suddenly burst into the light of day. Also, on either side of the road, they would set up market shops, brilliantly illuminated by streetlamps, surprisingly forming a bustling marketplace which does not suffer from the gloominess and obscurity of the long night. This is quite a novel idea.

The streets of London are all extremely level, well ordered, and clean. In places where an elevated roadway makes crossing the street on foot impossible, they construct a stone staircase with an iron railing on either side, to facilitate the passage of pedestrians. When they plant trees and flowers behind the balustrade, it especially helps travelers to feel refreshed. Westerners really like to plant trees. They say the benefits are fivefold: 1) they refresh the air, 2) they retain moisture so that the soil does not dry out, 3) fallen fruit is edible, 4) useful timber can be obtained from them, 5) they lead to more rain, so one does not need to worry about drought. So among the streets of London, there are numerous gardens and groves of trees. When people have a small area of open space, they all plant beautiful shade trees, which are especially flourishing in the suburbs. During the hottest period of summer, everyone can seek shade and take their rest under them.

It appears that the study of electricity began during the Ming dynasty in China, but although Chinese philosophers sought after its principles, rarely did they understand how to put it into practical use. Toward the end of the Daoguang reign [1820–1850], some people made a test run of a private electrical system, but the magical effects of electrical wires were first introduced to England, America, Germany, France, and other Western countries. Its benefits are immense, and the positive effects can be seen quickly. In areas where merchants are concentrated, a profusion of written correspondence speeds along telegraph wires. Even if the route traveled is lengthy or the time is pressing, the message still arrives in an instant, bringing relief and transmitting the news. Everyone feels it is a great convenience. Because the revenue generated by the telegraph companies was so enormous, in the seventh year of the Tongzhi reign [1868], the British Parliament prohibited all private systems and nationalized them under government control, levying a tax upon their use [Telegraph Act of 1868]. They established five bureaus throughout the country, with London as the main hub. There are domestic and foreign branch offices in 5,540 locations, and the tax revenue it brings in amounts to several tens of millions [of British pounds]. You could certainly say that it is a prosperous system. When I arrived in England, this nationalization had only occurred a few months previously.

* * * * * * * * * * * * * * * **BRIEF EXCERPT** * * * * * * * * * * * * * * * *

Manufacturing Marvelous Things

The minds of Englishmen are so clever and ingenious. For the manufacture of all sorts of implements and utensils, they concentrate on developing trade secrets which are extremely refined and insightful, and in many instances, it is because of this that some of them have become extraordinarily wealthy. This certainly manifests the essence of their motives, but the state also has means to encourage them and help bring their plans to fruition, relying on government officials to assist them.

According to English customary law, when a person invents something and does not want others to imitate it, he goes to a patent office, clearly describes the particular invention or method, and pays a fee to guarantee his exclusive patent for periods ranging from five to six years or as many as twenty. If there are people who imitate his design, regulations will not permit this. If someone violates the regulation, the inventor files a lawsuit with the officials, and the person will be fined.

Suppose that a poor man invents something, but he lacks the ability to apply for a patent or the funds to manufacture the item himself. He can relay the plans to a wealthy man and allow him to test the invention. If it is effective, then the wealthy industrialist can request to buy his designs for a price. Sometimes, for an initial investment of one or two times the value of the plans, the return can be as high as a hundred or a thousandfold.

When one first wants to manufacture things, he must exhaust his ideas, seek a broad base of knowledge, spare no expense, nor shirk from excessive work, nor calculate the total hours of his toil, draw on personal connections from a wide area, and repeatedly test his procedures. Then he will be sure to achieve his goal and can make his report to the patent official. If the results of the official's evaluation show the invention to be effective, a patent certificate will be issued protecting the invention for a number of years, and others will be prohibited from privately copying the design. Those who admire the invention and want to legally copy it will have to pay a fee to the inventor. It is also feared that those in foreign countries might copy the invention, so an announcement is circulated widely to neighboring countries, and the officials there will also enforce it. If there is someone who imitates the invention but does not pay a fee to the inventor, his fines will be doubled.

Therefore, once the thing is successfully manufactured, the profit can be calculated in the millions and billions. If this system were not in place, other people could discover the inventor's secret after repeated experiments and could sit back and enjoy success without lifting a finger, and the inventor would have no legal recourse. In that case, who would be willing to waste money and resources to invent anything new? If someone has not yet completed his invention but he is anxious [that someone is already copying it], he can also make this known to the officials. If the invention really is effective, but the patent official declares this is not the case, as well as when people have been prohibited from privately copying the invention, but the government is using it in secret, the inventor can bring a case to court. If there is a person who has a worthwhile skill, even the royal court cannot use its power to suppress him, so he has the courage to undertake his enterprise.

* *

A Dance Festival

In Western countries, men and women gather to dance, which is called *danchun* [dancing] in Western languages. Some say that it is a survival of the old Miao [southwestern Chinese ethnic group] custom of dancing in courtship under the moonlight. Japan and some other countries in the South China Sea still have such a custom. English people consider it one way to make merry and amuse themselves. Each year in June or July, large gatherings are held which are a sight to behold. They select one hundred, or as many as two or three hundred adolescent boys and girls, all with gleaming white teeth and beautiful features. The smallest are around 12 or 13 years of age, the oldest 15 or 16, and they partner them up according to age. Each of the styles of dance has particular steps or sequences and is given a specific name. Prior to this they have received guidance from a female instructor and must practice for several months before they have mastered the steps.

This year, the academy in Dollar, Scotland, held a great dance festival and invited me to come and watch. It was really filled with a variety of wonderful sights. The girls all came with made-up faces and dazzling clothes, and the boys were well groomed and attired smartly as well, all competing to be the best looking and most charming to earn bragging rights. At the outset, the dancers joined up then separated, suddenly moving forward then suddenly stepping back, advancing then retreating, wanting to approach then halting, sometimes coming close to one another, and sometimes moving far away. At times, they scattered unevenly, while at other times they danced in tidy formations. The boys might beckon the girls, or the girls beckon the boys. The boys might approach, and the girls pretend to avoid them, or the girls would draw near, and the boys would feign disinterest. When they came together, there was clasping of slender waists and a brushing up against fragrant shoulders. While forming pairs or in lines, spreading out or circling in a ring, moving and stopping, dancing quickly or slowly, each showed off their talent.

Each girl held a posy of red and white flowers in her hand whose fragrance could be sensed from far away. Their clothes were also made of scented gauze and patterned silk, with their upper shoulders left bare. When they danced with airy lightness in their rainbow skirts, they appeared like fairy flower maidens who had descended from heaven to earth.

Their way of dancing was entrancing in its unpredictability, moving in single file, or linked in a row, or like geese flying in a flock, or split apart like the tail of a swallow, or scattered in seeming disorder like the planets crossing the heavens, or bunched tightly like chess pieces in formation, or suddenly forming three lines like the character *chuan* 川, or abruptly joining up to make a formation like the character *yue* 日, or advancing in a single wall like the character *yi* 一, or forming a perfect circle or square. When formed into a circle, the dancers would face into the center, then suddenly turn to face the outside; then break into two circles, like joined links in a chain. Or the boys and girls might form two separate concentric circles, with the male circle surrounding the female one. Then, each girl in the inner ring would in turn break free and pass through the boys in the outer ring. Or there might be a ring of girls surrounding one of boys, and the boys in the inner ring would each break off and emerge through the female ring. If in square formation, they might double-up like the character *lü* 吕 or form three squares like the character *pin* 品.

It was such a bizarre and gaudily colorful spectacle, fantastic yet strange, that one couldn't help but watch it intently. At times, just the girls would perform a dance similar to the Central Asian twirling dance, with white silk ribbons more than seven feet long tied to each sleeve. They looked like white bats spreading their wings, giving the impression that they were gracefully flying up toward the heavens. All the girls tiptoed around on white leather shoes, and when they danced, they lifted off the ground as if weightless. It looked just like a thousand-petalled white lotus flower was springing up from the soil. Furthermore, this was accompanied by a variety of harmonious and mellifluous music played on stringed instruments. The spectator is so dazzled and confused that he becomes unaware of where he is. I was watching the show accompanied by Miss Mary [Legge], who asked me, "Wouldn't you consider that dancing like this is marvelous?" I clapped my hands together and sighed, "It was the most perfect thing I have ever seen!"[2]

Notes

1. Eliza Jane Gillett Bridgman, *Daughters of China* (Robert Carter & Brothers, 1853), vii–viii, 13, 20–27, 28–30, 33–35, 50–52, 58–65, 80–82, 126–127, 131–132, 157, 166–167, 200. Spelling, capitalization, and some punctuation has been modernized. Headings added by the author; dates are in the original.
2. Wang Tao 王韜, *Manyou suilu tuji* 漫游隨錄圖記 (Shandong huabao chubanshe, 2004) 91–93, 94–96, 104–107, 144–146. Translated by the author. Headings from the original.

CHAPTER 13

WITNESSES TO THE CONTRADICTIONS OF MODERNITY (1900–1936)

In part 1 of this chapter, we encounter a new genre of Chinese travel writing by two early 20th–century authors: Liang Qichao and Shan Shili. Like the embassy letters of Guo Songtao (chapter 11) or the travelogue of Wang Tao (chapter 12), these authors are still reporting on Western modernity and reflecting on Chinese backwardness, but their views are more critical, after encountering contradictions within Western modernity such as wealth inequality and racism. Hence, the authors are careful about which elements they would borrow to modernize China.

During the early 20th century, Chinese society felt the aftershocks of the end of imperial rule and also experienced the deep class contradictions brought about by uneven modernization and wealth inequality. Consequently, the second part of this chapter pairs the accounts of these Chinese authors with Edgar Snow's account of the Communist revolutionary forces in China.

Introduction to Liang Qichao

Liang Qichao's 梁啟超 (1873–1929) *Notes on Travels in the New World* (Xin Dalu youji 新大陸遊記) was the most influential account of early 20th–century America by a Chinese traveler. While several Chinese diplomats had written descriptions of the United States in the second half of the 19th century, making observations on technology, social customs, and the form of American government, none of these men possessed the incisive and searching mind of Liang, considered one of China's greatest intellectuals. His penetrating account, compiled and serialized from his extensive travel journals, has led him to be characterized as the "Chinese de Tocqueville."

The brilliant young Cantonese Liang Qichao had embarked on a typical Confucian scholarly career until he attached himself in 1890 to the reformer Kang Youwei 康有為 (1858–1927), an advocate of constitutional monarchy. The two men organized a protest against the humiliating treaty after China's defeat in the Sino-Japanese War (1894-95), then clamored incessantly for reform in newspapers and study societies. During the summer of 1898, Kang and Liang had the sponsorship of the Guangxu Emperor (r. 1875–1908) to propose major reforms, but a few months later, a palace coup led by Empress Dowager Cixi (1835–1908) condemned them to death, and they quickly fled to Japan. During his lengthy exile in Japan (1898–1912), Liang became deeply immersed in European history and sociology, looking for an answer to a difficult problem: how to modernize China's government and society to create a powerful nation that could survive in the modern world.

The ostensible reason for Liang's seven-month visit to North America in 1903 was to drum up support for the Protect the Emperor Society, the overseas Chinese society founded by his mentor Kang Youwei, which sought to restore the reform-minded Guangxu Emperor to the throne. His more important motivation for coming to America was to see the now century-old republic in action and evaluate its political model as a candidate for potential adoption in a post-imperial China.

Liang arrived in Vancouver in March 1903 then traveled by train across Canada, arriving in New York on May 12, 1903. Initially based in New York City, Liang took excursions to Boston, Philadelphia, Baltimore, and Washington, DC, traveled the Midwest through Pittsburgh and Cincinnati, and then went down to New Orleans. He subsequently crossed the country westward by rail through St. Louis, Chicago, Montana, and Idaho then down the West Coast from Seattle to Los Angeles, stopping over for a while in San Francisco, where the largest Chinatown in the New World was a special focus of his concern and his ultimate disappointment. He gave numerous speeches to Chinese associations and met briefly with President Theodore Roosevelt, Secretary of State John Hay, and the financier J.P. Morgan.

Though Liang was impressed and a bit alarmed by American power and dynamism, especially its new imperial designs on the Pacific, he developed serious misgivings about American-style democracy. He criticized the spoils system and the frequent elections that led to corruption and inefficiency. He disapproved that truly outstanding men would not go into politics in America and that most presidents since Washington were mediocre. He also noted the weak position of the president as head of the executive branch, except during wartime.

Most incisively, Liang argued that the American-style republic, which had developed organically out of the original colonies, could not be easily transported elsewhere. He pointed to the fragility of representative governments in Latin America and France, which had been installed after revolutions. He was decidedly pessimistic about the chances that direct democracy or a republican system could be transplanted to China, for he felt that his countrymen still had a "village mentality" and were too unruly and clannish for self-governance. When given near complete liberty in San Francisco and the chance to develop self-governing institutions, they failed miserably. He concluded that the Chinese needed several decades of tutelage under an enlightened despot before one could "give them works by John Jacques Rousseau to read and discuss with them the deeds of George Washington." After his return to Japan, Liang became more drawn to German-style statism than American liberal democracy.

Many of the problems with American politics, economy, and society that Liang pointed out still bedevil the country today. The wealth inequality that he condemned, where immigrants lived in squalid slums right next to glittering new buildings and where the top 0.25 percent of the population held 70 percent of the wealth, have worsened in recent decades. Surveying his critiques of anti-Black racism, gender inequality, and the danger of business monopolies makes one feel that things have not fundamentally changed in a century.

Liang was not resident in America long enough, however, to become adept at filtering the information he was receiving from newspapers and domestic critics. He often parrots the unfounded myths and stereotypes that suffused American media, such as anti-Semitic conspiracy theories about Jewish domination, the danger of dark-skinned "Latin" immigrants diluting the blood of lighter skinned northern European "Teutons," and the repugnant racial stereotype that Black men can't restrain their lust for white women.

PRIMARY SOURCE, ALTERNATE ENGLISH TRANSLATIONS, AND SELECTED STUDIES

Arkush, R. David, and Leo O. Lee, trans. *Land Without Ghosts: Chinese Impression of America from the Mid-Nineteenth Century to the Present.* University of California Press, 1989.

Chou, Yin-hwa. "Formal Features of Chinese Reportage and an Analysis of Liang Qichao's 'Memoirs of My Travels in the New World.'" *Modern Chinese Literature* 1 (1985): 201–215.

Grieder, Jerome B. "Liang Ch'i-ch'ao (1873–1929) and Hu Shih (1891–1962)." In *Abroad in America: Visitors to the New Nation, 1776–1914),* edited by Marc Pachter and Frances Wren. Addison-Wesley, 1976.

Levenson, Joseph. *Liang Ch'i-Ch'ao and the Mind of Modern China.* 2nd rev. ed. University of California Press, 1967.

Liang Qichao 梁啟超. *Xin Dalu youji* 新大陸遊記. Shehui kexue wenxian chubanshe, 2007.

Liang Qichao and Peter Zarrow, trans. *Thoughts from the Ice-Drinker's Studio: Essays on China and the World.* Penguin, 2023.

EXCERPTS FROM *NOTES ON TRAVELS IN THE NEW WORLD,* BY LIANG QICHAO (1903)

* * * * * * * * * * * * * * * **BRIEF EXCERPT** * * * * * * * * * * * * * * * *

New York

Uncivilized men live underground, partially enlightened men live at ground level, and civilized men live well above the ground. The most common housing is a dwelling of one or two floors. . . . It is not uncommon in New York for residential buildings to have from 10 to 20 stories. The tallest has 33 floors, which one could truly call living far above the ground! But residential buildings in major metropolitan cities in America usually have one or two basement levels, so you could say they have both aboveground and belowground dwellings.

In New York, what first strikes the eye as being pigeon coops placed everywhere are actually apartments; what strikes the eye as spider webs hanging all around are really power lines, and what strikes the eye (from high above in tall buildings) as being centipedes are really electric streetcars.

New York's Central Park, which goes from 71st street to 123rd street [actually from 59th to 110th], is roughly equal in area to the French and British Concessions in Shanghai. Every Sunday and holiday, the place is just packed with carriages bumping wheels and people brushing shoulders. It is situated right in the heart of the city. If it were converted to commercial real estate, the sales price of the land would be three or four times the entire annual revenue of the Chinese government. From the Chinese point of view, this is like throwing away money on wasted land. What a pity. The area of New York's Central Park is about 7,000 square *aaiga* [Chinese "acres" (*mu*); actual area, 843 English acres], making it the largest urban park in the world. London's [Royal Parks, including Hyde Park and Kensington Gardens] are second, with a combined area of 6,500 square *aaiga*.

Those who discuss city administration all say that for a thriving metropolis not to have a comparable urban park would be very harmful to public health and morals, and after arriving in New York, I can believe it. Even one day without visiting the park leaves one's spirit turbid and his mental capacity degraded.

All day long, streetcars, elevated trains, subways, horse-drawn carriages, electric automobiles, and bicycles go rumbling "*yin-yin*" above one's head, drumming "*peng-peng*" below one's feet, rattling "*lin-lin*" to one's left, galloping "*peng-peng*" to the right, thundering "*long-long*" in front, and jingling "*ding-ding*" behind, to the point that one's mind is dazed and his soul utterly shaken.

It is said that those who have lived in New York for a long time have quicker eyes than ordinary people. If this were not the case, when it came time to cross an intersection, they would probably stand frozen there all day, not daring to take a step. . . .

New York is the most prosperous city in the world, but probably also the grimmest. Let me tell you a little bit about New York's darker side. Those who want to eradicate the spread of Yellow Fever strongly condemn the Chinese people as being filthy, but from what I saw in New York, the Chinese aren't the dirtiest ones. In summer, along the streets where the Italians and Jews live, old women and young wives, boys and girls, each carrying a stool, sit or squat outdoors, blocking the street. Their clothing is ragged, and their appearance wretched. Electric streetcar routes don't access these parts of town, and even horse-drawn carriages rarely go there. Tourists often go to these neighborhoods just to gawk at the customs of the poor immigrants. Based on external appearances, they all seem to live in multistory residences, but each tenement actually houses dozens of families, and more than half of the apartments have no access to daylight or fresh air. Gas-lamps burn day and night, and when one enters the door, the stench of filth assaults the nose. In the whole city of New York, as many as 200,000–300,000 people live under these conditions.

According to statistics for the year 1888, on New York City's Houston and Mulberry streets, where most of the residents are Italians, with some Germans, Chinese, and Jews mixed in, the mortality rate was about 35 per 1,000 persons. The mortality rate for children under five was 139 per 1,000. Comparing that to the general mortality rate for the entire city, which was 26 per 1,000, you comprehend the desperate living conditions of the poor. It is said that these rates are probably brought about by the lack of fresh air and sunlight in the tenements where they live. Another statistician said that there are about 37,000 multifamily apartments in New York City, in which more than 1.2 million people live. This type of housing is not merely harmful to physical health but is also harmful to morals. Again, according to the statisticians, there is one building on a certain street in New York where 483 people live, and in just one year, 102 of them committed crimes. So one can see the negative influence of these living conditions is very great indeed!

A poem by Du Fu [712–770 CE] goes:

The vermillion gates of the wealthy reek of ale and meat,
While on the road just outside lie the frozen bones of the dead.
Flourishing and withering a mere foot apart.
It is so upsetting; I find it hard to describe it further.

I witnessed this for myself in New York. According to the statistics of the Socialists, 70 percent of the entire wealth of the United States belongs to the 200,000 richest individuals, and the other 30 percent is held by the remaining 79,800,000 poor people. Thus, the wealthy in America are truly rich, and this so-called wealthy class consists of only 0.25 percent of the population. For example, if there were $100, and 400 people split it, one person would get

$70 and the remaining 399 people would have to share $30, each not even getting a dime, only a little more than seven cents. Isn't that strange! Isn't that bizarre! All the civilized nations are like this, especially in the great metropolises, where the cases in New York and London are the most prominent. With wealth inequality reaching this extreme, and looking at the slums of New York, I heave a deep sigh and now realize that it is impossible to avoid socialism! . . .

* *

From Pittsburgh to Cincinnati and New Orleans

. . .

The rights of emancipated Black people are really rights in name only, for their situation is almost the same as before emancipation. Black people in New Orleans cannot move to another city without the permission of the city council. Most of the Southern states have similar restrictions. So whereas in the past, each Black person was owned by one private individual, they are now the collectively owned slaves of the city. So Chinese people [laborers] are not allowed to come, and Black people are not allowed to leave. What a monstrous absurdity! There are some that feel a strong attachment to their former masters and are reluctant to leave. It is not that they feel bound to them because of some lingering affection but that they fear them due to ongoing intimidation.

Here is the most ridiculous thing: the movement to emancipate the slaves was originally championed by the Republican Party. After emancipation, the former slaves now had the right of suffrage. The Republicans thought that since they had shown kindness toward the former slaves, they would certainly gain their political support, increasing their voting bloc by the millions and thereby subduing their political adversaries. Who would have thought that when it came time to vote, the Black people would all cast their ballots for the Confederate Party [the Southern Democrats]. Since the upper class of Southern society were their former masters and all members of the Confederate Party, they felt like they had to obey their directions. We can draw the general conclusion that their old servile nature could never be changed.

Americans have a type of private, extralegal punishment called *lihngjih* [lynching] that they apply to Black people. Such a practice is really inconceivable in a civilized country. It began with a farmer named Lynch [Charles Lynch, 1736–1796]. A Black man had offended him in some way, so he tied him up and suspended him from a tree, awaiting the arrival of the police. The Black man died before the authorities arrived. Later, people continued to use his name [to refer to this punishment]. Recently, the common practice has been to burn the person to death. When a Black person is considered guilty of something, they don't go through the courts, but a mob gathers straight away and burns him. If I had not personally traveled to America and someone had just told me that such a brutal and inhuman practice like this could happen in broad daylight in the 20th century, I would never have believed it.

In the 10 months I spent traveling throughout America, I had seen these strange reports in the newspapers on no less than 10 occasions. At first, I was appalled by this, but after growing accustomed to seeing these notices, I no longer considered it unusual. Looking into the statistics, since 1884, there has been an average of at least 157 of these lynchings each year. Oh my! When Russia kills a hundred or a few dozen Jews, the world considers this a brutal outrage, but I don't know which to choose as worse, America or Russia!

But there is something in the behavior of Black men that is sufficient to disgust people. If they can just get their hands on the luscious flesh of one white woman, they could die nine times without regret. They usually forcibly rape them in the middle of a dense forest at night, and when finished, they kill the woman to silence all witnesses. Nine out of 10 of the lynching incidents stem from this. While this is certainly something to be angry about, isn't there a judiciary to handle such cases? And why doesn't the government subject those who wantonly carry out these lynchings to an appropriate punishment? There is no reason other than prejudiced views concerning race.

The American Declaration of Independence states that all human beings are born free and equal. So is it just blacks that are not considered human beings? Alas! Now, I understand what currently passes for a "civilized society." . . .

"Shortcomings of Chinese People"

The first is that they are only qualified to be clan members and not citizens.

The organization of our Chinese society takes the extended-family lineage as the basic unit, not the individual person. This is borne out by the so-called "once the family is well ordered, then the country can be well governed" idea [that one sees in the Confucian Classic, *The Great Learning*]. Even though the formal characteristics of the lineage system of the Zhou dynasty [ca. 1045–256 BCE] were long ago discarded, the spirit of that institution remains with us. Discussing it at a superficial level, we see that the capacity for self-governance of the Aryan race of the West definitely developed the earliest, even though the ability of the Chinese people for local self-governance should be no weaker than theirs. So how is it that they can form a nation and we cannot? It is because what they have developed is a citizenship system of self-governance, while what we have developed is a clan system of self-governance. Try traveling to one of our country's villages, and you will see that the extent of their self-governance is something impossible to conceal. Take, for instance my hometown, whose population does not exceed a mere two or three thousand people. Its legislative and administrative organs are well ordered and do not overlap. People from other lineage-based villages would also say the same thing. If that is the case, these local village units should be able to serve as the primary foundation for the building of a modern nation state. However, if one then travels to a major metropolitan area, the situation is such a complete mess, it is absolutely inconceivable.

This is clear proof that Chinese people can serve as members of a clan but still cannot become proper citizens. Now that I have traveled to North America, I believe this even more strongly.

Once an overseas Chinese has left his home village, taking his individual self as his only qualification and coming to live in the freest of modern cities in America, then paradoxically, of all the capital he brings with him and all that he builds up, none of it will reside outside the confines of the lineage system. Moreover, it is only through upholding this that he remains a part of the social order; all depends on it. One can see that thousands of years of inherited tradition have formed extraordinarily deep roots, and that those who want to guide our Chinese people towards a modern nation, must always take this into serious consideration.

The second is that they have a village mentality and not a national mentality.

I heard Theodore Roosevelt's speech, in which he said that the greatest priority for today's American people should be to get rid of their village mentality, by which he meant the emotional connection of the residents of each state or city to their native state or hometown. If we analyze this from the perspective of historical development, the main reason that America was able to implement a republican form of government so thoroughly was by relying on this same village mentality as its very foundation, so it can't be completely faulted. However, when it is overdeveloped, village thinking becomes the greatest impediment to nation building. The criteria by which we can determine this boundary are not the most accurate standards, so it is difficult to measure. Now in the case of our China, this is clearly the perfect case of excessive development of village thinking. How could it just be the overseas Chinese in San Francisco who are like this? Domestically, in China, every place has the same issue. Even the worthiest scholars cannot avoid this sentiment. . . . But, if this barrier of village thinking is not demolished, China's desire to become a strong and stable empire will indeed be difficult to achieve.

***The third is that they can only be subjects of autocratic rule
and cannot enjoy the liberty of self-governance.***

Actually, this is a statement that treat's all the world's creatures like
straw dogs for sacrifice [i.e., it is an overgeneralization]. Although that
is the case, the real situation is quite close to this. Even if you wanted
to cover up the facts or avoid talking about it, you could not do so.
When I observe all the world's societies, none is as disorderly as the
overseas Chinese in San Francisco. Why is this? One can only say that it is
because of liberty. The character of domestic Chinese people is not really
superior to that of those who live in San Francisco, but inside China,
they are still under the control of officials and restrained by parents and
brothers. The situation of overseas Chinese living in Southeast Asia is
seemingly different from that in China, but England, Holland, France, and
the various other colonialist countries treat them with such harshness,
ordering the dispersing of assemblies of more than 10 people and stripping
away all their rights. Their harshness is even more severe than in China
itself, so the Chinese are quite docile and obedient there. Those overseas
Chinese who can really enjoy the same degree of liberty under the law as
Westerners are the ones who emigrated to live in America or Australia.
Even so, when they live in a sparsely populated town, the [negative]
potential of this liberty can't really be realized, so the drawbacks aren't
so obvious. It is when they live together in large numbers in a free city,
San Francisco being the prime example, that the social phenomenon of
disorder manifests like this.

There are locals who have further said that it was only previously,
when Mr. Zuo Geng was consul [in San Francisco; served 1889–1891],
that things were extremely tranquil there. No one would dare wield a
knife seeking revenge, none would dare form a mob to stir up trouble,
none would dare be idle or engage in nefarious activities, and each secret
society hid their tracks and held their breath [i.e., laid low]. There were
no alarms in the middle of the night, and the people worked diligently
at their occupations. Mr. Zuo quietly instructed the police in that city to
roughly seize Chinese miscreants and punish them severely. After Mr.

Zuo departed, the old state of affairs resumed. This case is clear proof that authoritarianism results in tranquility, whereas liberty leads to danger; autocracy is beneficial, but freedom is harmful. . . .

Now, complete liberty [i.e., democracy], constitutional [monarchy], and a republic are general terms for a government by the majority. But the bulk of the population in China, the greatest majority in fact, are just like these people [in America's Chinatowns]. So if we were to adopt a government by majority rule at this time, it would be like committing national suicide. Adopting a democracy, a constitutional monarchy, or a republic would be like wearing lightweight hempen clothing in the winter or a fur coat in summer; it isn't that they don't appear beautiful, but they are just wholly inappropriate for our situation. For now, we shouldn't be dazzled by fancy illusions; we shouldn't try to realize beautiful dreams. In a phrase, that is why I say that the people of China today can only be the subjects of autocratic rule; they cannot enjoy liberty. I pray and I yearn; I pray and yearn only that our country can have someone like Guan Zhong [statesman of the Spring and Autumn period], Lord Shang [the autocratic, reforming Qin minister of the Warring States period], Lycurgus [the 7th c. BCE Spartan lawgiver], or Oliver Cromwell alive today who could carry out reforms with the power of thunder and the speed of wind, and with iron and fire, mold and temper our countrymen's character over a period of 20, 30, or even 50 years. Then, we can give them works by John Jacques Rousseau to read and discuss with them the deeds of George Washington. . . .

The fourth is that they don't have noble goals.

This is a fundamental shortcoming of our Chinese people. Everyone in the world is a member of a country; sometimes it is a large or powerful country, but other times it is a small or weak country. Why is this? Every man occupies space in the physical world, but beyond the needs of one's body for clothing and food, there should be a greater purpose. Within the timespan of one's life, one needs a loftier goal, beyond just a short-term desire for wealth, glory, and honor. If this is the case, then a person can naturally progress and shine brilliantly in the light; otherwise, he will just stagnate or fall behind. This is true for the individual person, as well as for the collective body of persons who constitute the citizens of a nation.

In my estimation, the noble goals of Euro-Americans are varied, but these three are the most important. The first is a love for perfect beauty. When the Greeks discussed virtue, they considered the three qualities of truth, goodness, and perfect beauty to be the ultimate expression of this. We Chinese have often talked about goodness, but not so much about beauty. Confucius talked a little about the music of Shao [under the court of the sage Shun] being the perfection of beauty and goodness, and Mencius said that the desirable is "good," and to possess goodness fully in oneself is called "beautiful." These two notions are a bit at odds, but beyond these two references, other discussions of beauty in Chinese philosophy are rare. So talking about this from a comparative perspective, we could say that the Chinese are a people who do not cherish beauty. The second noble goal is a concern for one's social reputation, and the third is a notion within Western religions of a world to come. The development of civilization with a Western spirit considers these three principles to be nearly fundamental, but our China is greatly lacking in all of them. Thus, all their industriousness is focused on their own self, and all their diligence is directed towards the here and now. The reason for their stagnating and falling behind is to be found in this. This is not just true for overseas Chinese but for the entire country. It is only when I traveled overseas that this made such a deep impression upon me, and that is why I bring it up now. The reasons behind this national pathology are quite long and complex, so it is not possible to exhaust the topic today.

Beyond these, there are many other ways in which the character of Chinese people is inferior to that of Westerners

. . . Westerners work only eight hours a day, and on every Sunday, they rest. Chinese stores open their doors at 7 every morning and don't close until 10 or 11 at night. Chinese shopkeepers sit in their stores all day long and take no rest on Sundays, and yet they are not as wealthy as the Western businessmen. Moreover, they don't even accomplish as much work in all that time as their Western counterparts. Why? For working people, it is worst when they become chronically fatigued. And when people work at something all day, all year, they are certain to start loathing it. When they loathe their work, they are bound to become chronically fatigued, and then all other elements of their life fall apart. Rest is really an essential component of human life. One reason why the Chinese cannot pursue noble goals is to be blamed on their inability to rest.

American schools are, on average, only in session 140 days a year, and students attend only for an average of five or six hours a day, and yet Western schoolwork is superior to that of the Chinese, for this same reason.

A tiny little Chinese shop frequently employs several people, or sometimes even more than 10. A typical Western shop might employ only one or two persons. So, one of them can accomplish three times as much work as one of us. It is not that the Chinese are not hardworking, it is just that they don't work cleverly.

Sunday rest is indeed wonderful! After six days of toil, one has a sort of renewed spirit. A person's vigor and clarity of mind really rely on this. Chinese people are very muddle headed, so even if we don't adopt this day off for worship, we should at least implement the old system of "a day for hair washing" [i.e., one day of rest out of ten].

Try gathering more than a hundred Chinese in a single venue. Regardless of the extreme solemnity of the occasion, with no commotion allowed, you are bound to hear four types of noises: the most frequent is coughing, followed by stretching and yawning, sneezing, and nose blowing. Once, when I was attending a lecture for a Chinese audience, I listened attentively to them, and these four noises went on like a never-ending string of pearls. By contrast, when I have attended a speech at a Western lecture hall or theater, I listened with equal care, and even though more than a thousand were in attendance, I heard not a sound.

On trains and electric streetcars in the Far East, they always set up spittoons, but the passengers still spit everywhere, indiscriminately. In American passenger cabins, they rarely set up spittoons, and people hardly use them. On Far Eastern train journeys of more than two or three hours, at least half the passengers will doze off. On American trains, even if the trip occupies the whole day, not one person will fall asleep in his seat. Thus, you can see the superior physical constitution of the Western race compared to the Oriental. . . .

One is not permitted to spit or discard used paper and other litter on the sidewalks that line both sides of the streets in San Francisco. Vehicles travel in the middle. Violators are fined five silver dollars. It is not permitted to spit on New York streetcars, and violators there are fined 500 silver dollars. Americans value cleanliness so much that they are willing to restrict people's liberty to this extent. Since Chinese are such dirty and disorderly countrymen, no wonder Westerners are so disgusted by them.

When Westerners walk down the street, they all stand erect with their heads held high. We Chinese are quick to bow at one command, stoop over at a second, and prostrate ourselves meekly at a third. The contrast in attitude should certainly make us feel ashamed and inferior.

When Westerners are walking, they always seem in a hurry. One glance and you realize the city is full of people with matters to attend to, who act like they can't possibly accomplish everything. Chinese people walk around gracefully and poised, their jade ornaments all tinkling —truly detestable. When a Chinese person is approaching you on the street, you can make them out from hundreds of feet away, and not just because they are short or have a yellow complexion.

When a large number of Westerners are walking, they appear like an orderly formation of geese, whereas a large group of Chinese walking looks like a confused dispersal of ducks.

When a Westerner is addressing a single person, he makes sure that just one person can hear him; if he speaks with two people, he ensures that just two people can hear; if he speaks to 10 people, he ensures that all 10 can hear him; if he speaks to an audience of a hundred or more than a thousand people, he makes sure that hundreds or thousands can hear him. He modulates the loudness or softness of his voice as appropriate to the situation. When a few people sit down in a room to talk in China, their voices boom like thunder, but when thousands gather in a lecture hall, the speaker's voice is as soft as the buzzing of a mosquito.

When Westerners are conversing, until Person A has finished speaking, Person B will not interrupt him. When Chinese people are in a lecture hall, the voices are clamorous and chaotic. Some famous scholars in the capital will interrupt the speaker to indicate their own mastery of the topic. The scene can be described as extremely disorderly. . . . My friend Xu Junmian [1873–1945] once said, "Chinese people have not yet learned how to walk, not yet learned how to speak." This is not an exaggeration. Although these things might seem trivial matters, they speak to larger issues. . . .

Brief Commentary on American Customs [Women's Suffrage]

Westerners have a common saying that goes, "If you want to test how civilized or barbarous a particular country is, you should use the status of its women as the measure." Is that really the case or not? Everyone who travels to the United States all say that in American culture women are more honored than men. Even Americans themselves say that this is so. In my view, the reality of this phenomenon is so readily falsifiable that it isn't even up for debate. Although, if you told me that the status of American women is comparatively loftier than that of women in any other country, I would certainly believe that.

If you observe the external manifestations of this phenomenon, then you will see that in all hotels, trains, or in every class of entertainment facility, they always set aside a special room just for women, whose appointments and decoration far exceed those in the facilities reserved for men. When men encounter each other on the street, they will just nod their head slightly, but if they come across a woman, they will always remove their cap as a matter of courtesy. If a single woman enters a skyscraper's elevator cabin, then every man aboard will remove his hat. When all the seats on an electric streetcar are occupied and a woman comes aboard, the men will usually stand up and offer her their seat. In Eastern cities like New York and Boston this custom is not as prevalent. This is such elaborate etiquette, like one is greeting an important guest. These customs are not exclusive to the upper levels of society, though, for the same also goes for ordinary women. This is an indication that the ideal of egalitarianism is being put into practice.

In terms of actual substantial power, some women have made considerable progress. Within specialized higher occupations, women compete daily with men. One now sees female doctors, female lawyers, female newspaper editors and reporters, female pastors, and female orators, in ever increasing numbers. Especially in other occupations, like secretaries in government offices or businesses, or schoolteachers, women constitute the clear majority, with men falling far behind them in terms of numbers. Regarding legal rights, each state has slightly different laws, but their main features are not that far apart, and in general, women enjoy the same personal rights as men. Regardless of whether they are married or not, all women have the right to manage their own property and wealth. If her husband dies, each woman has the right to become a proxy manager of her children's inheritance. This is an area where the rights of American women exceed those of other countries.

> Appended note: before 1896, widows still did not have this right of legal guardianship over their children's inheritance.

There have been those who have advocated for women's suffrage since the founding of the country, and after the successful emancipation of the slaves, this movement has gained momentum. According to the principle of the Declaration of Independence, "All persons are created equal," and since the boundary between the black and white races has already been removed, then the barrier segregating men's and women's rights cannot but also be destroyed. This is the source of inspiration for the idealism [of the suffrage movement]. In recent years, especially in the states of the north and west, many citizens are passionate about this issue. They have frequently introduced bills for women's suffrage into the state legislatures and have repeatedly advocated for an amendment to the federal constitution. However, I fear such a constitutional amendment will be difficult to pass, since only the states of Wyoming [1869], Utah [1870], and Washington [1883] have so far approved women's right to vote in their state constitutions, and recently both Utah [1887] and Washington [1887] have repealed this approval. So among the states in America, only the women of Wyoming still have the right to vote.

Realistically, in today's society, there are few real benefits and many disadvantages to women getting involved in politics. Men like Johann Kaspar Bluntschli [1808–1881] have written detailed opinions on this issue, and it is fitting that the [suffrage] bill has not been able to achieve passage for a long time.

> Appended note: Australia's provinces of New Zealand [actually a separate colony since 1841], Western Australia, Adelaide [i.e., South Australia], and Tasmania all have women's suffrage. I heard that 10 years ago, a woman in New Zealand was elected head of a town council, but all the men obstructed her orders, so not long after, she resigned her post.

Regarding women's right to elect members of local school boards, or being elected to serve on one, 14 states in the United States have currently approved this, but most women have renounced this right and do not understand how valuable it is. I heard that in one city with a population of 200,000, when they opened elections for school board members, less than 200 or 300 women actually showed up to vote. Moreover, when the state of Massachusetts first implemented this policy, women first flocked to the polls, but each year after, the numbers have dwindled. From this perspective, women entering the political arena is not only something that they shouldn't do, but also something that they aren't able to do....[1]

INTRODUCTION TO SHAN SHILI

The travel accounts by the female gentry author Shan Shili 單士釐 (1858–1945) excerpted here are the first known travel accounts of Europe written by a Chinese woman. Shan Shili was born into a family of renowned scholars. She later married the Chinese progressive diplomat Qian Xun 錢恂 (1853–1927), who served as an envoy to Russia (1903) and later ambassador to Holland (1907) and Italy (1908–1909). Shan Shili accompanied Qian Xun on these postings, and he later encouraged her travel publications. It was unusually progressive for the primary wife of a Chinese diplomat to accompany him to his posting. A couple decades earlier, Guo Songtao (chapter 11) had only brought his concubine (Madame Guo, née Liang) to London and presented her as his wife, probably since a concubine would feel more comfortable at social events with mixed company than a traditionally-secluded primary wife.

Shan's *Travelogue of the Guimao Year* (Guimao lüxing ji 癸卯旅行記) documented the roughly eighty days of her journey (March 15–May 26, 1903) with her husband as they traveled through China, Japan, Korea, and Manchuria, finally crossing Russia on the newly completed Trans-Siberian Railway to arrive in St. Petersburg for his appointment. Her work builds on the tradition of embassy diaries like that of Guo Songtao (chapter 11) but has even more in common with the travel reportage style of Liang Qichao, whose observations of foreign places and peoples were influenced by a concern for strengthening China's position in the world. Shan remarks on China's lagging behind Japan in the areas of education and industry and advocated for improved education and greater physical mobility of a more modern Chinese woman, so they could serve as a bulwark of the nation.

In one of the last entries in her travelogue, Shan offers a fascinating ethnographic analysis of the cross section of Russian society through which she had just traveled by rail. Though gender decorum prevented her from ever alighting from her train coach, she was still a keen observer of the world outside her window. Summarizing a "Five Ages" theory of mankind, which was prevalent during the late 19th century in the West, she remarks that traveling across Eurasia was not just a journey across space but also through time, for she witnessed all the stages of mankind's social evolution, from primitive hunters in Manchuria to the Industrial Age of Western Europe.

Shan Shili's second book inspired by her travels, *Writings During My Husband's Retirement* (Gui qianji 歸潛記; 1910) was based on her time spent in Italy and other Western countries between 1907 and 1909. It is very different in character from her first travelogue, for it demonstrates her great capacity as a scholar of history, philology, and art history. It is arranged as a series of essays on topics like the art in the Vatican, Greco-Roman gods, and the history of Jews and Christians in China. In this work, she demonstrates her affinity with the best of Qing dynasty "evidential research" scholarship, a genre that had traditionally been only associated with elite male authors.

I have excerpted two sections from *Writings During My Husband's Retirement.* The first is her detailed history of the treatment of Jews in Rome. Shan Shili was able to walk about in Rome (usually chaperoned by her husband or son) and had observed the Jewish quarter in Rome and others like it in Europe. It is apparent, however, that she never interviewed any Roman Jews and that most of her account was based on library research. This female Chinese scholarly observer from the beginning of the awful 20th century reminds us of the deep history of anti-Semitism in Western Europe and prophetically warns her Chinese compatriots that a similar fate could befall them if they were to fall under the Western imperialist yoke.

The second excerpt is from her study of Marco Polo. Shan Shili employs her considerable scholarly talents to compare Marco's account of his service under Kublai Khan with Chinese historical records and points to contradictions, anachronisms, and implausibilities in his account that predate the critical works of Western Sinologists by decades.

PRIMARY SOURCE, ALTERNATE ENGLISH TRANSLATIONS, AND SELECTED STUDIES

Hu Ying. "Re-Configuring *Nei/Wai*: Writing the Woman Traveler in the Late Qing." *Late Imperial China* 18, no. 1 (1997): 72–99.

Hu Ying. "Would that I Were Marco Polo: The Travel Writing of Shan Shili (1856–1943)." In *Traditions of East Asian Travel*, edited by Joshua A. Fogel. Berghahn, 2006.

Qian Shan Shili 錢單士厘. *Guimao lüxing ji; Guiqian ji* 癸卯旅行記；歸潛記. Hunan renmin chubanshe, 1981.

Renditions Editorial Team, trans. "Shan Shili: Selections from *Travel in the Year Guimao*." *Renditions* 53–54 (Spring and Autumn, 2000): 214–218.

Widmer, Ellen. "Foreign Travel through a Woman's Eyes: Shan Shili's *Guimao lü xing ji* in Local and Global Perspective." *Journal of Asian Studies* 65, no. 4 (November, 2006):763–791.

EXCERPTS FROM *TRAVELOGUE OF THE GUIMAO YEAR* (1904) AND *WRITINGS DURING MY HUSBAND'S RETIREMENT* (1910), BY SHAN SHILI

Travelogue of the Guimao Year

Preface by the Author
. . . In the current *guimao* year [1903], my husband was to embark on a trip to European Russia on the Trans-Siberian Railway, and I happily

accompanied him. For the past decade, I have kept an unbroken diary, but as I look back upon those entries, they all record trivial matters, none that are really worth preserving. The only exception is this short travelogue, spanning eighty days and covering a journey of more than 20,000 Chinese miles [approx. 10,000 km] and four countries. It might considerably broaden the horizons of my readers' knowledge. I copied it out and handed it over to the printer to carve onto wooden blocks and called it *Travelogue of the Guimao Year*. Some of my fellow Chinese women will perhaps read this, and it will arouse a yearning for long-distance travel. I eagerly hope for this.

Qian Shan Shili, originally from Zhejiang. Written in St. Petersburg, capital of Russia. . . .

* * * * * * * * * * * * * * * **BRIEF EXCERPT** * * * * * * * * * * * * * * * *

May 13, [1903]

Chinese women are caged up at home in a single-room cloister, so they basically don't even know that the nation exists. Coming from Japan, I often heard that their women each take personal responsibility as a citizen of the nation, and I further considered that when a nation's foundation is strong and stable, this is especially connected to its women. I involuntarily felt a surge of patriotic sentiment. Because I was crossing the national border [into Russia], I couldn't help but sigh with emotion when I said this. . . .

The 25th day [of the Chinese lunar calendar]; May 21, [1903, on the Gregorian calendar]. Normally, one would cross the Ural Mountain range only during daytime. But because we encountered mishaps on one or two occasions, we were delayed until nightfall. So in the end I was not able to see the widely known stone monument that marks the boundary between Europe and Asia [i.e., the Europe-Asia Obelisk in Ekaterinburg]. The monument is said to have a triangular, almost pagoda-like shape, and is surrounded by an iron railing. On one face is written the word "Asia," and on the other is written "Europe." It was erected in 1845.

After passing Miyasi [Myza?] Station, the light was still dim, and it was hard to distinguish anything. I could only faintly make out the shadows of dense pine trees and hear the gurgling brook beneath the train tracks. I was born 46 years ago. Today, I entered Europe for the first time from Asia. Baron Fukushima Yasumasa [1852–1919; in his *Long Solo Journey on Horseback,* 1894] wrote, "[Eurasia] is one big, undifferentiated continent. How can Europe and Asia exist separately within it? Moreover, all of humanity's eyes are placed horizontally, and their noses run vertically, and there are no real qualitative differences in the nature of the soul or mind. The only variations are in spoken languages and skin complexion."

So true!

. . .

May 23, [1903]

When discussing the principle of human societal evolution, from primitive society and arriving at civilization, in general, one can divide it into a sequence of five stages. At the very beginning, all of man's existence was just trying to avoid hunger and cold. When he encountered a body of water, he went fishing, when he strode through the hills, he was hunting. People ate only meat and slept on animal pelts, and that is all. This was the so-called Age of Hunting and Fishing. After man long realized that wild game could not be relied upon for permanent subsistence, he began to breed and herd domesticated animals for food. This was the so-called Age of Animal Husbandry. After a long while, man realized that moving about to follow herds, with no settled place, was insufficient to provide fixed property and a steady livelihood. Beyond clothing and food, he also sought somewhere to reside; in addition to blood and flesh, he also had a taste for plant products. Thereupon, he began the work of cultivating the land, ushering in the so-called Age of Agriculture. After a long period, man realized that it was very inconvenient for each person, who relied upon his own leftover grain and cloth, to be without the means to exchange it for what he lacked. Thereupon, he organized trading networks and credit instruments as a medium for mutual economic assistance with such surpluses and shortages. This was the so-called Age of Commerce. At this time, far and near were in constant communication, so people's cleverness and ingenuity became increasingly advanced, and their tastes and predilections became ever more complex. So each group deployed their ingenuity to refine manufacturing, and each made these refined products to cater to people's tastes and desires, thus leading to the Industrial Age.

Each of these five ages has its proper place in the sequence and cannot just be surpassed in one leap. The speed with which one society progresses depends on the level of intelligence of its people and on the presence or lack of education. When contemplating this, we are talking about a span of thousands of years to progress from the lowest to the highest stages of evolution, not just a span of tens of thousands of miles of travel. But unexpectedly, I personally witnessed all of them in my 30 days and 20,000 Chinese miles of travel.

Penetrating the mountains west of Vladivostok, one enters the frontier of Ningguta [present-day Ning'an City in southeast Heilongjiang]. Three hundred years ago, this was the auspicious birthplace [of the Manchu dynasty]. The customs of the so-called forest dwelling people or the so-called hunting Jurchens of the Butha-Ula tribe mentioned in old historical texts are still preserved here. Isn't this what is called the Age of Hunting and Fishing? Going farther west and emerging onto the Mongolian frontier, passing north of the Yin Mountains, the fertile soil has not yet been cultivated, and oxen, sheep, camels, and horses all propagate extremely well. Wouldn't this be called the Age of Animal Husbandry? Farther west still, entering the western frontier of Siberia, the people's customs are plain and unaffected, and the granaries are overflowing. Wouldn't you say they were living in the Age of Agriculture? (Their wheat harvest is exported to Germany, Austria, and other countries.) When one crosses the

Ural Mountains and enters Moscow, the transportation becomes much more convenient and there are prosperous commercial districts. Although its industry is not yet known to the world, this area has certainly galloped headlong into the Age of Commerce. Then, when we talk again about Germany, France, England, and the various countries in America, aren't we looking upon the so-called Industrial Age!

* *

Writings During My Husband's Retirement (1910)
. . .

Record of the Hebrew Religion in China
This piece brings together into one chapter those minor facts and anecdotes that I learned about the Jewish quarter, the Ghetto, when I was in Rome. On the one hand, I want to trace the shared origins of Christianity and Judaism, but on the other hand, I express the opinion that Christianity and Judaism are difficult to reconcile. Moreover, I want to show the tragic situation of these stateless refugees under the rule of the white race, and their relative freedom living under the control of the yellow race [in China]. . . .

The Jewish Quarter in Rome—The "Ghetto"
Traveling by rail and carriage between the capitals and great cities of Europe, the commercial districts that one witnesses are largely identical, yet the people who reside within them are not without some slight differences in terms of their facial features, and their customs also deviate slightly from the norm. When one inquires, these invariably turn out to be districts of Jewish people.

They are the remnants of a stateless people who lost their original homeland, a diaspora scattered among various other countries. The oppression and abuse that they have suffered is almost beyond description, particularly in the case of Russian practices.

Even though one might be a great scholar or a successful capitalist, most of the heads of the "red shop-signs" [the illustrious banking houses] that control finance in Europe and America are Jews, and one certainly can't say that Caucasians don't depend upon them, but in the end, Jewish people are unable to arouse their sympathy or dissolve their prejudiced opinions about "us" vs. "them." Some say that this is due to a difference in religion between Judaism and Christianity. Although you could call what Moses did in reaction to polytheism a true "religious reformation," you could not say that what Jesus changed about Mosaic Judaism was equally a religious reformation. So although there was no fundamental change in the religious basis, yet opinions are so deep and hardened to this extent. Well! If this is not then because of racial differences, then what else could it be?

The Jewish quarter in Rome is called the "Ghetto." During the middle of the 19th century, it was still a separate, walled area. After the districts of Rome were reorganized in 1885, large portions of this were destroyed [in 1888], so now only the broad outlines are discernable. Some say that the word "ghetto" is derived from the Hebrew *chat*, which means "to destroy" or "to abandon." One sees this word in Hebrew literature such as in Isaiah (chapters 14 and 15), Jeremiah (chapter 48), or Zechariah (chapter 11). In general, those Jews who are foreign [to Italy] all follow this etymology. But actually, the word derives from an abbreviation of the Italian term *borghetto*, which means "little walled town." Thus, when Jewish sojourners were made subjects of Rome, they were segregated by a walled enclosure to control where they could live.

This arrival of the Jews in Rome was not for commerce, nor was it for colonization. It was Pompey the Great, the Roman general, who besieged Jerusalem [63 BCE] and forced his way into the "Holy of Holies" of Solomon's Temple. The Holy of Holies is a room in the temple. No one dares enter except the sacrificial priest. Any others who enter commit great disrespect. The Jews made a big uproar, and Pompey captured some of them and returned to Rome. This was the beginning of the presence of Jews in Rome.

According to ancient Roman custom, anyone who arrived as a captive was made a slave. When King Herod I of Judea [r. 37–4 BCE], and after him, Herod Agrippa [r. 41–44 CE] came to Rome, they resided in the imperial palace and were afforded the ceremonial etiquette suited to kings. Nevertheless, they were still the final rulers of a vanquished kingdom; so how were they much different than war captives? The old kings of Israel ended with the Herodian dynasty.

During the imperial era, the Jews of Rome resided near the banks of the Tiber River, and there was no so-called Ghetto yet. According to Christian legend, when St. Peter was in Rome, he rented a house together with Aqila [Aquila] and his wife Bolisila [Priscilla] [famous early converts and supporters of Christianity] at the foot of the Aventine Hill. Peter had come to preach the faith and would not live in areas with a preexisting Jewish population.

The two emperors Julius Caesar and Augustus Caesar both looked favorably upon the Jews, but the emperor Caligula [r. 37–41 CE] treated them very cruelly. Caligula wanted to erect a statue of himself in the Holy of Holies in Solomon's Temple, but the Jewish people would not permit it. Caligula became enraged and treated the Jews with particular brutality. Despite the brutal treatment, the Jews themselves had still not interfered. That meddling began with internal disputes within Judea. Ever since the emergence of the one called John the Baptist, groups of Jewish believers in the coming Messiah had formed. But nonbelievers denounced these groups as heretics. After angry disputes had gone on for some time, they shifted their target towards the Christians and petitioned to the [Roman] law courts for their ban. And so Roman law was then implemented upon the Jewish race. Thus, Roman interference in Judea was summoned by the residents of Judea themselves.

When the future Emperor Titus [r. 79–81] destroyed Jerusalem [70 CE] and captured thousands of Jews as slaves, he put them to work on the Colosseum project in Rome, the ancient stone amphitheater. Even his unintentional brutality was still brutal. His father, the Emperor Vespasian [r. 69–79], had permitted the Jews to pay a tax of two drachma [Roman silver coins] per person [the *fiscus Judaicus*], to allow them to continue to freely practice their religion. Previously, the Jews were compelled to support Roman polytheism. The revenue from the Jewish tax was used to support the rebuilding and maintenance of the Temple of Jupiter Capitolinus in Rome. But by paying the tax, they were permitted to freely practice their traditional monotheism. This two drachmae tax was the origin of the Capitoline New Year's Tax. During the reign of the Emperor Domitian [r. 81–96], because unrest among the Jews of Rome was not yet quelled, he drove them into the Aiqila [Aquila] Valley, near the Caelian Hill (one of the seven hills of Rome). The Jews then made their own livelihood, as diviners, love fortune tellers, which was a popular custom at the time, street magicians, highly skilled physicians, and the like. After that point, the Jews lived separately and undisturbed, in a period without incident for almost a thousand years.

During the 12th century according to the Christian calendar, there was the Antipope Anacletus II [ca. 1130–1138 CE], who came from the Pierleoni family. The family was known for its wealth and had been patricians in Rome for a long time. But the family forebearer was a converted Jew who later received baptism. While Anacletus was on the papal throne, he was particularly lenient on the Jews, steadfastly promoted friendly relations with them, and made good use of Jewish scholars. At that time, it was only the Israelis [i.e., Jewish scholars] who could recognize ancient scripts, and many could read ancient books that white people were still unable to consult. The success of Christianity had not been brought about by just empty talk, so they could not avoid employing Israelis to assist in literary [i.e., biblical] studies. This trend continued into the papacy of Pope Martin V [1417–1431], when the papal palace still employed many such people.

The pope who was the most hostile enemy of the Jews was Pope Eugene IV [1431–1447], who forbade Christians from engaging in commerce with Jews, as well as eating together or residing together with them. He prohibited Jews from building new synagogues and completely excluded them from undertaking any official work. Pope Paul II [1464–1471], moreover, forced the Jews [starting in 1468] to participate in a race on Carnival [the so-called Race of the Jews, Palio degli Ebrei] in which they were jeered at by the crowds, like some sort of horse race. This cruel practice went on for 200 years before it was halted. In the race, donkeys were driven at the fore, and the Jews chased behind the donkeys. They were permitted to wear only a loincloth, and their four limbs were exposed. Behind the Jews were water buffaloes, and after them were wild horses (precisely, African zebras), generally showing that the Jews were not regarded as human beings. The Jews endured this humiliation and did not dare to disobey. During the papacy of Clement IX [in 1668], the Jews were allowed to remit an annual redemption fee [of 300 *scudi*] instead, and the race was abolished.

On the first Saturday of Carnival, the rabbi and other leaders of the Jewish community in Rome, as a rule, went to the Capitoline, near the Arch of Titus, which depicted the triumphal procession following the destruction of Jerusalem, to appear before a city councilor [*caporione*], an official title. The representative for the Jews had to kneel and present a tribute of 20 *scudi* as well as a wreath of flowers, requesting that the wreath be placed in the Piazza del Campidoglio to decorate the Palazzo Senatorio. Then the Jewish representative went to the Senate and knelt in homage as before, and following ancient precedent, requested permission for the Jews to reside in Rome. The senator raised his foot and struck the forehead of the Jewish person, replying using the formulaic response, "Jews are not allowed to reside in Rome, and it is only with our pardon that they are permitted continued residence." Today, although this example is no longer followed, the Jewish people still go to the Capitoline on the first Saturday of Carnival to salute a saddle, in commemoration of these past events, and to thank the horse for amusing the Roman people in their place.

On New Year's Day (September 9, 1553; the Jewish New Year [Rosh Hashanah]), all copies of the Talmud and many other scriptures were confiscated and burned in public. Originally, Jews had gathered to live on the opposite bank of the Tiber River, close to the Basilica of Saint Cecilia in Trastevere. After the 12th century, they were driven to reside on the right bank of the Tiber, in what would become the territory of the later Jewish Ghetto. Then, starting from the papacy of the Dominican madman, Pope Paul IV [1555–1559], they were locked up within the encircling walls of the Ghetto, and it was also ordered that males who were not wearing a distinctive yellow colored hat and women who did not wear a yellow veil would be unable to leave the Ghetto for any reason.

The Ghetto was originally called Jewish Street, and it was surrounded by walls, from the Ponte dei Quattro Capito [Ponte Fabricio] to the Piazza of Tears [Piazza Giudea, with its Fountain of Tears]. The name "Piazza of Tears" was given to this place, because on July 25, 1556, the Jewish people were forced into their collective prison cell, and from this point onward, they were subjected to a tragedy of infinite troubles.

The Jewish people of the Ghetto had to give their property rights to other people and could no longer own any property themselves. The houses in this area were basically all owned by Romans and included housing where at one time anyone could live. After people moved, it was easy to transfer ownership of the house. Jewish people could only lease property to engage in minor occupations. However, if a Jewish person wanted long-term residence on this street, there needed to be a permanent Jewish land lease contract. This could prevent two types of dangers: first, if the owner went bankrupt or lost ownership, the land lease to the Jewish person could not be revoked or cancelled. Second, the owner was not able to raise the rent on the property. Thus, a law was enacted in which the property ownership right to all land on which Jews resided must revert to a Roman citizen. Outside of the fixed land lease payment, the Roman citizen could not levy labor services on the borrower and must permanently lease the land to them. So if a Jewish person happened to go bankrupt, they would not have to lose their rented land on account of this, and in addition to paying the yearly fixed payment, there would be no risk of increasing rental amounts, and the Jewish person was able to expand or increase the height of his residence as he wished. This special privilege was called the Eastern Rights in the laws (Jus Gazzaga; a Greek term. Gazzaga was the name of a place in the East, thus it temporarily came to stand in for "the East" in translating this term). By applying these Eastern Rights, Jews could inherit the leases to land, and they could buy, sell, sublet, or even pass down through inheritance the rights to land among other Jews. Jewish women used this right to form their dowries when they married, and people all welcomed this. Under these financial rights, Jewish people, within a certain limit, could almost call the places where they lived as their own houses.

Pope Sixtus V [1585–1590] treated the Jews with particular tolerance, for he considered that Jesus had been born from among their tribe. He permitted the Jews to engage in different trades and to have social inter-actions with Christians. He also had housing, libraries, and synagogues built for them. Pope Clement VIII [1592–1605] completely revoked the privileges that were given to the Jews by Sixtus V and codified this into law. Under the papacy of Innocent XIII [1721–1724], the restrictions became even more severe, and Jews were not permitted to engage in any trade outside of scrap iron and textiles. Under the papacy of Benedict XIV [1740–1758], this was relaxed slightly, as Jews were allowed to enter the carpet trade, and this is still a flourishing Jewish business today.

Pope Gregory XIII [1572–1585] forced Jews to go to a Christian church each week to listen to a sermon, first at the Church of St. Benedict [just across the Tiber from the Ghetto] and later at the Church of St. Mary of the Angels. Every sabbath (the Jewish day of worship), he sent policemen to the Ghetto to drive men and women, young and old, to the Christian church with horse whips. If there were any who were indolent or disrespectful in the church, they were also whipped or beaten. This sermon at mass was not something that the Jews wanted to hear; they were certainly forced to listen to it as a means of humiliation. The sermons were filled with parables, with the general theme that Christianity was merciful and compassionate and that everyone could be saved, even dogs who were starving to death, just like the Hebrews who had come to live in Rome, who were all blind and muddled with barely the appearance of human beings. Nevertheless, they had to be beckoned and compelled to come, if they were to receive the grace of the Kingdom of Heaven, and so on. This custom of calling upon the Jews to listen to Christian sermons reappeared during the papacy of Leo XII [1823]. During the papacy of Pius IX [1846–1878], the Duke of Sermoneta, Michelangelo Caetani [1804–1882], made a petition to the pope in 1848 to abolish the policy of forcing the Jews to listen to sermons.

Before the time of Pius IX, the gates to the Ghetto were locked tightly at night from the outside. Pius IX changed the restrictions regarding the Ghetto walls and abolished all the laws that were unfavorable to the Jews. Pope Pius IX's tolerance and goodwill toward this long-oppressed and humiliated race could actually begin to be seen in his actions. One day, he personally gave a generous sum of money to a beggar. His attendant suddenly said to him, "This is a Jew." Pius IX rebuked him, saying, "What's wrong with him being a Jew? Aren't Jews human beings?" Nevertheless, in later years, he was misled by some Jesuits and reinstated the troublesome restrictions against the Jews. During Pius IX's funeral, Jewish people threw stones at his honor guard and wounded some. They were grateful for what he had done earlier in his papacy but regretted what he had done later. But you could say that Pius IX had brought this upon himself.

Opposite the gates of the Ghetto, near the Ponte dei Quattro Capito (the name of the bridge has since been changed), they built a Christian church for converted Jews. On the outer wall there is an image of the cross, and whenever Jews emerged from the Ghetto, they had to look at it. Below the image of the cross, using Hebrew and Latin characters, they wrote in large script a verse from Isaiah (65:2). The text said, "I have spread out my hands all the day unto a rebellious people, which walketh in a way that was not good, after their own thoughts."

The low-lying streets in the Ghetto, near the banks of the Tiber River, are constantly flooded with water during the period of spring rains and melting snow, giving rise to endless misery. The Jews are known for having large families, and they had to reside in such narrow alleys and damp places, so one would expect them to be susceptible to epidemic diseases. And yet diseases like cholera and smallpox were less prevalent in the Ghetto than in other areas of Rome. This was just due to the merit of the Jews always cleaning their residences during each festival and their custom of eating hygienically prepared food. Rome had no Jewish hospital, so if a Jew wanted to be admitted to a regular hospital, this would not be permitted unless he hung a crucifix from the end of his hospital bed to embarrass him.

At the very center of the Jewish quarter is the Portico of Octavia. After the two emperors Vespasian and Titus destroyed Jerusalem, they held a triumphal procession, and the march began from this spot. It was already unbearable enough for the Jews to endure a triumphal procession after the destruction of Jerusalem, but the Roman emperors intentionally staged this march in the area of the future Ghetto. The insults they heaped upon a subjugated and stateless people were as terrible as this. On the walls of the cramped streets inside the Ghetto, one can still see the paintings of menorah everywhere. Today, this is the symbol of their religion.

There were many shops inside the Ghetto. They sold everything from gemstones to buttons to sundry domestic articles. They also had embroidered cloth from Algeria and Constantinople, and striped fabrics from Spain. However, all the sale items are hidden inside the shop and not put on display. When the shopkeeper sees someone passing near the store, he always asks him what he desires, and tries to entice him to purchase. What each shop has for sale is largely the same. On Friday evening, each shop closes its doors and begins to bake the bread for the next day's Sabbath. All the sale goods are locked away inside and the Jewish people all go to the synagogue. On their return, they would exchange greetings of "Shabbat shalom!"

There are five schools in the Jewish quarter, but they are all located in the same building, namely the Temple School, Catalan School, Castilian School, Sicilian School, and New School. They are all schools for Hebrew. These five schools divided the Ghetto into five school districts. Even though people from each district were all Jews, they belonged to different ethnic communities. As for the Jews who lived in Rome at this point, most were indigenous Roman Jews. But there are also some who came from Spain, and some who came from Sicily. As for those in the Temple School district, tradition says that they are the descendants of the Jews from the time of Emperor Titus. On the school grounds there was a large synagogue. Before the New Synagogue [Tempio Maggiore di Roma] near the Portico of Octavia was completed in 1903, this used to be the largest and most magnificent one. Now, that is not the case, but it is still famous for its carvings and gilded decoration. On the facade of the building and along the roofbeam, there are carvings of holy menorahs, the harp of King David, and Miriam's hand drum. The interior decoration is even finer. On festival days, they cover the walls with tapestries. The carvings on the roofbeams all depict Solomon's Temple in Jerusalem and its sacred objects. There is a circular window on the north wall, divided into 12 segments and inlaid with different colors of stained glass, with each color representing one of the tribes of Israel. On the west side is a platform for the choir, where they also set up a wooden pipe organ for the cantor and musicians to use. Opposite this, on the east wall, is the Holy of Holies. On a Corinthian column (one of the three Greek orders of columns), there is a high relief panel carrying the Ten Commandments. They are displayed on two tablets, in the same style as the stone tablets received by Moses. This is veiled by a curtain that is covered with embroidered patterns of the commandments and rose flowers. This Arabic style is known to have imitated the design of

Solomon's Temple, and all the embroidery has used gold thread. At the top-center, there is an embroidered menorah. Within the Holy of Holies, they placed a copy of the Torah under seal. It is written on sheepskin parchment in one giant scroll. This is the so-called Holy Scroll. On every festival day, a person would carry this scripture around the interior of the temple. After one circuit was complete, they would place it atop the organ in the choir area, so that all the people in the temple could see it. The Jews would then raise their hands and chant loudly.

Postscript:
As for this account of the Ghetto, I hope that my readers shall attentively savor it, for several hundred years from now, our [Chinese] people may comprehend it only too well. . . .

Matters Concerning Marco Polo

Twenty years ago, when my traveling husband returned from Western Europe for the first time and told me about the Venetian Marco Polo and how he had served China during the reign of Emperor Shizu of the Yuan dynasty [Kublai Khan], I was immediately envious of Marco as a person. After the passage of nineteen years, I personally was able to visit Venice and sought out Marco's former residence and admired a stone statue I saw of him. So, after recording my own travels, I append a summary of the evidence for the arrival in China of Marco [Polo], father and son, uncle and nephew, and their activities there. . . .

Marco Polo was unsurpassably intelligent and fluent in the Persian and Arabic spoken languages and written scripts. Once he went to the Orient, he also became fluent in the Chinese and Mongolian spoken and written languages. Kublai Khan loved and trusted him and appointed him as his personal attendant, and although Marco had no specific duties, he was quite involved in government affairs. The book he wrote describes Chinese affairs in great detail, and much of what he noticed or carried out can be corroborated in the official *History of the Yuan Dynasty.*

Thus, Kublai Khan frequently sent him to each province or route-command to check on financial affairs. Such envoys were sent often during the reign of Kublai Khan, and Marco Polo was repeatedly selected as one of them. He also sent him as an envoy to Champa [central and southern Vietnam], Ceylon and other countries. His book describes Champa affairs in great detail. In 1277, he was sent overseas for a journey of six months. Some have conjectured that he was serving as an envoy to the kingdom of Annam [Dai Viet kingdom in present-day northern Vietnam]. In that year, the retired emperor of Annam, Trần Thái Tông [r. 1226–1258, 1258–1277] died, and his son Trần Thánh Tông [r. 1258–1278, 1279–1290] was put on the throne, so an envoy would have been sent to invite him to the Yuan dynasty court. In the following year [1278], the grand secretary of the Ministry of Rites, Chai Chun, and others were sent there as envoys. Moreover, at one time Marco was a route command official of Yangzhou circuit, perhaps the *darughachi* ["overseer," a Mongol appointed co-official alongside a regular Chinese official]. When an English scholar said that Marco once served as a major provincial official in China, he was probably pointing to this. That route command [of Yangzhou] comprised 27 cities during the period of 1277 to 1280.

On April 10, 1282, the chief minister of the left, Aḥmad, was assassinated by the battalion head from Yidu, Wang Zhu. Marco Polo investigated the case and soon exposed the particularly heinous crimes of Aḥmad. Westerners only learned about this in 1559 when they obtained a complete version of Marco Polo's book [the printed Ramusio edition]. Perhaps, Marco did not want the world to know about this, so he had not spread this story previously. Western scholars say that according to the *History of the Yuan Dynasty*, the one who investigated this case and revealed Aḥmad's crimes was a deputy military affairs commissioner called Boluo 博羅/孛羅 in the text. So they concluded that Marco Polo had once served as deputy military affairs commissioner. However, the "Basic Annals of Kublai Khan" in the *History of the Yuan Dynasty* records that "in the second lunar month of 1277, the minister of revenue, censor-in-chief, pacification office commissioner, and concurrently imperial attendant,

Boluo 孛羅 was made deputy military affairs commissioner." Previously, it records, "in the fourth lunar month of 1275, the minister of revenue and vice censor-in-chief, Boluo, was made censor-in-chief." And before that, "in the 12th lunar month of 1270, the vice censor-in-chief, Boluo, was concurrently made minister of revenue." This is clearly the same individual. In 1270, Marco Polo had not yet arrived in the Orient, so how could he have served in the concurrent positions of vice censor-in-chief and minister of revenue? Western scholars suspect that the *History of the Yuan Dynasty* is full of errors. It is uncertain how many men are named Boluo in the *History of the Yuan Dynasty* or if some other Boluo has been confused or conflated with Marco Polo. There is no concrete evidence to prove it one way or the other. In the seventh lunar month of 1282, a Mongolian named Boluo was put in charge of panning for gold in Hubei and other provinces. And in the ninth month of that year, Boluo was named overseer of gold mining. No one knows who this person is either.

Marco Polo said that the Yuan [Mongol] forces captured the city of Xiangyang [the garrison city of the Southern Song northern frontier] with the help of counterweight trebuchets, which are machines that hurl large boulders, and that his father and uncle actually contributed the design for these trebuchets. His account is as vivid and detailed as a painting.

The Mongols attacked the city of Xiangyang in the fifth year of the Zhiyuan reign period [1268 CE], and the siege lasted for five years. It wasn't until the 10th year of the Zhiyuan reign [1273 CE] that they conquered it. According to Chinese historical records, the ones who constructed the trebuchets were a Labudan ['Alā al-Dīn] from Maosali [Mosul, Iraq] in the Western Regions and one Yisimayin [Ismā'īl] from Shila [Herat, Afghanistan], also in the Western Regions.

Western scholars say that these two men should be the ones who accompanied Nicolò Polo when he first went to Mongolia and are the same as the trebuchet engineers in their entourage mentioned by Marco Polo, one being a Nestorian Christian and the other a German. Supposedly, Yisimayin sounds a little like a German surname. However, Chinese historical records say that both men ['Alā al-Dīn and Ismā'īl] were sent by the Il-Khan of Persia, Abakha [r. 1265–1282] to the Mongol court in the capital in response to a summons in the eighth year of the Zhiyuan reign [1271 CE], which does not accord with the statements of Marco Polo. . . .[2]

INTRODUCTION TO EDGAR SNOW

Failed attempts at state building and uneven modernization after the fall of the Qing dynasty exposed deep contradictions within Chinese society. In some ways, these paralleled the sort of inequality and class tensions witnessed by Liang Qichao in the fully "modernized" America of the late Gilded Age. These fissures, along with nationalist discontent after the Versailles Conference, and the success of the Russian Revolution eventually fostered the rise of the Chinese Communist Party in 1921. After some initial successes in urban areas like Shanghai, the Communist Party was violently purged in 1927 by the Nationalist Party under Chiang Kai-shek and driven into rural areas, where they reoriented their revolution toward land reform and peasant mobilization. This signaled the start of China's long civil war between the Communists and Nationalists (1927–1949).

American journalist Edgar Snow's (1905–1972) trip behind the lines of "Red China" in the summer and fall of 1936 to interview Mao Zedong and other leaders of the Communist Party marks the final parallel journey in our anthology. It has been called the greatest journalistic scoop of the century. The remnants of the Red Army had recently arrived in the northwest after the perilous Long March (October 1934–October 1935) and were being blockaded by the troops of Zhang Xueliang (1901–2001). The exploits of Mao and his comrades Zhou Enlai, Lin Biao, and Zhu De were known in the West, but no reporter had interviewed them extensively or discovered what happened during the Long March from their perspective.

Edgar Snow had been living in China for seven years working as a journalist. He had covered famines, student movements, and Japanese aggression. He had heard about the exploits of Red Army fighters in the face of overwhelming opposition but wanted to understand what gave rise to the Communist movement and what enabled these men to survive such adversity and garner popular support.

Snow's visit behind the lines did not just happen by good fortune. He had essentially been hand selected by the Communists to help get their message out to the world and arrived at their stronghold through the introduction of Sun Yat-sen's widow and with Zhang Xueliang's cooperation. Once he was in the Red stronghold of Bao'an (near the later "Red capital" of Yan'an), he conducted hundreds of hours of interviews with Mao and the other leaders and went on fact-finding missions to other areas. The Communists translated the transcripts of these interviews into Chinese and edited them, before they were translated back into English, to ensure that they were being represented positively.

In his character portrait of Mao, Snow manages to capture the central paradox of the man's personality: his "native shrewdness" that combined cultured sophistication with simple crudeness. At one moment, Mao could be discussing Western philosophy or writing poetry, and the next, he might suddenly drop his pants in front of visitors and search for lice. Snow felt a "certain force of destiny" propelling this man who would later control China for decades. In hindsight, Snow's portrait is prophetic, especially when he remarks, "There is as yet, at least, no ritual of hero worship built up around him."

When Edgar Snow returned to Beijing in early November 1936, the pace of events accelerated, and he was hard pressed to get his story out before the whirlwind could sweep away the relevance of his coup. Shortly after he returned from Xi'an, Chiang Kai-shek was kidnapped by Zhang Xueliang's troops in the Xi'an Incident (December 1936) and forced to join the Second United Front. The Marco Polo Bridge Incident (July 7–9, 1937) and Japan's invasion of North China followed, leading to the Rape of Nanjing (December 1937). Snow was able to incorporate the Xi'an Incident into his book, relying on other informants, but the book went into press in the United Kingdom (1937) and the United States (1938) as the other events were transpiring. His wife, Helen Foster Snow (writing under the penname Nym Wales), was also an intrepid journalist. While Edgar was completing his draft of *Red Star,* she returned alone to Yan'an to interview Mao and other leaders further, helping Edgar fill in some of the holes in his narrative.

Red Star over China became a sensation upon its publication and continued to have a lasting impact around the world for decades. It was soon translated into Chinese and became one of the few ways people in China could learn about their Communist leaders. It even led some in the Chinese diasporic community to trek to the northwest to join them.

Red Star's humanizing and glowing portrait of the Chinese Communists as democratic patriots gained greater sympathy for the movement in the West but also landed Snow in trouble during the McCarthy Era, compelling him to move to Switzerland. Snow revised his book in 1944 and again in 1968, adding an epilogue and extensive biographical notes. A telling alteration in the 1968 edition was changing verbs from present to past tense. Going from "Mao *appears* to be quite free from symptoms of megalomania" to "*appeared* to be" seem especially potent, given the historical context of the Cultural Revolution, then in progress.

PRIMARY SOURCE AND SELECTED STUDIES

Lovell, Julia. "The Red Star." In *Maoism: A Global History.* Knopf, 2019.

Snow, Edgar. *Red Star over China.* Random House, 1938; revised and enlarged edition, Grove, 1968.

Thomas, S. Bernard. *Season of High Adventure: Edgar Snow in China.* University of California Press, 1996.

EXCERPTS FROM THE CHAPTER "SOVIET STRONGMAN" IN *RED STAR OVER CHINA* (1938), BY EDGAR SNOW

Small villages are numerous in the North-west, but towns of any size are infrequent. Except for the industries begun by the Reds it is agrarian and in places semi-pastoral country. Thus, it was quite breathtaking to ride out suddenly on the brow of the wrinkled hills and see stretched out below me in a green valley the ancient walls of Pao An [Bao'an], which means "Defended Peace." . . .

Here at last I found the Red leader whom Nanking [Nanjing] had been fighting for ten years—Mao Tse-tung [Mao Zedong], chairman of the "Chinese People's Soviet Republic," to employ the official title which had recently been adopted. The old cognomen, "Chinese Workers' and Peasants' Soviet Republic," was dropped when the Reds began their new policy of struggle for a "United Front."

Chou En-lai's [Zhou Enlai] radiogram had been received and I was expected. A room was provided for me in the "Foreign Office," and I became temporarily a guest of the Soviet State. My arrival resulted in a phenomenal increase of the foreign population of Pao An. The other Occidental resident was a German known as Li Teh T'ung-chih—the "Virtuous Comrade Li" [Li De Tongzhi, i.e., Otto Braun]. Of Li Teh, a former high official in the German Army and the only foreign adviser ever with the Red Army (to Herr Hitler's extreme vexation), more later.

I met Mao soon after my arrival: a gaunt, rather Lincolnesque figure, above average height for a Chinese, somewhat stooped, with a head of thick black hair grown very long, and with large, searching eyes, a high-bridged nose and prominent cheekbones. My fleeting impression was of an intellectual face of great shrewdness, but I had no opportunity to verify this for several days. Next time I saw him, Mao was walking hatless along the street at dusk, talking with two young peasants and gesticulating earnestly. I did not recognize him until he was pointed out to me moving along unconcernedly with the rest of the strollers, despite the $250,000 which Nanking had hung over his head.

I could have written a book about Mao Tse-tung alone. I talked with him many nights, on a wide range of subjects, and I heard dozens of stories about him from soldiers and Communists. My written interviews with him totaled about twenty thousand words. He told me of his childhood and youth, how he became a leader in the Kuomintang [Guomindang, KMT] and the Nationalist Revolution, why he became a Communist, and how the Red Army grew. He described the Long March to the Northwest and wrote a classical poem about it for me. He told me stories of many other famous Reds, from Chu Teh [Zhu De] down to the youth who carried on his shoulders for over 7,000 miles the two iron dispatch boxes that held the archives of the Soviet Government.

How can I select, from all the wealth of unexploited, unknown material, a few hundred words to tell you about this peasant-born intellectual turned revolutionary? I do not propose to attempt such a condensation. The story of Mao's life is a rich cross-section of a whole generation, an important guide to understanding the sources of action in China, and farther on I shall include that full exciting record of personal history, just as he told it to me. But here I want to try to convey some subjective impressions, and a few facts of interest about him.

* * * * * * * * * * * * * * **BRIEF EXCERPT** * * * * * * * * * * * * * * * *

Do not suppose, first of all, that Mao Tse-tung could be the "saviour" of all of China. Nonsense. There will never be any one "saviour" of China. Yet undeniably you feel a certain force of destiny in him. It is nothing quick or flashy, but a kind of solid elemental vitality. You feel that whatever extraordinary there is in this man grows out of the uncanny degree to which he synthesizes and expresses the urgent demands of millions of Chinese, and especially the peasantry—those impoverished, underfed, exploited, illiterate, but kind, generous, courageous, and just now rather rebellious human beings who are the vast majority of the Chinese people. If these demands and the movement which is pressing them forward are the dynamics which can regenerate China, then in this deeply historical sense Mao Tse-tung may possibly become a very great man.

But I do not intent to pronounce the verdicts of history. Meanwhile, Mao is of interest as a personality, apart from his political life, because, although his name is as familiar to many Chinese as that of Chiang Kai-shek, very little was known about him, and all sorts of strange legends existed about him. I was the first foreign newspaperman to interview him.

Mao has the reputation of a charmed life. He had been repeatedly pronounced dead by Nanking, only to return to the news columns a few days later, as active as ever. The Kuomintang has also officially "killed" and buried Chu Teh many times, assisted by occasional corroborations from clairvoyant missionaries. Numerous deaths of the two famous men, nevertheless, did not prevent them from being involved in many spectacular exploits, including the Long March. Mao was indeed in one of his periods of newspaper demise when I visited Red China, but I found him quite substantially alive. There seems to be some basis for the legend of his charmed life, however, in the fact that, although he has been in scores of battles, was once captured by enemy troops and escaped, and has the world's highest reward on his head, during all these years he has never once been wounded.

I happened to be in his house one evening when he was given a complete physical examination by a Red surgeon—a returned student from Europe who knew his business—and pronounced in excellent health. He has never had tuberculosis or any "incurable disease," as has been rumored by some romantic travelers. His lungs are completely sound, although, unlike most Red commanders, he is an inordinate cigarette-smoker. During the Long March, Mao and Li Teh (another heavy smoker) carried on original botanical research by testing out various kinds of leaves as tobacco substitutes.

Ho Tze-nien [He Zizhen], Mao's present wife, a former school-teacher and a Communist organizer herself, has been less fortunate than her husband. She has more than a dozen wounds, caused by splinters from an air bomb, but all of them were superficial. Just before I left Pao An the Maos were proud parents of a new baby girl. He had two other children by his former wife, Yang K'ai-hui [Yang Kaihui], the daughter of a noted Chinese professor. She was killed by Ho Chien [He Jian] several years ago.

Mao Tse-tung is now (1937) forty-four years old. He was elected chairman of the provisional Central Soviet Government at the Second All-China Soviet Congress, attended by delegates representing approximately 9,000,000 people then living under Red laws. . . .

The influence of Mao Tse-tung throughout the Communist world of China is probably greater than that of anyone else. He is a member of nearly everything—the revolutionary military committee, the political bureau of the Central Committee, the finance commission, the organization committee, the public health commission, and others. His real influence is asserted through his domination of the political bureau, which has decisive power in the policies of the Party, the Government, and the Army. Yet, while everyone knows and respects him, there is—as yet, at least—no ritual of hero worship built up around him. I never met a Chinese Red who driveled "our-great-leader" phrases, I did not hear Mao's name used as a synonym for the Chinese people, but still I never met one who did not like "the Chairman"—as everyone called him—and admire him. The rôle of his personality in the movement was clearly immense.

Mao seemed to me a very interesting and complex man. He had the simplicity and naturalness of the Chinese peasant, with a lively sense of humor and a love of rustic laughter. His laughter was even active on the subject of himself and the shortcomings of the Soviets—a boyish sort of laughter which never in the least shook his inner faith in his purpose. He is plain-speaking and plain-living, and some people might think him rather coarse and vulgar. Yet he combines curious qualities of naïveté with incisive wit and worldly sophistication.

I think my first impression—dominantly one of native shrewd-ness—was probably correct. And yet Mao is an accomplished scholar of Classical Chinese, an omnivorous reader, a deep student of philosophy and history, a good speaker, a man with an unusual memory and extraordinary powers of concentration, an able writer, careless in his personal habits and appearance but astonishingly meticulous about details of duty, a man of tireless energy, and a military and political strategist of considerable genius. It is an interesting fact that many Japanese regard him as the ablest Chinese strategist alive.

* *

The Reds were putting up some new buildings in Pao An, but accommodations were very primitive while I was there. Mao lived with his wife in a two-room *yao-fang* [loess cave house] with bare, poor, map-covered walls. He had known much worse, and as the son of a "rich" peasant in Hunan he had also known better. The chief luxury they boasted was a mosquito net. Otherwise Mao lived very much like the rank and file of the Red Army. After ten years of leadership of the Reds, after hundreds of confiscations of property of landlords, officials, and tax collectors, he owned only his blankets and a few personal belongings, including two cotton uniforms. Although he is a Red Army commander as well as chairman, he wore on his coat collar only the two red bars that are the insignia of the ordinary Red soldier.

I went with Mao several times to mass meetings of the villagers and the Red cadets, and to the Red theater. He sat inconspicuously in the midst of the crowd and enjoyed himself hugely. I remember once, between acts at the Anti-Japanese Theater, there was a general demand for a duet by Mao Tse-tung and Lin Piao [Lin Biao], the twenty-eight-year-old president of the Red Academy, and formerly a famed young cadet on Chiang Kai-shek's staff. Lin blushed like a schoolboy and got them out of the "command performance" by a graceful speech, calling upon the women Communists for a song instead.

Mao's food was the same as everybody's, but being a Hunanese he had the southerner's *ai-la*, or "love of pepper." He even had pepper cooked into his bread. Except for this passion, he scarcely seemed to notice what he ate. One night at dinner I heard him expand on a theory of pepper-loving peoples being revolutionaries. He first submitted his own province, Hunan, famous for the revolutionaries it has produced. Then he listed Spain, Mexico, Russia, and France to support his contention, but laughingly had to admit defeat when somebody mentioned the well-known Italian love of red pepper and garlic, in refutation of his theory. One of the most amusing songs of the "bandits," incidentally, is a ditty called "The Hot Red Pepper." It told of the disgust of the pepper with his pointless vegetable existence, waiting to be eaten, and how he ridiculed the contentment of the cabbages, spinach, and beans with their invertebrate careers. He ends up by leading a vegetable insurrection. "The Hot Red Pepper" was a great favorite with Chairman Mao.

He appears to be quite free from symptoms of megalomania, but he has a deep sense of personal dignity, and something about him suggests a power of ruthless decision when he deems it necessary. I never saw him angry, but I heard from others that on occasions he has been roused to an intense and withering fury. At such times his command of irony and invective is said to be classic and lethal.

I found him surprisingly well informed on current world politics. Even on the Long March, it seems, the Reds received news broadcasts by radio, and in the North-west they publish their own newspapers. Mao was exceptionally well read in world history and had a realistic conception of European social and political conditions. He was very interested in the Labour Party of England, and questioned me intensely about its present policies, soon exhausting all my information. It seemed to me that he found it difficult fully to understand why, in a country where workers are enfranchised, there is still no workers' government. I am afraid my answers did not satisfy him. He expressed profound contempt for Ramsay MacDonald, whom he designated as a *han-chien* [Han jian]—an arch-traitor of the British people.

His opinion of President Roosevelt was rather interesting. He believed him to be anti-Fascist, and thought China could cooperate with such a man. He asked innumerable questions about the New Deal, and Roosevelt's foreign policy. The questioning showed a remarkably clear conception of the objectives of both. He regarded Mussolini and Hitler as mountebanks, but considered Mussolini a much abler man, a real Machiavellian, with a knowledge of history, while Hitler was a mere will-less puppet of the reactionary capitalists.

Mao had read a number of books about India and had some definite opinions on that country. Chief among these was that Indian independence would never be realized without an agrarian revolution. He questioned me about Gandhi, Jawaharlal Nehru, Suhasini Chattopadhyay, and other Indian leaders I had known. He knew something about the Negro question in America, and unfavorably compared the treatment of Negroes and American Indians with policies in the Soviet Union toward national minorities. However, he was also interested when I pointed out certain great differences in the historical and psychological background of the Negro in America and that of minor races of Russia. Interested—but he did not agree with me.

Mao is an ardent student of philosophy. Once when I was having nightly interviews with him on Communist history, a visitor brought him several new books on philosophy, and Mao asked me to postpone our engagements. He consumed these books in three or four nights of intensive reading, during which he seemed oblivious to everything else. He had not confined his reading to Marxist philosophers, but had read something of the ancient Greeks, of Spinoza, Kant, Goethe, Hegel, and Rousseau, of course, and others.

I often wondered about Mao's own sense of responsibility over the question of force, violence, and the "necessity of killing." He had had in his youth strongly liberal and humanistic tendencies, and the transition from idealism to realism could only have been made philosophically. Although he was peasant-born, he did not as a youth personally suffer much from oppression of the landlords, as did many Reds, and, although Marxism is the core of his thought, I deduced that class hatred is for him probably fundamentally a mechanism in the bulwark of his philosophy, rather than a basic impulse to action.

There seemed to be nothing in him that might be called religious feeling; his judgments were reached, I believe, on the basis of reason and necessity. Because of this I think he has probably on the whole been a moderating influence in the Communist movement where life and death are concerned. It seemed to me that he tried to make his philosophy, the dialectics of "the long view," his criterion in any large course of action, and in that range of thought the preciousness of human life is only relative. This is distinctly unusual among Chinese leaders, who historically have always placed expediency above ethics.

Mao works thirteen or fourteen hours a day, often until very late at night, frequently retiring at two or three. He seems to have an iron constitution. This he traces to a youth spent in hard work on his father's farm, and to an austere period in his schooldays when he formed a kind of Spartan club with some comrades. They used to fast, go on long hikes in the wooded hills of South China, swim in the coldest weather, walk shirtless in the rain and sleet—all this to toughen themselves. They intuitively knew that the years ahead in China would demand the capacity for withstanding great hardship and suffering.

Mao once spent a summer tramping all over Hunan, his native province. He earned his bread by working from farm to farm, and sometimes by begging. Another time, for days he ate nothing but hard beans and water —again a process of "toughening" his stomach. The friendships he made on these country rambles in his early youth were of great value to him when, some ten years later, he began to organize thousands of farmers

in Hunan into the famous peasant unions which became the first base of
the Soviets, after the Kuomintang broke with the Communists in 1927.

Mao impressed me as a man of considerable depth of feeling. I remember
that his eyes moistened once or twice when speaking of dead comrades,
or recalling incidents in his youth, during the rice riots and famines of
Hunan, when some starving peasants were beheaded in his province for
demanding food from the yamen. One soldier told me of seeing Mao give
his coat away to a wounded man at the front. They said that he refused
to wear shoes when the Red warriors had none.

Yet I doubt very much if he would ever command great respect from
the intellectual *élite* of China, perhaps not entirely because he has an
extraordinary mind, but because he has the personal habits of a peasant.
The Chinese disciples of Pareto might think him uncouth. I remember,
when talking with Mao one day, seeing him absent-mindedly turn down
the belt of his trousers and search for some guests—but then it is just
possible that Pareto might have done a little searching himself if he lived
in similar circumstances. But I am sure that Pareto would never take off
his trousers in the presence of the president of the Red Academy—as
Mao did once when I was interviewing Lin Piao. It was extremely hot
inside the little room. Mao lay down on the bed, pulled off his pants,
and for twenty minutes carefully studied a military map on the wall
-- interrupted occasionally by Lin Piao, who asked for confirmation of
dates and names, which Mao invariably knew. His nonchalant habits
fitted with his complete indifference to personal appearance, although
the means were at hand to fix himself up like a chocolate-box general
or a politician's picture in *Who's Who in China*.

Except for a few weeks when he was ill, he walked most of the 6,000
miles of the Long March, like the rank and file. At any time in recent
years he could have achieved high office and riches by "betraying" to the
Kuomintang, and this applies to most Red commanders. The tenacity with
which these Communists for ten years clung to their principles cannot be

fully evaluated unless you know the history of "silver bullets" in China, by means of which other rebels have customarily been bought off.

He seemed to me sincere, honest, and truthful in his statements. I was able to check up on many of his assertions, and usually found them to be correct. He subjected me to mild doses of political propaganda, but nothing compared to what I have received in non-bandit quarters, and he never imposed any censorship on me, either in my writing or my photography, courtesies for which I was grateful. He did his best to see that I got facts to explain various aspects of Soviet life.

Because of their tremendous importance in the political scene of China today, his main declarations of Communist policies are worth serious consideration. For it is quite possible now—since the whole North-west and other large sections of the armed and unarmed Chinese people seem to be in sympathy with many of these policies— that they may become vital instruments of fundamental changes in the destiny of China.[3]

NOTES

1. Liang Qichao 梁啟超, *Xin Dalu youji* 新大陸遊記 (Shehui kexue wenxian chubanshe, 2007), 49–50, 51–52, 109–110, 154–159, 180–181. Translation by the author. Headings from the original.
2. Qian Shan Shili 錢單士厘, *Guimao lüxing ji; Guiqian ji* 癸卯旅行記；歸潛記, edited by Yang Jian 楊堅 (Hunan renmin chubanshe, 1981), 22, 74, 88–89, 91, 196, 204–211, 223–229. Translation by the author. Headings from the original.
3. Excerpts from *Red Star over China* by Edgar Snow, copyright © 1938 (pages 65–74), 1944 by Random House, Inc, © 1961 by John K. Fairbank, © 1968 by Edgar Snow. Used by permission of Grove/Atlantic, Inc. Any third party use of this material, outside of this publication, is prohibited. As required by the publisher, this excerpt reproduces the exact text of the 1938 edition, including the vestigial Britishisms in the first American printing and the Wade-Giles romanization of Chinese names. Notes in square brackets are not in the original text but have been added by the author to modernize the romanization and correct the spelling of some names, on first appearance only.

INDEX

www.ingramcontent.com/pod-product-compliance
Lightning Source LLC
Chambersburg PA
CBHW070740220326
41598CB00026B/3717